Electrochemistry of Organic Compounds

Electrochemistry of Organic Compounds

Contributors :
Johan Lindstrom,
Pontus Svens, *et al.*

KOROS PRESS LIMITED
London, UK

Electrochemistry of Organic Compounds
Contributors : Johan Lindstrom *and* Pontus Svens, *et al.*

Published by Koros Press Limited

www.korospress.com

United Kingdom

Copyright 2016

Printed in 2017 for Sale in the Indian Subcontinent

Electrochemistry of Organic Compounds

ISBN: 978-1-78163-555-1

British Library Cataloguing in Publication Data
A CIP record for this book is available from the British Library

Exclusively distributed by CBS Publishers & Distributors Pvt. Ltd.

Sales & Distribution Rights only for India, Pakistan, Bangladesh, Sri Lanka, Nepal and Bhutan.This book is not to be sold outside these territories.

PREFACE

Electrochemistry can offer a clean and efficient method for the selective oxidation or reduction of organic molecules and can accomplish transformations that are quite different from those realized by chemical reagents. Moreover the use of electricity means that the use of environmentally undesirable redox species based on Cr, Os, Mn, *etc* is avoided; it is green as well as powerful. Industrial applications of electro-organic synthesis are increasingly appealing but only a few are well-established, the most notable being the Monsanto synthesis of nylon-66. At the same time of course the huge commercial use of electrolysis in the formation of chlorine, sodium hydroxide and aluminium is noteworthy.

This is a comprehensive, exclusive and exhaustive work on the subject. It is an asset for all researchers and scholars.

This page left intentionally blank.

CONTENTS

This page left intentionally blank.

LIST OF CONTRIBUTORS

Pontus Svens

Scania CV AB, SE-151 87, Sodertalje, Sweden; E-Mails: johan.lindstrom@scania.com (J.L.); olle.gelin@scania.com (O.G.)

and

School of Chemical Science and Engineering, Department of Chemical Engineering and Technology, Applied Electrochemistry, KTH Royal Institute of Technology, SE-100 44, Stockholm, Sweden; E-Mails: behm@kth.se (M.B.); gnli@kth.se (G.L.)

Johan Lindstrom

Scania CV AB, SE-151 87, Sodertalje, Sweden; E-Mails: johan.lindstrom@scania.com (J.L.); olle.gelin@scania.com (O.G.)

Olle Gelin

Scania CV AB, SE-151 87, Sodertalje, Sweden; E-Mails: johan.lindstrom@scania.com (J.L.); olle.gelin@scania.com (O.G.)

Marten Behm

School of Chemical Science and Engineering, Department of Chemical Engineering and Technology, Applied Electrochemistry, KTH Royal Institute of Technology, SE-100 44, Stockholm, Sweden; E-Mails: behm@kth.se (M.B.); gnli@kth.se (G.L.)

Goran Lindbergh

School of Chemical Science and Engineering, Department of Chemical Engineering and Technology, Applied Electrochemistry, KTH Royal Institute of Technology, SE-100 44, Stockholm, Sweden; E-Mails: behm@kth.se (M.B.); gnli@kth.se (G.L.)

This page left intentionally blank.

Chapter 1

ELECTRO-CHEMISTRY

Electro-chemistry is a branch of chemistry that studies chemical reactions which take place in a solution at the interface of an electron conductor (the electrode : a metal or a semi-conductor) and an ionic conductor (the electrolyte). These reactions involve electron transfer between the electrode and the electrolyte or species in solution. Thus electro-chemistry deals with interactions between electrical energy and chemical change and *vice versa*.

If a chemical reaction is driven by an externally applied voltage, as in electrolysis, or if a voltage is created by a chemical reaction as in a battery, it is an *electrochemical* reaction. In contrast, chemical reactions where electrons are transferred between molecules are called oxidation-reduction (redox) reactions. In general, electro-chemistry deals with situations where redox reactions are separated in space or time, connected by an external electric circuit.

HISTORY OF ELECTRO-CHEMISTRY

Electro-chemistry, a branch of *chemistry*, went through several changes during its evolution from early principles related to *magnets* in the early 16th and 17th centuries, to complex theories involving *conductivity*, *electrical charge* and mathematical methods. The term *electro-chemistry* was used to describe electrical phenomena in the late 19th and 20th centuries. In recent decades, *electro-chemistry* has become an area of current research, including research in *batteries* and *fuel cells*, preventing *corrosion* of metals, and improving techniques in *refining* chemicals with *electrolysis* and *electrophoresis*.

Background and Dawn of Electro-chemistry

The 16th century marked the beginning of scientific understanding of electricity and magnetism that culminated with the production of electric power and the industrial revolution in the late 19th century.

In the 1550s, English scientist William Gilbert spent 17 years experimenting with magnetism and, to a lesser extent, electricity. For his work on magnets, Gilbert became known as "The Father of Magnetism." His book *De Magnete* quickly became the standard work throughout Europe on electrical and magnetic phenomena. He made the first clear distinction between magnetism and what was then called the "amber effect" (static electricity).

In 1663, German physicist Otto von Guericke created the first electrostatic generator, which produced static electricity by applying friction. The generator was made of a large sulfur ball inside a glass globe, mounted on a shaft. The ball was rotated by means of a crank and a static electric spark was produced when a pad was rubbed against the ball as it rotated. The globe could be removed and used as an electrical source for experiments with electricity. Von Guericke used his generator to show that like charges repelled each other.

The 18th Century and Birth of Electro-chemistry

In 1709, Francis Hauksbee at the Royal Society in London discovered that by putting a small amount of mercury in the glass of Von Guericke's generator and evacuating the air from it, it would glow whenever the ball built up a charge and his hand was touching the globe. He had created the first gas-discharge lamp.

Between 1729 and 1736, two English scientists, Stephen Gray and Jean Desaguliers, performed a series of experiments which showed that a cork or other object as far away as 800 or 900 feet (245–275 m) could be electrified by connecting it via a charged glass tube to materials such as metal wires or hempen string. They found that other materials, such as silk, would not convey the effect.

By the mid-18th century, French chemist Charles François de Cisternay Du Fay had discovered two forms of static electricity, and that like charges repel each other while unlike charges attract. Du Fay announced that electricity consisted of two fluids : *vitreous* (from the Latin for "glass"), or positive, electricity; and *resinous*, or negative, electricity. This was the "two-fluid theory" of electricity, which was opposed by Benjamin Franklin's "one-fluid theory" later in the century.

In 1745, Jean-Antoine Nollet developed a theory of electrical attraction and repulsion that supposed the existence of a continuous flow of electrical matter between charged bodies. Nollet's theory at first gained wide acceptance, but met resistance in 1752 with the translation of Franklin's *Experiments and Observations on Electricity* into French. Franklin and Nollet debated the nature of electricity, with Franklin supporting action at a distance and two qualitatively opposing types of electricity, and Nollet advocating mechanical action and a single type of electrical fluid. Franklin's argument eventually won and Nollet's theory was abandoned.

In 1748, Nollet invented one of the first electrometers, the electroscope, which showed electric charge using electrostatic attraction and repulsion. Nollet is reputed to be the first to apply the name "Leyden jar" to the first device for storing electricity. Nollet's invention was replaced by Horace-Bénédict de Saussure's electrometer in 1766.

By the 1740s, William Watson had conducted several experiments to determine the speed of electricity. The general belief at the time was that electricity was faster than sound, but no accurate test been devised to measure the velocity of a current. Watson, in the fields north of London, laid out a line of wire supported by dry sticks and silk which ran for 12,276 feet (3.7 km). Even at this length, the velocity of electricity seemed instantaneous. Resistance in the wire was also noticed but apparently not fully understood, as Watson related that "we observed again, that although the electrical compositions were very severe to those who held the wires, the report of the Explosion at the prime Conductor was little, in comparison of that which is heard when the Circuit is short." Watson eventually decided not to pursue his electrical experiments, concentrating instead upon his medical career.

By the 1750s, as the study of electricity became popular, efficient ways of producing electricity were sought. The generator developed by Jesse Ramsden was among the first electrostatic generators invented. Electricity produced by such generators was used to treat paralysis, muscle spasms, and to control heart rates. Other medical uses of electricity included filling the body with electricity, drawing sparks from the body, and applying sparks from the generator to the body.

Charles-Augustin de Coulomb developed the law of electrostatic attraction in 1781 as an outgrowth of his attempt to investigate the law of electrical repulsions as stated by Joseph Priestley in England. To this end, he invented a sensitive apparatus to measure the electrical forces involved in Priestley's law. He also established the inverse square law of attraction and repulsion magnetic poles, which became the basis for the mathematical theory of magnetic forces developed by Siméon Denis Poisson. Coulomb wrote seven important works on electricity and magnetism which he submitted to the Académie des Sciences between 1785 and 1791, in which he reported having developed a theory of attraction and repulsion between charged bodies, and went on to search for perfect conductors and dielectrics. He suggested that there was no perfect dielectric, proposing that every substance has a limit, above which it will conduct electricity. The SI unit of charge is called a coulomb in his honour.

In 1789, Franz Aepinus developed a device with the properties of a "condenser" (now known as a capacitor.) The Aepinus condenser was the first capacitor developed after the Leyden jar, and was used to demonstrate conduction and induction. The device was constructed so that the space between two plates could be adjusted, and the glass dielectric separating the two plates could be removed or replaced with other materials.

Electricitatis in Motu Musculari Commentarius (Commentary on the Effect of Electricity on Muscular Motion), where he proposed a "nerveo-electrical substance" in life forms.

In his essay, Galvani concluded that animal tissue contained a before-unknown innate, vital force, which he termed "animal electricity," which activated muscle when placed between two metal probes. He believed that this was evidence of a new form of electricity, separate from the "natural" form that is produced by lightning and the "artificial" form that is produced by friction (static electricity). He considered the brain to be the most important organ for the secretion of this "electric fluid" and that the nerves conducted the fluid to the muscles. He believed the tissues acted similarly to the outer and inner surfaces of Leyden jars. The flow of this electric fluid provided a stimulus to the muscle fibres.

Galvani's scientific colleagues generally accepted his views, but Alessandro Volta, the outstanding professor of physics at the University of Pavia, was not convinced by the analogy between muscles and Leyden jars. Deciding that the frogs' legs used in Galvani's experiments served only as an electroscope, he held that the contact of dissimilar metals was the true source of stimulation. He referred to the electricity so generated as "metallic electricity" and decided that the muscle, by contracting when touched by metal, resembled the action of an electroscope. Furthermore, Volta claimed that if two dissimilar metals in contact with each other also touched a muscle, agitation would also occur and increase with the dissimilarity of the metals. Galvani refuted this by obtaining muscular action using two pieces of similar metal. Volta's name was later used for the unit of electrical potential, the volt.

Rise of Electro-chemistry as Branch of Chemistry

In 1800, English chemists *William Nicholson* and *Johann Wilhelm Ritter* succeeded in separating water into *hydrogen* and *oxygen* by *electrolysis*. Soon thereafter, Ritter discovered the process of *electroplating*. He also observed that the amount of metal deposited and the amount of oxygen produced during an electrolytic process depended on the distance between the *electrodes*. By 1801 Ritter had observed thermoelectric currents, which anticipated the discovery of *thermo-electricity* by *Thomas Johann Seebeck*.

In 1802, *William Cruickshank* designed the first electric battery capable of mass production. Like Volta, Cruickshank arranged square copper plates, which he soldered at their ends, together with plates of zinc of equal size. These plates were placed into a long rectangular wooden box which was sealed with cement. Grooves inside the box held the metal plates in position. The box was then filled with an electrolyte of *brine*, or watered down acid. This flooded design had the advantage of not drying out with use and provided more energy than Volta's arrangement, which used brine-soaked papers between the plates.

In the quest for a better production of *platinum* metals, two scientists, *William Hyde Wollaston* and *Smithson Tennant*, worked together to design an efficient electro-chemical technique to refine or purify platinum. Tennant ended up discovering the elements *iridium* and *osmium*. Wollaston's effort, in turn, led him to the discovery of the metals *palladium* in 1803 and *rhodium* in 1804.

Wollaston made improvements to the galvanic battery (named after Galvani) in the 1810s. In Wollaston's battery, the wooden box was replaced with an earthenware vessel, and a copper plate was bent into a U-shape, with a single plate of zinc placed in the center of the bent copper. The zinc plate was prevented from making contact with the copper by dowels (pieces) of cork or wood. In his single cell design, the U-shaped copper plate was welded to a horizontal handle for lifting the copper and zinc plates out of the electrolyte when the battery was not in use.

In 1809, *Samuel Thomas von Soemmering* developed the first *telegraph*. He used a device with 26 wires (1 wire for each letter of the *German alphabet*) terminating in a container of acid. At the sending station, a key, which completed a circuit with a battery, was connected as required to each of the line wires. The passage of current caused the acid to decompose chemically, and the message was read by observing at which of the terminals the bubbles of gas appeared. This is how he was able to send messages, one letter at a time.

Humphry Davy's work with electrolysis led to conclusion that the production of electricity in simple *electrolytic cells* resulted from chemical reactions between the electrolyte and the metals, and occurred between substances of opposite charge. He reasoned that the interactions of electrical currents with chemicals offered the most likely means of *decomposing* all substances to their basic elements. These views were explained in 1806 in his lecture *On Some Chemical Agencies of Electricity*, for which he received the *Napoleon Prize* from the *Institut de France* in 1807 (despite the fact that England and France were at war at the time). This work led directly to the isolation of sodium and potassium from their common compounds and of the *alkaline earth metals* from theirs in 1808.

Hans Christian Ørsted's discovery of the magnetic effect of electrical currents in 1820 was immediately recognised as an important advance, although he left further work on *electromagnetism* to others. *André-Marie Ampère* quickly repeated Ørsted's experiment, and formulated them mathematically (which became *Ampère's law*). Ørsted also discovered that not only is a magnetic needle deflected by the electric current, but that the live electric wire is also deflected in a magnetic field, thus laying the foundation for the construction of an electric motor. Ørsted's discovery of *piperine*, one of the pungent components of pepper, was an important contribution to chemistry, as was his preparation of *aluminium* in 1825.

During the 1820s, *Robert Hare* developed the *Deflagrator*, a form of voltaic battery having large plates used for producing rapid and powerful *combustion*. A modified form of this apparatus was employed in 1823 in volatilising and fusing *carbon*. It was with these batteries that the first use of voltaic electricity for blasting under water was made in 1831.

In 1821, the *Estonian*-German physicist, Thomas Johann Seebeck, demonstrated the electrical potential in the juncture points of two dissimilar metals when there is a temperature difference between the joints. He joined a copper wire with a *bismuth* wire to form a loop or circuit. Two junctions were formed by connecting the ends of the wires to each other. He then accidentally discovered that if he heated one junction to a high temperature, and the other junction remained at room temperature, a magnetic field was observed around the circuit.

He did not recognise that an electrical current was being generated when heat was applied to a bi-metal junction. He used the term "thermomagnetic currents" or "thermomagnetism" to express his discovery. Over the following two years, he reported on his continuing observations to the *Prussian Academy of Sciences*, where he described his observation as "the magnetic polarization of metals and ores produced by a temperature difference." This *Seebeck effect* became the basis of the *thermocouple*, which is still considered the most accurate measurement of temperature today. The converse *Peltier effect* was seen over a decade later when a current was run through a circuit with two dissimilar metals, resulting in a temperature difference between the metals.

In 1827 German scientist *Georg Ohm* expressed his *law* in his famous book *Die galvanische Kette, mathematisch bearbeitet* (The Galvanic Circuit Investigated Mathematically) in which he gave his complete theory of electricity.

In 1829 *Antoine-César Becquerel* developed the "constant current" cell, forerunner of the well-known *Daniell cell*. When this acid-alkali cell was monitored by a *galvanometer*, current was found to be constant for an hour, the first instance of "constant current". He applied the results of his study of thermo-electricity to the construction of an electric thermometer, and measured the temperatures of the interior of animals, of the soil at different depths, and of the atmosphere at different heights. He helped validate *Faraday's laws* and conducted extensive investigations on the *electroplating* of metals with applications for metal finishing and *metallurgy*. *Solar cell* technology dates to 1839 when Becquerel observed that shining light on an electrode submerged in a conductive solution would create an electric current.

Michael Faraday began, in 1832, what promised to be a rather tedious attempt to prove that all electricities had precisely the same properties and caused precisely the same effects. The key effect was electro-chemical decomposition. Voltaic and electro-magnetic electricity posed no problems, but static electricity did. As Faraday delved deeper into the problem, he made two startling discoveries. First, electrical force did not, as had long been supposed, act at a distance upon molecules to cause them to dissociate. It was the passage of electricity through a conducting liquid medium that caused the molecules to dissociate, even when the electricity merely discharged into the air and did not pass through a "pole" or "center of action" in a voltaic cell. Second, the amount of the decomposition was found to be related directly to the amount of electricity passing through the solution.

These findings led Faraday to a new theory of electro-chemistry. The electric force, he argued, threw the molecules of a solution into a state of tension. When the force was strong enough to distort the *forces* that held the molecules together so as to permit the interaction with neighbouring particles, the tension was relieved by the migration of particles along the lines of tension, the different parts of atoms migrating in opposite directions. The amount of electricity that passed, then, was clearly related to the chemical affinities of the substances in solution. These experiments led directly to Faraday's two laws of electro-chemistry which state :

• The amount of a substance deposited on each electrode of an electrolytic cell is directly proportional to the amount of electricity passing through the cell.

- The quantities of different elements deposited by a given amount of electricity are in the ratio of their chemical *equivalent weights*.

William Sturgeon built an electric motor in 1832 and invented the *commutator*, a ring of metal-bristled brushes which allow the spinning *armature* to maintain contact with the electrical current and changed the *alternating current* to a pulsating *direct current*. He also improved the voltaic battery and worked on the theory of thermo-electricity.

Hippolyte Pixii, a French instrument maker, constructed the first *dynamo* in 1832 and later built a direct current dynamo using the commutator. This was the first practical mechanical generator of electrical current that used concepts demonstrated by Faraday.

John Daniell began experiments in 1835 in an attempt to improve the voltaic battery with its problems of being unsteady and a weak source of electrical current. His experiments soon led to remarkable results. In 1836, he invented a primary cell in which hydrogen was eliminated in the generation of the electricity. Daniell had solved the problem of polarization. In his laboratory he had learned to alloy the amalgamated zinc of Sturgeon with mercury. His version was the first of the two-fluid class battery and the first battery that produced a constant reliable source of electrical current over a long period of time.

William Grove produced the first fuel cell in 1839. He based his experiment on the fact that sending an electric current through water splits the water into its component parts of hydrogen and oxygen. So, Grove tried reversing the reaction — combining hydrogen and oxygen to produce electricity and water. Eventually the term *fuel cell* was coined in 1889 by Ludwig Mond and Charles Langer, who attempted to build the first practical device using air and industrial coal gas. He also introduced a powerful battery at the annual meeting of the British Association for the Advancement of Science in 1839. Grove's first cell consisted of zinc in diluted sulfuric acid and platinum in concentrated nitric acid, separated by a porous pot. The cell was able to generate about 12 amperes of current at about 1.8 volts. This cell had nearly double the voltage of the first Daniell cell. Grove's nitric acid cell was the favourite battery of the early American telegraph (1840–1860), because it offered strong current output.

As telegraph traffic increased, it was found that the Grove cell discharged poisonous nitrogen dioxide gas. As telegraphs became more complex, the need for a constant voltage became critical and the Grove device was limited (as the cell discharged, nitric acid was depleted and voltage was reduced). By the time of the American Civil War, Grove's battery had been replaced by the Daniell battery. In 1841 Robert Bunsen replaced the expensive platinum electrode used in Grove's battery with a carbon electrode. This led to large scale use of the "Bunsen battery" in the production of arc-lighting and in electroplating.

Wilhelm Weber developed, in 1846, the electro-dynamometer, in which a current causes a coil suspended within another coil to turn when a current is passed through both. In 1852, Weber defined the absolute unit of electrical resistance (which was named the ohm after Georg Ohm). Weber's name is now used as a unit name to describe magnetic flux, the weber.

German physicist Johann Hittorf concluded that *ion movement* caused electric current. In 1853 Hittorf noticed that some ions travelled more rapidly than others. This observation led to the concept of transport number, the rate at which particular ions carried the electric current. Hittorf measured the changes in the concentration of electrolysed solutions, computed from these the transport numbers (relative carrying capacities) of many ions, and, in 1869, published his findings governing the migration of ions.

In 1866, Georges Leclanché patented a new battery system, which was immediately successful. Leclanché's original cell was assembled in a porous pot. The positive electrode (the cathode) consisted of crushed manganese dioxide with a little carbon mixed in. The negative pole (anode) was a zinc rod. The cathode was packed into the pot, and a carbon rod was inserted to act as a current collector. The anode and the pot were then immersed in an ammonium chloride solution. The liquid acted as the electrolyte, readily seeping through the porous pot and making contact with the cathode material. Leclanché's "wet" cell became the forerunner to the world's first widely used battery, the zinc-carbon cell.

Late 19th Century Advances and the Advent of Electro-chemical Societies

In 1869 Zénobe Gramme devised his first clean direct current dynamo. His generator featured a ring armature wound with many individual coils of wire.

Svante August Arrhenius published his thesis in 1884, *Recherches sur la conductibilité galvanique des électrolytes* (Investigations on the galvanic conductivity of electrolytes). From the results of his experiments, the author concluded that electrolytes, when dissolved in water, become to varying degrees split or dissociated into positive and negative ions. The degree to which this dissociation occurred depended above all on the nature of the substance and its concentration in the solution, being more developed the greater the dilution. The ions were supposed to be the carriers of not only the electric current, as in electrolysis, but also of the chemical activity. The relation between the actual number of ions and their number at great dilution (when all the molecules were dissociated) gave a quantity of special interest ("activity constant").

The race for the commercially viable production of aluminium was won in 1886 by Paul Héroult and Charles M. Hall. The problem many researchers had with extracting aluminium was that electrolysis of an aluminium salt dissolved in water yields aluminium hydroxide. Both Hall and Héroult avoided this problem by dissolving aluminium oxide in a new solvent— fused cryolite (Na_3AlF_6).

Wilhelm Ostwald, 1909 Nobel Laureate, started his experimental work in 1875, with an investigation on the law of mass action of water in relation to the problems of chemical affinity, with special emphasis on electro-chemistry and chemical dynamics. In 1894 he gave the first modern definition of a catalyst and turned his attention to catalytic reactions. Ostwald is especially known for his contributions to the field of electro-chemistry, including important studies of the electrical conductivity and electrolytic dissociation of organic acids.

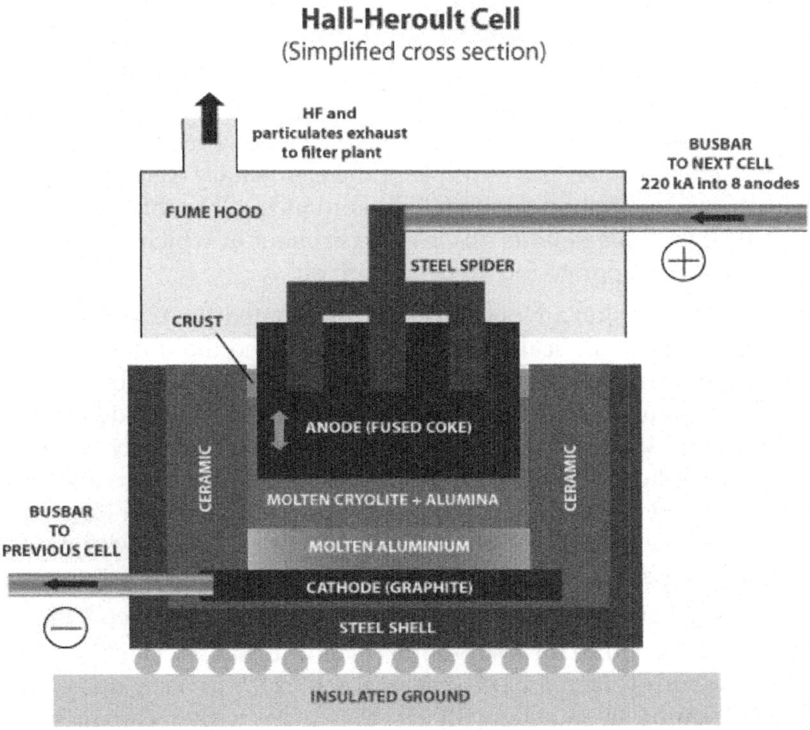

Hall-Heroult Cell
(Simplified cross section)

Fig. : A Hall-Héroult industrial cell.

Hermann Nernst developed the theory of the electromotive force of the voltaic cell in 1888. He developed methods for measuring dielectric constants and was the first to show that solvents of high dielectric constants promote the ionization of substances. Nernst's early studies in electro-chemistry were inspired by Arrhenius' dissociation theory which first recognised the importance of ions in solution. In 1889, Nernst elucidated the theory of galvanic cells by assuming an "electrolytic pressure of dissolution," which forces ions from electrodes into solution and which was opposed to the osmotic pressure of the dissolved ions. He applied the principles of thermodynamics to the chemical reactions proceeding in a battery. In that same year he showed how the characteristics of the current produced could be used to calculate the free energy change in the chemical reaction producing the current. He constructed an equation, known as Nernst Equation, which describes the relation of a battery cell's voltage to its properties.

In 1898 Fritz Haber published his textbook, *Electro-chemistry : Grundriss der technischen Elektrochemie auf theoretischer Grundlage* (The Theoretical Basis of Technical Electro-chemistry), which was based on the lectures he gave at Karlsruhe. In the preface to his book he expressed his intention to relate chemical research to industrial processes and in the same year he reported the results of his work on electrolytic oxidation and reduction, in which he showed that definite reduction products can result if the voltage at the cathode is kept constant. In 1898 he explained the reduction of nitrobenzene in stages at the cathode and this became the model for other similar reduction processes.

In 1909, Robert Andrews Millikan began a series of experiments to determine the electric charge carried by a single electron. He began by measuring the course of charged water droplets in an electrical field. The results suggested that the charge on the droplets is a multiple of the elementary electric charge, but the experiment was not accurate enough to be convincing. He obtained more precise results in 1910 with his famous oil-drop experiment in which he replaced water (which tended to evaporate too quickly) with oil.

Jaroslav Heyrovský, a Nobel laureate, eliminated the tedious weighing required by previous analytical techniques, which used the differential precipitation of mercury by measuring drop-time. In the previous method, a voltage was applied to a dropping mercury electrode and a reference electrode was immersed in a test solution. After 50 drops of mercury were collected, they were dried and weighed. The applied voltage was varied and the experiment repeated. Measured weight was plotted *versus* applied voltage to obtain the curve. In 1921, Heyrovský had the idea of measuring the current flowing through the cell instead of just studying drop-time.

On February 10, 1922, the "polarograph" was born as Heyrovský recorded the current-voltage curve for a solution of 1 mol/L NaOH. Heyrovský correctly interpreted the current increase between −1.9 and −2.0 V as being due to the deposit of Na^+ ions, forming an amalgam. Shortly thereafter, with his Japanese colleague Masuzo Shikata, he constructed the first instrument for the automatic recording of polarographic curves, which became world famous later as the polarograph.

In 1923, Johannes Nicolaus Brønsted and Thomas Martin Lowry published essentially the same theory about how acids and bases behave using electrochemical basis.

The International Society of Electro-chemistry (ISE) was founded in 1949, and some years later the first sophisticated electrophoretic apparatus was developed in 1937 by Arne Tiselius, who was awarded the 1948 Nobel prize for his work in protein electrophoresis. He developed the "moving boundary," which later would become known as *zone electrophoresis*, and used it to separate serum proteins in solution. Electrophoresis became widely developed in the 1940s and 1950s when the technique was applied to molecules ranging from the largest proteins to amino acids and even inorganic ions.

During the 1960s and 1970s quantum electro-chemistry was developed by Revaz Dogonadze and his pupils.

Chapter 2

REDOX

Redox (reduction–oxidation) reactions include all *chemical reactions* in which atoms have their *oxidation state* changed; in general, redox reactions involve the transfer of *electrons* between *species*.

This can be either a simple redox process, such as the oxidation of *carbon* to yield *carbon dioxide* (CO_2) or the reduction of carbon by *hydrogen* to yield *methane* (CH_4), or a complex process such as the oxidation of *glucose* ($C_6H_{12}O_6$) in the human body through a series of complex *electron transfer* processes.

The term "redox" comes from two concepts involved with electron transfer: reduction and oxidation. It can be explained in simple terms :

- **Oxidation** is the *loss* of *electrons* or an *increase* in oxidation state by a *molecule, atom,* or *ion.*
- **Reduction** is the *gain* of electrons or a *decrease* in oxidation state by a molecule, atom, or ion.

Although oxidation reactions are commonly associated with the formation of oxides from oxygen molecules, these are only specific examples of a more general concept of reactions involving electron transfer.

Redox reactions, or oxidation–reduction reactions, have a number of similarities to *acid–base reactions*. Like acid–base reactions, redox reactions are a matched set, that is, there cannot be an oxidation reaction without a reduction reaction happening simultaneously. The oxidation alone and the reduction alone are each called a *half–reaction*, because two half-reactions always occur together to form a whole reaction. When writing half–reactions, the gained or lost electrons are typically included explicitly in order that the *half–reaction be balanced* with respect to electric charge.

Though sufficient for many purposes, these descriptions are not precisely correct. Oxidation and reduction properly refer to *a change in oxidation state* — the actual transfer of electrons may never occur. Thus, oxidation is better defined as an *increase in oxidation state*, and reduction as a *decrease in oxidation state*. In practice, the transfer of electrons will always cause a change in oxidation state, but there are many reactions that are classed as "redox" even though no electron transfer occurs (such as those involving *covalent* bonds).

$$\text{Reduction}$$
$$\text{Oxidant} + e^- \longrightarrow \text{Product}$$
(Electrons **gained**; oxidation number **decreases**)

$$\text{Oxidation}$$
$$\text{Reductant} \longrightarrow \text{Product} + e^-$$
(Electrons **lost**; oxidation number **increases**)

Fig. : The two parts of a redox reaction.

ETYMOLOGY

"Redox" is a portmanteau of "reduction" and "oxidation".

The word *oxidation* originally implied reaction with oxygen to form an oxide, since (di)oxygen was historically the first recognized oxidizing agent. Later, the term was expanded to encompass oxygen-like substances that accomplished parallel chemical reactions. Ultimately, the meaning was generalized to include all processes involving loss of electrons.

The word *reduction* originally referred to the loss in weight upon heating a metallic ore such as a metal oxide to extract the metal. In other words, ore was "reduced" to metal. Antoine Lavoisier (1743-1794) showed that this loss of weight was due to the loss of oxygen as a gas. Later, scientists realized that the metal atom gains electrons in this process. The meaning of *reduction* then became generalized to include all processes involving gain of electrons. Even though "reduction" seems counter-intuitive when speaking of the **gain** of electrons, it might help to think of reduction as the loss of oxygen, which was its historical meaning.

The electro-chemist John Bockris has used the words *electronation* and *deelectronation* to describe reduction and oxidation processes respectively when they occur at electrodes. These words are analogous to protonation and deprotonation, but they have not been widely adopted by chemists.

The term "hydrogenation" could be used instead of reduction, since hydrogen is the reducing agent in a large number of reactions, especially in organic chemistry and bio-chemistry. But, unlike oxidation, which has been generalized beyond its root element, hydrogenation has maintained its specific connection to reactions that *add* hydrogen to another substance (*e.g.*, the hydrogenation of unsaturated fats in saturated fats, $R-CH=CH-R + H_2 \rightarrow R-CH_2-CH_2-R$).

OXIDIZING AND REDUCING AGENTS

In redox processes, the reductant transfers electrons to the oxidant. Thus, in the reaction, the reductant or *reducing agent* loses electrons and is oxidized, and the oxidant or *oxidizing agent* gains electrons and is reduced. The pair of an oxidizing and reducing agent that are involved in a particular reaction is called a **redox pair**. A **redox couple** is a reducing species and its corresponding oxidized form, *e.g.*, Fe^{2+}/Fe^{3+}.

Oxidizing Agent

An **oxidizing agent** (also **oxidant, oxidizer** or **oxidiser**) is the element or compound in an oxidation–reduction (redox) reaction that accepts an electron from another species. Because the oxidizing agent is gaining electrons (and is thus often called an **electron acceptor**), it is said to have been reduced.

The oxidizing agent itself is reduced, as it is taking electrons onto itself, but the reactant is oxidized by having its electrons taken away by the oxidizing agent. Oxygen is the prime (and eponymous) example among the varied types of oxidizing agents.

Overview

• The oxidizing agent takes electrons from another species, and thus itself is *reduced*.

• The reducing agent gives electrons to another species, and thus itself is *oxidized*.

• All atoms in a molecule can be assigned an *oxidation number*. This number changes when an oxidant acts on a substrate.

• Redox reactions occur when *oxidation states* of the reactants change.

Example of Oxidation

The formation of iron (III) oxide (rust) through the oxidation of iron;

$$4Fe + 3O_2 \rightarrow 2Fe_2O_3$$

In the above equation, the iron (Fe) has an oxidation number of 0 before and 3+ after the reaction. For oxygen (O) the oxidation number began as 0 and decreased to 2−. These changes can be viewed as two (balanced) "half–reactions" that occur concurrently :

1. Oxidation half reaction : $4Fe \rightarrow 4Fe^{3+} + 12e^-$
2. Reduction half reaction : $3O_2 + 12e^- \rightarrow 6O^{2-}$

Iron (Fe) has become oxidised because its oxidation number increased and was the reducing agent because it gave electrons to the oxygen (O). Oxygen (O) has been reduced because the oxidation number has decreased and was the oxidising agent because it took electrons from iron (Fe).

Electron Acceptor

Because the process of oxidation is so widespread (fire, explosives, chemical synthesis, corrosion), the term *oxidising agent* has acquired multiple meanings.

In one definition, an oxidising agent accepts–or gains–electrons. In this context, the reducing agent is called an **electron donor**. A classic oxidising agent is the ferrocenium ion $[Fe(C_5H_5)_2]^+$, which accepts an electron to form $Fe(C_5H_5)_2$. Of great interest to chemists are the details of the electron transfer event, which can be described as inner sphere or outer sphere.

In more colloquial usage, an oxidising agent transfers oxygen atoms to the substrate. In this context, the oxidising agent can be called an oxygenation reagent or oxygen–atom transfer agent. Examples include $[MnO_4]^-$ (permanganate), $[CrO_4]^{2-}$ (chromate), OsO_4 (osmium tetroxide), and especially $[ClO_4]^-$ (perchlorate). Notice that these species are all oxides, and are in fact polyoxides. In some cases, these oxides can also serve as electron acceptors, as illustrated by the conversion of $[MnO_4]^-$ to $[MnO_4]^{2-}$, manganate.

Dangerous Materials Definition

The dangerous materials definition of an **oxidizing agent** is a substance that is not necessarily combustible, but may, generally by yielding oxygen, cause or contribute to the combustion of other material. By this definition some materials that are classified as oxidising agents by analytical chemists are not classified as oxidising agents in a dangerous materials sense. An example is potassium dichromate, which does not pass the dangerous goods test of an oxidising agent.

The U.S. Department of Transportation defines Oxidizing agent specifically. There are two definitions for oxidizing agents governed under DOT regulations. These two are Class 5; Division 5.1 and Class 5; Division 5.2. Division 5.1 "means a material that may, generally by yielding oxygen, cause or enhance the combustion of other materials." Division 5.1 of the DOT code applies to solid oxidizers "if, when tested in accordance with the UN Manual of Tests and Criteria, its mean burning time is less than or equal to the burning time of a 3 :7 potassium bromate/ cellulose mixture." 5.1 of the DOT code applies to liquid oxidizers "if, when tested in accordance with the UN Manual of Tests and Criteria, it spontaneously ignites or its mean time for a pressure rise from 690 kPa to 2070 kPa gauge is less then the time of a 1 :1 nitric acid (65 per cent)/cellulose mixture."

Common Oxidizing Agents

- Oxygen (O_2)
- Ozone (O_3)
- Hydrogen peroxide (H_2O_2) and other inorganic peroxides
- Fluorine (F_2), chlorine (Cl_2), and other halogens
- Nitric acid (HNO_3) and nitrate compounds
- Sulfuric acid (H_2SO_4)
- Peroxydisulfuric acid $(H_2S_2O_8)$
- Peroxymonosulfuric acid (H_2SO_5)
- Chlorite, chlorate, perchlorate, and other analogous halogen compounds

- Hypochlorite and other hypohalite compounds, including household bleach (NaClO)
- Hexavalent chromium compounds such as chromic and dichromic acids and chromium trioxide, pyridinium chlorochromate (PCC), and chromate/dichromate compounds
- Permanganate compounds such as potassium permanganate
- Sodium perborate
- Nitrous oxide (N_2O)
- Silver oxide (Ag_2O)
- Osmium tetroxide (OsO_4)
- Potassium nitrate (KNO_3), the oxidizer in black powder
- Tollens' reagent
- 2,2'–Dipyridyldisulfide (DPS).

Common Oxidizing Agents and Their Products

Agent	Product(s)
O_2 oxygen	Various, including the oxides H_2O and CO_2
O_3 ozone	Various, including ketones, aldehydes, and H_2O;
F_2 fluorine	F^-
Cl_2 chlorine	Cl^-
Br_2 bromine	Br^-
I_2 iodine	I^-, I_3^-
ClO^- hypochlorite	Cl^-, H_2O
ClO_3^- chlorate	Cl^-, H_2O
HNO_3 nitric acid	NO nitric oxide NO_2 nitrogen dioxide
Hexavalent chromium CrO_3 chromium trioxide CrO_4^{2-} chromate $Cr_2O_7^{2-}$ dichromate	Cr^{3+}, H_2O
MnO_4^- permanganate MnO_4^{2-} manganate	Mn^{2+} (acidic) or MnO_2 (basic)
H_2O_2, other peroxides	Various, including oxides and H_2O

Reducing Agent

A **reducing agent** (also called a **reductant** or **reducer**) is an element or compound that loses (or "donates") an *electron* to another *chemical species* in a *redox* chemical reaction. Since the reducing agent is losing electrons, it is said to have been oxidized.

If any chemical is an electron donor (reducing agent), another must be an electron recipient (*oxidizing agent*). A reducing agent is oxidized because it loses electrons in the redox reaction. Thus reducers are "oxidized" by oxidizers and oxidizers are "reduced" by reducers; reducers are by themselves reduced (have more electrons) and oxidizers are by themselves oxidized (have fewer electrons). A reducing agent typically is in one of its lower possible *oxidation states* and is known as the electron donor. Examples of reducing agents include the earth metals, *formic acid,* and *sulfite* compounds.

For example, consider the overall reaction for aerobic *cellular respiration* :

$$C_6H_{12}O_6(s) + 6O_2(g) \rightarrow 6CO_2(g) + 6H_2O(l)$$

The *oxygen* (O_2) is being reduced, so it is the oxidizing agent. The *glucose* ($C_6H_{12}O_6$) is being oxidized, so it is the reducing agent.

In *organic chemistry,* reduction more specifically refers to the addition of hydrogen to a molecule, though the aforementioned definition still applies. For example, *benzene* is reduced to *cyclohexane* in the presence of a platinum *catalyst* :

$$C_6H_6 + 3 H_2 \rightarrow C_6H_{12}$$

In organic chemistry, good reducing agents are reagents that deliver H_2.

Characteristics of Reducing Agents

Consider the following reaction :

$$2 [Fe(CN)_6]^{4-} + Cl$$

$$2 \rightarrow 2 [Fe(CN)_6]^{3-} + 2 Cl-$$

The reducing agent in this reaction is ferrocyanide ($[Fe(CN)_6]^{4-}$). It donates an electron, becoming oxidized to ferricyanide ($[Fe(CN)_6]^{3-}$). Simultaneously, the oxidizer chlorine is reduced to chloride.

Strong reducing agents easily lose (or donate) electrons. An atom with a relatively large atomic radius tends to be a better reductant. In such species, the distance from the nucleus to the valence electrons is so long that these electrons are not strongly attracted. These elements tend to be strong reducing agents. Good reducing agents tend to consist of atoms with a low electronegativity, the ability of an atom or molecule to attract bonding electrons, and species with relatively small ionization energies serve as good reducing agents too. "The measure of a material to oxidize or lose electrons is known as its oxidation potential". The table below shows a few reduction potentials that could easily be changed to oxidation potential by simply reversing the sign. Reducing agents can be ranked by increasing strength by ranking their oxidation potentials. The reducing agent is stronger when it has a more positive oxidation potential and weaker when it has a negative oxidation potential. The following table provides the reduction potentials of the indicated reducing agent at 25°C.

Oxidizing agent	Reducing agent	Reduction potential (V)
$Li^+ + e^- =$	Li	−3.04
$Na^+ + e^- =$	Na	−2.71

(Contd...)

(Contd...)

Oxidizing agent	Reducing agent	Reduction potential (V)
$Mg^{2+} + 2e^- =$	Mg	−2.38
$Al^{3+} + 3e^- =$	Al	−1.66
$2H_2O_{(l)} + 2e^- =$	$H_{2(g)} + 2OH^-$	−0.83
$Cr^{3+} + 3e^- =$	Cr	−0.74
$Fe^{2+} + 2e^- =$	Fe	−0.44
$2H^+ + 2e^- =$	H_2	0.00
$Sn^{4+} + 2e^- =$	Sn^{2+}	+0.15
$Cu^{2+} + e^- =$	Cu^+	+0.16
$Ag^+ + e^- =$	Ag	+0.80
$Br_2 + 2e^- =$	$2Br^-$	+1.07
$Cl_2 + 2e^- =$	$2Cl^-$	+1.36
$MnO_4^- + 8H^+ + 5e^- =$	$Mn^{2+} + 4H_2O$	+1.49
$F_2 + 2e^- =$	$2F^-$	+2.87

To tell which is the strongest reducing agent, one can change the sign of its respective reduction potential to make it oxidation potential. The bigger the number, the stronger the reducing agent. For example, among Na, Cr, Cu and Cl^-, Na is the strongest reducing agent and Cl^- is the weakest one.

Common reducing agents include metals potassium, calcium, barium, sodium and magnesium, and also compounds that contain the H^- ion, those being NaH, LiH, $LiAlH_4$ and CaH_2.

Some elements and compounds can be both reducing or oxidizing agents. Hydrogen gas is a reducing agent when it reacts with non–metals and an oxidizing agent when it reacts with metals.

$$2\,Li_{(s)} + H_2(g) \rightarrow 2\,LiH_{(s)}$$

Hydrogen acts as an oxidizing agent because it accepts an electron donation from lithium, which causes Li to be oxidized.

Half reactions : $2\,Li^0_{(s)} \rightarrow 2\,Li^+_{(s)} + 2\,e^- ::::: H^0_{2\,(g)} + 2\,e^- \rightarrow 2\,H^-_{(g)}$

$$H_2(g) + F_2(g) \rightarrow 2\,HF(g)$$

Hydrogen acts as a reducing agent because it donates its electrons to fluorine, which allows fluorine to be reduced.

Half reactions : $H^0_{2\,(g)} \rightarrow 2\,H^+_{(g)} + 2\,e^- ::::: F^0_{2\,(g)} + 2\,e^- \rightarrow 2\,F^-_{(g)}$

Importance of Reducing and Oxidizing Agents

Reducing agents and oxidizing agents are the ones responsible for corrosion, which is the "degradation of metals as a result of electro-chemical activity". Corrosion requires an anode and cathode to take place. The anode is an element that loses electrons (reducing agent), thus oxidation always occurs in the anode, and the cathode is an element that gains electrons (oxidizing agent), thus reduction always occurs in the cathode. Corrosion occurs whenever there's a difference in oxidation potential. When this is present, the anode metal begins deteriorating, given there is an electrical connection and the presence of an electrolyte.

Example of Redox Reaction

The formation of iron(III) oxide;

$$4Fe + 3O_2 \rightarrow 2Fe_2^{3+}O_3^{6-}$$

In the above equation, the Iron (Fe) has an oxidation number of 0 before and 3+ after the reaction. For oxygen (O) the oxidation number began as 0 and decreased to 2–. These changes can be viewed as two "half–reactions" that occur concurrently :

1. Oxidation half reaction : $Fe^0 \rightarrow Fe^{3+} + 3e^-$
2. Reduction half reaction : $O_2 + 4e^- \rightarrow 2\ O^{2-}$

Iron (Fe) has been oxidized because the oxidation number increased. Iron is the reducing agent because it gave electrons to the oxygen (O_2). Oxygen (O_2) has been reduced because the oxidation number has decreased and is the oxidizing agent because it took electrons from iron (Fe). ferric

Common Reducing Agents

- Lithium aluminum hydride ($LiAlH_4$)
- Nascent (atomic) hydrogen
- Sodium amalgam
- Diborane
- Sodium borohydride ($NaBH_4$)
- Compounds containing the Sn^{2+} ion, such as tin (II) chloride
- Sulfite compounds
- Hydrazine (Wolff–Kishner reduction)
- Zinc–mercury amalgam ($Zn(Hg)$) (Clemmensen reduction)
- Diisobutylaluminum hydride (DIBAL-H)
- Lindlar catalyst
- Oxalic acid ($C_2H_2O_4$)
- Formic acid (HCOOH)
- Ascorbic acid ($C_6H_8O_6$)
- Phosphites, hypophosphites, and phosphorous acid
- Dithiothreitol (DTT)–used in bio-chemistry labs to avoid S–S bonds
- Compounds containing the Fe^{2+} ion, such as iron (II) sulfate
- Carbon monoxide(CO)
- Carbon (C)

Several Common Reducing Agents and their Products

Agent	Product
Hydrogen	H^+, H_2O
NADH	NAD^+
Metals	metal ions
Hydrocarbons	CO_2 carbon dioxide, H_2O

STANDARD ELECTRODE POTENTIALS (REDUCTION POTENTIALS)

Each half–reaction has a *standard electrode potential* (E^0_{cell}), which is equal to the potential difference (or voltage) (E^0_{cell}) at equilibrium under standard conditions of an electro-chemical cell in which the cathode reaction is the half–reaction considered, and the anode is a standard hydrogen electrode where hydrogen is oxidized : $\frac{1}{2} H_2 \rightarrow H^+ + e^-$.

The electrode potential of each half–reaction is also known as its *reduction potential* E^0_{red}, or potential when the half–reaction takes place at a cathode. The reduction potential is a measure of the tendency of the oxidizing agent to be re-duced. Its value is zero for $H^+ + e^- \rightarrow \frac{1}{2} H_2$ by definition, positive for oxidizing agents stronger than H^+ (*e.g.,* +2.866 V for F_2) and negative for oxidizing agents that are weaker than H^+ (*e.g.,*–0.763 V for Zn^{2+}).

For a redox reaction that takes place in a cell, the potential difference $E^0_{cell} = E^0_{cathode} - E^0_{anode}$

However, the potential of the reaction at the anode was sometimes expressed as an *oxidation potential,* $E^0_{ox} = -E^0$. The oxidation potential is a measure of the ten-dency of the reducing agent to be oxidized, but does not represent the physical potential at an electrode. With this notation, the cell voltage equation is written with a plus sign $E^0_{cell} = E^0_{cathode} + E^0_{ox (anode)}$.

Examples of Redox Reactions

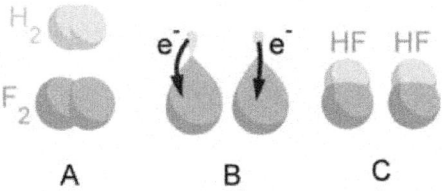

Fig. : Illustration of a redox reaction.

A good example is the reaction between hydrogen and fluorine in which hydrogen is being oxidized and fluorine is being reduced :

$H_2 + F_2 \rightarrow 2 HF$

We can write this overall reaction as two half–reactions :

the oxidation reaction :

$$H_2 \rightarrow 2\,H^+ + 2\,e^-$$

and the reduction reaction :

$$F_2 + 2\,e^- \rightarrow 2\,F^-$$

Analyzing each half–reaction in isolation can often make the overall chemical process clearer. Because there is no net change in charge during a redox reaction, the number of electrons in excess in the oxidation reaction must equal the number consumed by the reduction reaction (as shown above).

Elements, even in molecular form, always have an oxidation state of zero. In the first half–reaction, hydrogen is oxidized from an oxidation state of zero to an oxidation state of +1. In the second half–reaction, fluorine is reduced from an oxidation state of zero to an oxidation state of −1.

When adding the reactions together the electrons are canceled :

$$H_2 \rightarrow 2\,H^+ + 2\,e^-$$

$$F_2 + 2\,e^- \rightarrow 2\,F^-$$

$$H_2 + F2 \rightarrow 2\,H^+ + 2\,F^-$$

And the ions combine to form hydrogen fluoride :

$$2\,H^+ + 2\,F^- \rightarrow 2\,HF$$

The overall reaction is :

$$H_2 + F_2 \rightarrow 2\,HF$$

Metal Displacement

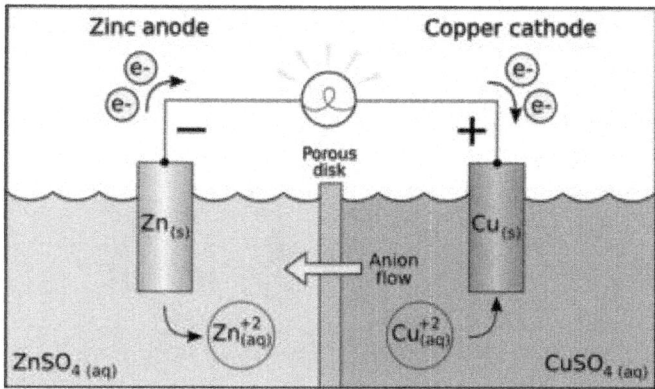

Fig. : A redox reaction is the force behind an electro–chemical cell like the Galvanic cell pictured. The battery is made out of a zinc electrode in a $ZnSO_4$ solution connected with a wire and a porous disk to a copper electrode in a $CuSO_4$ solution.

In this type of reaction, a metal atom in a compound (or in a solution) is replaced by an atom of another metal. For example, copper is deposited when zinc metal is placed in a copper (II) sulfate solution :

$Zn(s) + CuSO_4(aq) \rightarrow ZnSO_4(aq) + Cu(s)$

In the above reaction, zinc metal displaces the copper (II) ion from copper sulfate solution and thus liberates free copper metal.

The ionic equation for this reaction is :

$Zn + Cu^{2+} \rightarrow Zn^{2+} + Cu$

As two half-reactions, it is seen that the zinc is oxidized :

$Zn \rightarrow Zn^{2+} + 2\,e^-$

And the copper is reduced :

$Cu^{2+} + 2\,e^- \rightarrow Cu.$

Other Examples

- The reduction of nitrate to nitrogen in the presence of an acid (denitrification) :

  $2\,NO_3^- + 10\,e^- + 12\,H^+ \rightarrow N_2 + 6\,H_2O$

- The combustion of hydrocarbons, such as in an internal combustion engine, which produces water, carbon dioxide, some partially oxidized forms such as carbon monoxide, and heat energy. Complete oxidation of materials containing carbon produces carbon dioxide.

- In organic chemistry, the stepwise oxidation of a hydrocarbon by oxygen produces water and, successively, an alcohol, an aldehyde or a ketone, a carboxylic acid, and then a peroxide.

Corrosion and Rusting

- The term *corrosion* refers to the electro-chemical oxidation of metals in reaction with an oxidant such as oxygen. *Rusting*, the formation of *iron oxides*, is a well-known example of electro-chemical corrosion; it forms as a result of the oxidation of *iron* metal. Common rust often refers to *iron (III) oxide*, formed in the following chemical reaction :

  $4Fe + 3O_2 \rightarrow 2Fe_2O_3$

- The oxidation of iron (II) to iron (III) by *hydrogen peroxide* in the presence of an acid :

  $Fe^{2+} \rightarrow Fe^{3+} + e^-$

  $H_2O_2 + 2\,e^- \rightarrow 2\,OH^-$

  Overall equation :

  $2\,Fe^{2+} + H_2O_2 + 2\,H^+ \rightarrow 2\,Fe^{3+} + 2\,H_2O.$

REDOX REACTIONS IN INDUSTRY

Cathodic protection is a technique used to control the corrosion of a metal surface by making it the cathode of an electro-chemical cell. A simple method of protection connects protected metal to a more easily corroded "sacrificial anode" to act as the anode. The sacrificial metal instead of the protected metal, then, corrodes. A common application of cathodic protection is in galvanized steel, in which a sacrificial coating of zinc on steel parts protects them from rust.

The primary process of reducing ore at high temperature to produce metals is known as smelting.

Oxidation is used in a wide variety of industries such as in the production of cleaning products and oxidizing ammonia to produce nitric acid, which is used in most fertilizers.

Redox reactions are the foundation of electro-chemical cells.

The process of electroplating uses redox reactions to coat objects with a thin layer of a material, as in chrome–plated automotive parts, silver plating cutlery, and gold–plated jewelry.

The production of compact discs depends on a redox reaction, which coats the disc with a thin layer of metal film.

REDOX REACTIONS IN BIOLOGY

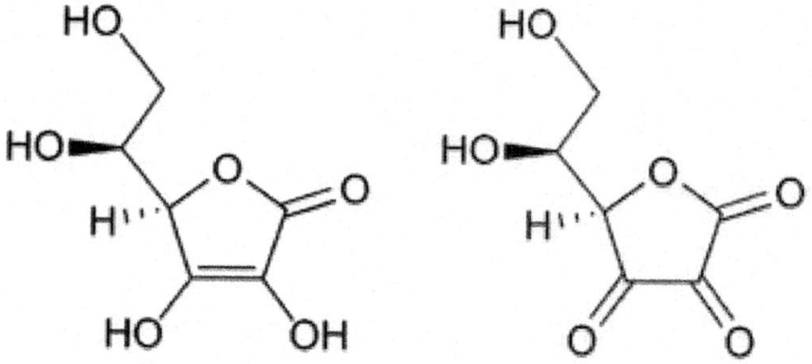

Top : ascorbic acid (reduced form of Vitamin C)

Bottom : dehydroascorbic acid (oxidized form of Vitamin C).

Many important biological processes involve redox reactions.

Cellular respiration, for instance, is the oxidation of glucose ($C_6H_{12}O_6$) to CO_2 and the reduction of oxygen to water. The summary equation for cell respiration is :

$$C_6H_{12}O_6 + 6\,O_2 \rightarrow 6\,CO_2 + 6\,H_2O$$

The process of cell respiration also depends heavily on the reduction of NAD^+ to NADH and the reverse reaction (the oxidation of NADH to NAD^+). Photo-synthesis and cellular respiration are complementary, but photo-synthesis is not the reverse of the redox reaction in cell respiration :

$$6\,CO_2 + 6\,H_2O + \text{light energy} \rightarrow C_6H_{12}O_6 + 6\,O_2$$

Biological energy is frequently stored and released by means of redox reactions. Photo-synthesis involves the reduction of carbon dioxide into sugars and the oxidation of water into molecular oxygen. The reverse reaction, respiration, oxidizes sugars to produce carbon dioxide and water. As intermediate steps, the reduced carbon compounds are used to reduce nicotinamide adenine dinucleotide (NAD^+), which then contributes to the creation of a proton gradient, which drives the synthesis of adenosine triphosphate (ATP) and is maintained by the reduction of oxygen. In animal cells, mitochondria perform similar functions.

Free radical reactions are redox reactions that occur as a part of homeostasis and killing micro-organisms, where an electron detaches from a molecule and then reattaches almost instantaneously. Free radicals are a part of redox molecules and can become harmful to the human body if they do not reattach to the redox molecule or an anti-oxidant. Unsatisfied free radicals can spur the mutation of cells they encounter and are, thus, causes of cancer.

The term **redox state** is often used to describe the balance of GSH/GSSG, NAD^+/NADH and $NADP^+$/NADPH in a biological system such as a cell or organ. The redox state is reflected in the balance of several sets of metabolites (*e.g.*, lactate and pyruvate, beta-hydroxybutyrate, and acetoacetate), whose interconversion is dependent on these ratios. An abnormal redox state can develop in a variety of deleterious situations, such as hypoxia, shock, and sepsis. Redox mechanism also control some cellular processes. Redox proteins and their genes must be co-located for redox regulation according to the CoRR hypothesis for the function of DNA in mitochondria and chloroplasts.

Redox cycling

A wide variety of aromatic compounds are enzymatically reduced to form free radicals that contain one more electron than their parent compounds. In general, the electron donor is any of a wide variety of flavoenzymes and their coenzymes. Once formed, these anion free radicals reduce molecular oxygen to superoxide, and regenerate the unchanged parent compound. The net reaction is the oxidation of the flavoenzyme's coenzymes and the reduction of molecular oxygen to form superoxide. This catalytic behaviour has been described as futile cycle or redox cycling.

Examples of redox cycling-inducing molecules are the herbicide paraquat and other viologens and quinones such as menadione.

REDOX REACTIONS IN GEOLOGY

In geology, redox is important to both the formation of minerals and the mobilization of minerals, and is also important in some depositional environments. In general, the redox state of most rocks can be seen in the colour of the rock. The rock forms in oxidizing conditions, giving it a red colour. It is then "bleached" to a green—or sometimes white—form when a reducing fluid passes through the rock. The reduced fluid can also carry uranium-bearing minerals. Famous examples of redox conditions affecting geological processes include uranium deposits and Moqui marbles.

BALANCING REDOX REACTIONS

Describing the overall electro-chemical reaction for a redox process requires a *balancing* of the component half–reactions for oxidation and reduction. In general, for reactions in aqueous solution, this involves adding H^+, OH^-, H_2O, and electrons to compensate for the oxidation changes.

Acidic Media

In acidic media, H+

ions and water are added to half–reactions to balance the overall reaction.

For instance, when manganese (II) reacts with sodium bismuthate :

Unbalanced reaction : Mn2+

(aq) + NaBiO

3(s) → Bi3+

(aq) + MnO

4⁻ (aq)

Oxidation :   4 H

2O(l) + Mn2+

(aq) → MnO–

4(aq) + 8 H+

(aq) + 5 e⁻

Reduction :   2 e⁻ + 6 H+

+ BiO–

3(s) → Bi3+

(aq) + 3 H

2O(l)

The reaction is balanced by scaling the two half–cell reactions to involve the same number of electrons (multiplying the oxidation reaction by the number of electrons in the reduction step and *vice versa*) :

8 H

2O(l) + 2 Mn2+

(aq) → 2 MnO–

4(aq) + 16 H+

(aq) + 10 e⁻

10 e⁻ + 30 H+

+ 5 BiO–

3(s) → 5 Bi3+

(aq) + 15 H

2O(l)

Adding these two reactions eliminates the electrons terms and yields the balanced reaction :

14 H+

(aq) + 2 Mn2+

(aq) + 5 NaBiO

3(s) → 7 H

2O(l) + 2 MnO−

4(aq) + 5 Bi3+

(aq) + 5 Na+

(aq)

Basic Media

In basic media, OH⁻ ions and water are added to half-reactions to balance the overall reaction.

For example, in the reaction between potassium permanganate and sodium sulfite :

Unbalanced reaction : KMnO

4 + Na

2SO

3 + H

2O → MnO

2 + Na

2SO

4 + KOH

Reduction : 3 e⁻ + 2 H

2O + MnO

4⁻ → MnO

2 + 4 OH⁻

Oxidation : 2 OH⁻ + SO

3²⁻ → SO

4²⁻ + H

2O + 2 e⁻

Balancing the number of electrons in the two half–cell reactions gives :

$6\,e^- + 4\,H$

$2O + 2\,MnO$

$4^- \rightarrow 2\,MnO$

$2 + 8\,OH^-$

$6\,OH^- + 3\,SO$

$3^{2-} \rightarrow 3\,SO$

$4^{2-} + 3\,H$

$2O + 6\,e^-$

Adding these two half-cell reactions together gives the balanced equation :

$2\,KMnO$

$4 + 3\,Na$

$2SO$

$3 + H$

$2O \rightarrow 2\,MnO$

$2 + 3\,Na$

$2SO$

$4 + 2\,KOH$

MEMORY AIDS

The key terms involved in redox are often confusing to students. For example, an element that is oxidized loses electrons; however, that element is referred to as the reducing agent. Likewise, an element that is reduced gains electrons and is referred to as the oxidizing agent. Acronyms or mnemonics are commonly used to help remember what is happening :

- "OIL RIG" — Oxidation Is Loss of electrons, Reduction Is Gain of electrons.
- "LEO the lion says GER" — Loss of Electrons is Oxidation, Gain of Electrons is Reduction.
- "LEORA says GEROA" — Loss of Electrons is Oxidation (Reducing Agent) and Gain of Electrons is Reduced (Oxidizing Agent).
- "RED CAT" and "AN OX" — Reduction occurs at the Cathode and the Anode is for Oxidation.

BALANCING REDOX REACTIONS

Chemical Equation

A **chemical equation** is the symbolic representation of a chemical reaction wherein the reactant entities are given on the left-hand side and the product entities on the right-hand side. The coefficients next to the symbols and formulae of entities are the absolute values of the stoichiometric numbers. The first chemical equation was diagrammed by Jean Beguin in 1615.

Form

A chemical equation consists of the chemical formulas of the reactants (the starting substances) and the chemical formula of the products (substances formed in the chemical reaction). The two are separated by an arrow symbol ($\rightarrow$, usually read as "yields") and each individual substance's chemical formula is separated from others by a plus sign.

As an example, the equation for the reaction of hydrochloric acid with sodium can be denoted :

$2\,HCl + 2\,Na \rightarrow 2\,NaCl + H_2$

This equation would be read as "two HCl plus two Na yields two NaCl and H two." But, for equations involving complex chemicals, rather than reading the letter and its subscript, the chemical formulas are read using IUPAC nomenclature. Using IUPAC nomenclature, this equation would be read as "hydrochloric acid plus sodium yields sodium chloride and hydrogen gas."

This equation indicates that sodium and HCl react to form NaCl and H_2. It also indicates that two sodium molecules are required for every two hydrochloric acid molecules and the reaction will form two sodium chloride molecules and one diatomic molecule of hydrogen gas molecule for every two hydrochloric acid and two sodium molecules that react. The stoichiometric coefficients (the numbers in front of the chemical formulas) result from the law of conservation of mass and the law of conservation of charge.

Common Symbols

Symbols are used to differentiate between different types of reactions. To denote the type of reaction :

- "=" symbol is used to denote a stoichiometric relation.
- "$\rightarrow$" symbol is used to denote a net forward reaction.
- "$\rightleftharpoons$" symbol is used to denote a reaction in both directions.
- "$\rightleftharpoons$" symbol is used to denote an equilibrium.

Physical state of chemicals is also very commonly stated in parentheses after the chemical symbol, especially for ionic reactions. When stating physical state, (s) denotes a solid, (l) denotes a liquid, (g) denotes a gas and (aq) denotes an aqueous solution.

If the reaction requires energy, it is indicated above the arrow. A capital Greek letter delta (Δ) is put on the reaction arrow to show that energy in the form of heat is added to the reaction. *hv* is used if the energy is added in the form of light.

Balancing Chemical Equations

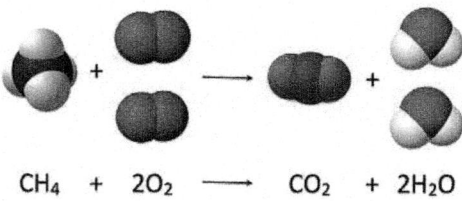

$$CH_4 \;+\; 2O_2 \;\longrightarrow\; CO_2 \;+\; 2H_2O$$

Fig. : As seen from the equation CH.

4 + 2 O

2 → CO

2 + 2 H

2O, a coefficient of 2 must be placed before the oxygen gas on the reactants side and before the water on the products side in order for, as per the law of conservation of mass, the quantity of each element does not change during the reaction

The law of conservation of mass dictates that the quantity of each element does not change in a chemical reaction. Thus, each side of the chemical equation must represent the same quantity of any particular element. Likewise, the charge is conserved in a chemical reaction. Therefore, the same charge must be present on both sides of the balanced equation.

One balances a chemical equation by changing the scalar number for each chemical formula. Simple chemical equations can be balanced by inspection, that is, by trial and error. Another technique involves solving a system of linear equations.

Balanced equations are written with smallest whole–number coefficients. If there is no coefficient before a chemical formula, the coefficient 1 is understood.

The method of inspection can be outlined as putting a coefficient of 1 in front of the most complex chemical formula and putting the other coefficients before everything else such that both sides of the arrows have the same number of each atom. If any fractional coefficient exist, multiply every coefficient with the smallest number required to make them whole, typically the denominator of the fractional coefficient for a reaction with a single fractional coefficient.

As an example, seen in the above image, the burning of methane would be balanced by putting a coefficient of 1 before the CH_4 :

$$1\,CH_4 + O_2 \rightarrow CO_2 + H_2O$$

Since there is one carbon on each side of the arrow, the first atom (carbon) is balanced.

Looking at the next atom (hydrogen), the right–hand side has two atoms, while the left–hand side has four. To balance the hydrogens, 2 goes in front of the H_2O, which yields :

$$1\,CH_4 + O_2 \rightarrow CO_2 + 2\,H_2O$$

Inspection of the last atom to be balanced (oxygen) shows that the right-hand side has four atoms, while the left-hand side has two. It can be balanced by putting a 2 before O_2, giving the balanced equation :

$$CH_4 + 2 O_2 \rightarrow CO_2 + 2 H_2O$$

This equation does not have any coefficients in front of CH_4 and CO_2, since a coefficient of 1 is dropped.

Ionic Equations

An ionic equation is a chemical equation in which electrolytes are written as dissociated ions. Ionic equations are used for single and double displacement reactions that occur in aqueous solutions. For example in the following precipitation reaction :

$$CaCl_2(aq) + 2 AgNO_3(aq) \rightarrow Ca(NO_3)_2(aq) + 2 AgCl(s)$$

the full ionic equation is :

$$Ca^{2+}(aq) + 2 Cl^-(aq) + 2 Ag^+(aq) + 2 NO_3^-(aq) \rightarrow Ca^{2+}(aq) + 2 NO_3^-(aq) + 2 AgCl(s)$$

In this reaction, the Ca^{2+} and the NO_3^- ions remain in solution and are not part of the reaction. That is, these ions are identical on both the reactant and product side of the chemical equation. Because such ions do not participate in the reaction, they are called spectator ions. A *net ionic* equation is the full ionic equation from which the spectator ions have been removed. The net ionic equation of the proceeding reactions is :

$$2 Cl^-(aq) + 2 Ag^+(aq) \rightarrow 2 AgCl(s)$$

or, in *reduced* balanced form,

$$Ag^+(aq) + Cl^-(aq) \rightarrow AgCl(s)$$

In a neutralization or acid/base reaction, the net ionic equation will usually be :

$$H^+(aq) + OH^-(aq) \rightarrow H_2O(l)$$

There are a few acid/base reactions that produce a precipitate in addition to the water molecule shown above. An example is the reaction of barium hydroxide with phosphoric acid, which produces not only water but also the insoluble salt barium phosphate. In this reaction, there are no spectator ions, so the net ionic equation is the same as the full ionic equation.

$$3 Ba(OH)_2(aq) + 2 H_3PO_4(aq) \rightarrow 6 H_2O(l) + Ba_3(PO_4)_2(s)$$
$$3 Ba^{2+}(aq) + 6 OH^-(aq) + 6 H^+(aq) + 2 PO_4^{3-}(aq) \rightarrow 6 H_2O(l) + Ba_3(PO_4)_2(s)$$

Double displacement reactions that feature a carbonate reacting with an acid have the net ionic equation :

$$2 H^+(aq) + CO_3^{2-}(aq) \rightarrow H_2O(l) + CO_2(g)$$

If every ion is a "spectator ion" then there was no reaction, and the net ionic equation is null.

ELECTRO-CHEMICAL CELL

An electro-chemical cell is a device capable of either deriving electrical energy from chemical reactions or facilitating chemical reactions through the introduction of electrical energy. A common example of an electro-chemical cell is a standard 1.5-volt "battery". (Actually a single "Galvanic cell"; a battery properly consists of multiple cells, connected in either parallel or series pattern.)

Half-cells

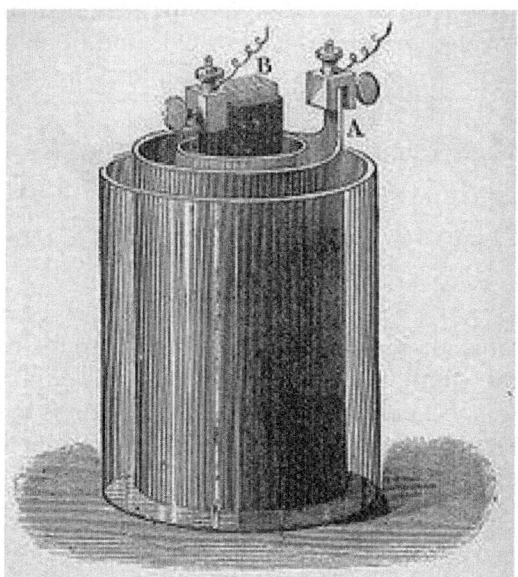

Fig. : The Bunsen cell, invented by Robert Bunsen.

An electro-chemical cell consists of two half-cells. Each *half-cell* consists of an electrode and an electrolyte. The two half-cells may use the same electrolyte, or they may use different electrolytes. The chemical reactions in the cell may involve the electrolyte, the electrodes, or an external substance (as in fuel cells that may use hydrogen gas as a reactant). In a full electro-chemical cell, species from one half-cell lose electrons (oxidation) to their electrode while species from the other half-cell gain electrons (reduction) from their electrode. A *salt bridge* (*e.g.,* filter paper soaked in KNO_3) is often employed to provide ionic contact between two half-cells with different electrolytes, to prevent the solutions from mixing and causing unwanted side reactions. As electrons flow from one half-cell to the other, a difference in charge is established. If no salt bridge were used, this charge difference would prevent further flow of electrons. A salt bridge allows the flow of ions to maintain a balance in charge between the oxidation and reduction vessels while keeping the contents of each separate. Other devices for achieving separation of solutions are porous pots and gelled solutions. A porous pot is used in the Bunsen cell (right).

Equilibrium Reaction

Each half-cell has a characteristic voltage. Different choices of substances for each half-cell give different potential differences. Each reaction is undergoing an equilibrium reaction between different oxidation states of the ions : When equilibrium is reached, the cell cannot provide further voltage. In the half-cell that is undergoing oxidation, the closer the equilibrium lies to the ion/atom with the more positive oxidation state the more potential this reaction will provide. Likewise, in the reduction reaction, the closer the equilibrium lies to the ion/atom with the more *negative* oxidation state the higher the potential.

Cell Potential

The cell potential can be predicted through the use of electrode potentials (the voltages of each half-cell). These half-cell potentials are derived from the assignment of 0 volts to the standard hydrogen electrode (SHE). The difference in voltage between electrode potentials gives a prediction for the potential measured When calculating the difference in voltage, one must first manipulate the half-cell reactions to obtain a balanced oxidation–reduction equation.

1. Reverse the reduction reaction with the smallest potential (to create an oxidation reaction/overall positive cell potential)
2. Half-reactions must be multiplied by integers to achieve electron balance.

An important note with this is that the cell potential does not change when the reaction is multiplied.

Cell potentials have a possible range of about zero to 6 volts. Cells using water-based electrolytes are usually limited to cell potentials less than about 2.5 volts, because the very powerful oxidizing and reducing agents that would be required to produce a higher cell potential tend to react with the water.

STANDARD ELECTRODE POTENTIAL

In *electro-chemistry*, the **standard electrode potential**, abbreviated $E°$ or $E^{\ominus}$ (with a superscript *plimsoll* character, pronounced "standard" or "*nought*"), is the measure of *individual potential* of a reversible electrode at *standard state*, which is with solutes at an effective concentration of 1 mol dm^{-3}, and gases at a pressure of 1 atm. The reduction potential is an *intensive property*. The values are most often tabulated at 25°C. The basis for an *electro-chemical cell* such as the *galvanic cell* is always a *redox reaction* which can be broken down into two *half-reactions* : oxidation at anode (loss of electron) and *reduction* at cathode (gain of electron). *Electricity* is generated due to *electric potential* difference between two electrodes. This potential difference is created as a result of the difference between individual potentials of the two metal *electrodes* with respect to the *electrolyte*. (Reversible electrode An electrode that owes its potential to unit charges of a reversible nature, in contrast to electrodes used in electroplating and destroyed during their use).

Although the overall potential of a cell can be measured, there is no simple way to accurately measure the *electrode/electrolyte potentials* in isolation. The electric potential also varies with temperature, concentration and pressure. Since the oxidation potential of a half–reaction is the negative of the reduction potential in a redox reaction, it is sufficient to calculate either one of the potentials. Therefore, standard electrode potential is commonly written as standard reduction potential.

Calculation of Standard Electrode Potentials

The electrode potential may not be obtained empirically. The galvanic cell potential results from a *pair* of electrodes. Thus, only one empirical value is available in a pair of electrodes and it is not possible to determine the value for each electrode in the pair using the empirically obtained galvanic cell potential. A reference electrode, standard hydrogen electrode (SHE), for which the potential is *defined* or agreed upon by convention, needed to be established. In this case SHE is set to 0.00 V and any electrode, for which the electrode potential is not yet known, can be paired with SHE — to form a galvanic cell — and the galvanic cell potential gives the unknown electrode's potential. Using this process, any electrode with an unknown potential can be paired with either the SHE or another electrode for which the potential has already been derived and that unknown value can be established.

Since the electrode potentials are conventionally defined as reduction potentials, the sign of the potential for the metal electrode being oxidized must be reversed when calculating the overall cell potential. Note that the electrode potentials are independent of the number of electrons transferred — that is, they are set to one mole of electrons transferred — and so the two electrode potentials can be simply combined to give the overall *cell* potential even if different numbers of electrons are involved in the two electrode reactions.

For practical measurements, the electrode in question is connected to the positive terminal of the electrometer, while SHE is connected to the negative terminal.

Standard Reduction Potential Table

The larger the value of the standard reduction potentials, the easier it is for the element to be reduced (accept electrons); in other words, they are better oxidizing agents. For example, F_2 has 2.87 V and Li^+ has −3.05 V. F reduces easily and is therefore a good oxidizing agent. In contrast, $Li_{(s)}$ would rather undergo oxidation (hence a good reducing agent). Thus Zn^{2+} whose standard reduction potential is −0.76 V can be oxidized by any other electrode whose standard reduction potential is greater than −0.76 V (*e.g.* $H^+(0\ V)$, $Cu^{2+}(0.16\ V)$, $F_2(2.87\ V)$) and can be reduced by any electrode with standard reduction potential less than −0.76 V (*e.g.* $H_2(−2.23\ V)$, $Na^+(−2.71\ V)$, $Li^+(−3.05\ V)$).

In a galvanic cell, where a spontaneous redox reaction drives the cell to produce an electric potential, Gibbs free energy $\Delta G°$ must be negative, in accordance with the following equation :

$$\Delta G^\circ_{cell} = -nFE^\circ_{cell}$$

where n is number of moles of electrons per mole of products and F is the Faraday constant, ~96485 C/mol. As such, the following rules apply :

If $E^\circ_{cell} > 0$, then the process is spontaneous (galvanic cell)

If $E^\circ_{cell} < 0$, then the process is non-spontaneous (electrolytic cell)

Thus in order to have a spontaneous reaction ($\Delta G^\circ < 0$), E°_{cell} must be positive, where :

$$E^\circ_{cell} = E^\circ_{cathode} - E^\circ_{anode}$$

where E°_{anode} is the standard potential at the anode and $E^\circ_{cathode}$ is the standard potential at the cathode as given in the table of standard electrode potential.

Influence of the Neutron Flux

The neutron flux applied to the standard hydrogen electrode modifies the value of the potential as reported by Bagotski etall.

SPONTANEOUS PROCESS

A **spontaneous process** is the time–evolution of a system in which it releases free energy (usually as heat) and moves to a lower, more thermodynamically stable energy state. The sign convention of changes in free energy follows the general convention for thermodynamic measurements, in which a release of free energy from the system corresponds to a negative change in free energy, but a positive change for the surroundings.

Depending on the nature of the process, the free energy is determined differently. For example, the Gibbs free energy is used when considering processes that occur under constant pressure and temperature conditions whereas the Helmholtz free energy is used when considering processes that occur under constant volume and temperature conditions.

A spontaneous process is capable of proceeding in a given direction, as written or described, without needing to be driven by an outside source of energy. The term is used to refer to macro processes in which entropy increases; such as a smell diffusing in a room, ice melting in lukewarm water, salt dissolving in water, and iron rusting.

The laws of thermodynamics govern the direction of a spontaneous process, ensuring that if a sufficiently large number of individual interactions (like atoms colliding) are involved then the direction will always be in the direction of increased entropy (since entropy increase is a statistical phenomenon).

Overview

For a reaction at constant temperature and pressure, ΔG in the Gibbs free energy is :

$$\Delta G = \Delta H - T\Delta S$$

The sign of ΔG depends on the signs of the changes in enthalpy (ΔH) and entropy (ΔS), as well as on the absolute temperature (T, in kelvins). ΔG changes from positive to negative (or *vice versa*) where $T = \Delta H / \Delta S$.

For heterogeneous systems where all of the species of the reaction are in different phases and can be mechanically separated, the following is true.

When ΔG is negative, a process or chemical reaction proceeds spontaneously in the forward direction.

When ΔG is positive, the process proceeds spontaneously in reverse.

When ΔG is zero, the process is already in equilibrium, with no net change taking place over time.

We can further distinguish four cases within the above rule just by examining the signs of the two terms on the right side of the equation.

When ΔS is positive and ΔH is negative, a process is always spontaneous

When ΔS is positive and ΔH is positive, a process is spontaneous at high temperatures, where exothermicity plays a small role in the balance.

When ΔS is negative and ΔH is negative, a process is spontaneous at low temperatures, where exothermicity *is* important.

When ΔS is negative and ΔH is positive, a process is not spontaneous at any temperature, but the reverse process is spontaneous.

For Homogeneous systems where all of the species of the reaction are in the same phase, ΔG cannot accurately predict reaction spontaneity.

The second law of thermodynamics states that for any spontaneous process the overall ΔS must be greater than or equal to zero, yet a spontaneous chemical reaction can result in a negative change in entropy. This does not contradict the second law, however, since such a reaction must have a sufficiently large negative change in enthalpy (heat energy) that the increase in temperature of the reaction surroundings (considered to be part of the system in thermodynamic terms) results in a sufficiently large increase in entropy that overall the change in entropy is positive. That is, the ΔS of the *surroundings* increases enough because of the exothermicity of the reaction that it over-compensates for the negative ΔS of the system, and since the overall $\Delta S = \Delta S_{surroundings} + \Delta S_{system}$, the overall change in entropy is still positive.

Another way to view the fact that some spontaneous chemical reactions can lead to products with lower entropy is to realize that the second law states that entropy of an **isolated** system must increase (or remain constant). Since a negative enthalpy change in a reaction means that energy is being released to the surroundings, then the 'isolated' system includes the chemical reaction plus its surroundings. This means that the heat release of the chemical reaction sufficiently increases the entropy of the surroundings such that the overall entropy of the isolated system increases in accordance with the second law of thermodynamics.

Spontaneity does not imply that the reaction proceeds with great speed. For example, the decay of diamonds into graphite is a spontaneous process occurs very slowly, taking millions of years. The *rate* of a reaction is independent of its spontaneity, and instead depends on the chemical kinetics of the reaction. Every reactant in a spontaneous process has a tendency to form the corresponding product. This tendency is related to stability. Stability is gained by a substance if it is in a minimum energy state or is in maximum randomness. Only one of these can be applied at a time. *e.g.* Water converting to ice is a spontaneous process because ice is more stable since it is of lower energy. However, the formation of water is also a spontaneous process as water is the more random state.

Chapter 3

CELL EMF DEPENDENCY ON CHANGES IN CONCENTRATION

NERNST EQUATION

In electro-chemistry, the **Nernst equation** is an equation that relates the equilibrium reduction potential of a half–cell in an electro-chemical cell (or the total voltage (electro-motive force) for a full cell) to the standard electrode potential, temperature, activity, and reaction quotient of the underlying reactions and species used. It is named after the German physical chemist who first formulated it, Walther Nernst.

The Nernst equation gives a formula that relates the numerical values of the concentration gradient to the electric gradient that balances it. For example, if a concentration gradient was established by dissolving KCl in half of a divided vessel that was originally full of H_2O, and then a membrane permeable to K^+ ions was introduced between the two halves — empirically, an equilibrium situation would arise where the chemical concentration gradient (that would normally cause ions to move from the region of high concentration to the region of low concentration) could be balanced by an electrical gradient that opposes the movement of charge.

Expression

The two (ultimately equivalent) equations for these two cases (half–cell, full cell) are as follows :

$$E_{red} = E_{red}^{\ominus} - \frac{RT}{zF} \ln \frac{a_{Red}}{a_{Ox}} \text{ (half–cell reduction potential)}$$

$$E_{cell} = E_{cell}^{\ominus} - \frac{RT}{zF} \ln Q \text{ (total cell potential)}$$

where :

- E_{red} is the half–cell reduction potential at the temperature of interest
- E°_{red} is the *standard* half–cell reduction potential
- E_{cell} is the cell potential (electro-motive force) at the temperature of interest
- E°_{cell} is the *standard* cell potential
- R is the universal gas constant : $R = 8.314\,472(15)\ J\,K^{-1}\,mol^{-1}$
- T is the absolute temperature
- a is the chemical activity for the relevant species, where a_{Red} is the reductant and a_{Ox} is the oxidant. $a_X = \gamma_X c_X$, where γ_X is the activity coefficient of species X. (Since activity coefficients tend to unity at low concentrations, activities in the Nernst equation are frequently replaced by simple concentrations.)
- F is the Faraday constant, the number of coulombs per mole of electrons : $F = 9.648\,533\,99(24)\times10^4\ C\,mol^{-1}$
- z is the number of moles of electrons transferred in the cell reaction or half-reaction
- Q is the reaction quotient.

At room temperature (25°C), RT/F may be treated like a constant and replaced by 25.693 mV for cells.

The Nernst equation is frequently expressed in terms of base 10 logarithms (*i.e.*, common logarithms) rather than natural logarithms, in which case it is written, *for a cell at 25°C* :

$$E = E^0 - \frac{0.05916\ V}{z} \log_{10} \frac{a_{Red}}{a_{Ox}}.$$

The Nernst equation is used in physiology for finding the electric potential of a cell membrane with respect to one type of ion.

REVERSAL POTENTIAL

In a biological membrane, the **reversal potential** (also known as the **Nernst potential**) of an ion is the membrane potential at which there is no net (overall) flow of that particular ion from one side of the membrane to the other. In the case of post–synaptic neurons, the reversal potential is the membrane potential at which a given neurotransmitter causes no net current flow of ions through that neurotransmitter receptor's ion channel.

In a single–ion system, *reversal potential* is synonymous with **equilibrium potential**; their numerical values are identical. The two terms refer to different aspects of the difference in membrane potential. *Equilibrium* refers to the fact that the net ion flux at a particular voltage is zero. That is, the outward and inward rates of ion movement are the same; the ion flux is in equilibrium. *Reversal* refers to the fact that a change of membrane potential on either side of the equilibrium potential reverses the overall direction of ion flux.

The reversal potential is often called the "Nernst potential", as it can be calculated from the Nernst equation. Ion channels conduct most of the flow of simple ions in and out of cells. When a channel type that is selective to one species of ion dominates within the membrane of a cell (because other ion channels are closed, for example) then the voltage inside the cell will equilibrate (*i.e.* become equal) to the reversal potential for that ion (assuming the outside of the cell is at 0 volts). For example, the resting potential of most cells is close to the K^+ (potassium ion) reversal potential. This is because at resting potential, potassium conductance dominates. During a typical action potential, the small resting ion conductance mediated by potassium channels is overwhelmed by the opening of a large number of Na^+ (sodium ion) channels, which brings the membrane potential close to the reversal potential of sodium.

The relationship between the terms "reversal potential" and "equilibrium potential" only holds true for single–ion systems. In multi–ion systems, there are areas of the cell membrane where the summed currents of the multiple ions will equal zero. While this is a reversal potential in the sense that membrane current reverses direction, it is not an equilibrium potential because not all (and in some cases, none) of the ions are in equilibrium and thus have net fluxes across the membrane. When a cell has significant permeabilities to more than one ion, the cell potential can be calculated from the Goldman–Hodgkin–Katz equation rather than the Nernst equation.

Mathematical Models

The term *driving force* is related to equilibrium potential, and is likewise useful in understanding the current in biological membranes. Driving force refers to the difference between the actual membrane potential and an ion's equilibrium potential. It is defined by the following equation :

$I_{ion} = g_{ion} (V_m - E_{ion})$ where $V_m - E_{ion}$ is the Driving Force.

In words, this equation says that : the ionic current (I_{ion}) is equal to that ion's conductance (g_{ion}) multiplied by the driving force, which is represented by the difference between the membrane potential and the ion's equilibrium potential (*i.e.* $V_m - E_{ion}$). Note that the ionic current will be zero if the membrane is impermeable ($g_{ion} = 0$) to the ion in question, regardless of the size of the driving force.

A related equation (which is derived from the more general equation above) determines the magnitude of an end plate current (EPC), at a given membrane potential, in the neuromuscular junction :

$EPC = g_{ACh}(V_m - E_{rev})$

where EPC is the end plate current, g_{ACh} is the ionic conductance activated by acetylcholine, V_m is the membrane potential, and E_{rev} is the reversal potential. When the membrane potential is equal to the reversal potential, $V_m - E_{rev}$ is equal to 0 and there is no driving force on the ions involved.

Use in Research

When V_m is at the reversal potential ($V_m - E_{rev}$ is equal to 0), the identity of the ions that flow during an EPC can be deduced by comparing the reversal potential of the

EPC to the equilibrium potential for various ions. For instance several excitatory ionotropic ligand–gated neurotransmitter receptors including glutamate receptors (AMPA, NMDA, and kainate), nicotinic acetylcholine (nACh), and serotonin (5–HT$_3$) receptors are non-selective cation channels that pass Na$^+$ and K$^+$ in nearly equal proportions, giving an equilibrium potential close to 0 mV. The inhibitory ionotropic ligand–gated neurotransmitter receptors that carry Cl$^-$, such as GABA$_A$ and glycine receptors, have equilibrium potentials close to the resting potential (approximately–70 mV) in neurons.

This line of reasoning led to the development of experiments (by Akira Takeuchi and Noriko Takeuchi in 1960) that proved that acetylcholine–activated ion channels are approximately equally permeable to Na$^+$ and K$^+$ ions. The experiment was performed by lowering the external Na$^+$ concentration, which lowers (more negative) the Na$^+$ equilibrium potential and produces a negative shift in reversal potential. Conversely, increasing the external K$^+$ concentration raises (more positive) the K$^+$ equilibrium potential and produces a positive shift in reversal potential.

Derivation

Using Boltzmann Factors

For simplicity, we will consider a solution of redox–active molecules that undergo a one–electron reversible reaction :

Ox + e$^-$ $\rightleftharpoons$ Red

and that have a standard potential of zero. The chemical potential μ_c of this solution is the difference between the energy barriers for taking electrons from and for giving electrons to the working electrode that is setting the solution's electrochemical potential.

The ratio of oxidized to reduced molecules, [Ox]/[Red], is equivalent to the probability of being oxidized (giving electrons) over the probability of being reduced (taking electrons), which we can write in terms of the Boltzmann factor for these processes :

$$\frac{[Ox]}{[Red]} = \frac{\exp(-[\text{barrier for losing an electron}]/kT)}{\exp(-[\text{barrier for gaining an electron}]/kT)} = \exp(\mu_c/kT).$$

Taking the natural logarithm of both sides gives :

$$\mu_c = kT \ln \frac{[Ox]}{[Red]}.$$

If $\mu_c \neq 0$ at [Ox]/[Red] = 1, we need to add in this additional constant :

$$\mu_c = \mu_c^0 + kT \ln \frac{[Ox]}{[Red]}.$$

Dividing the equation by e to convert from chemical potentials to electrode potentials, and remembering that $kT/e = RT/F$, we obtain the Nernst equation for the one–electron process $Ox + e^- \rightarrow Red$:

$$E = E^0 + \frac{kT}{e} \ln \frac{[Ox]}{[Red]} = E^0 - \frac{RT}{F} \ln \frac{[Red]}{[Ox]}.$$

Using Thermodynamics (Chemical Potential)

Quantities here are given per molecule, not per mole, and so Boltzmann constant k and the electron charge e are used instead of the gas constant R and Faraday's constant F. To convert to the molar quantities given in most chemistry textbooks, it is simply necessary to multiply by Avogadro's number : $R = kN_A$ and $F = eN_A$.

The entropy of a molecule is defined as :

$S \underline{def} k \ln \Omega,$

where Ω is the number of states available to the molecule. The number of states must vary linearly with the volume V of the system, which is inversely proportional to the concentration c, so we can also write the entropy as :

$S = k \ln (\text{constant} \times V) = -k \ln (\text{constant} \times c).$

The change in entropy from some state 1 to another state 2 is therefore,

$$\Delta S = S_2 - S_1 = -k \ln \frac{c_2}{c_1},$$

so that the entropy of state 2 is

$$S_2 = S_1 - k \ln \frac{c_2}{c_1}.$$

If state 1 is at standard conditions, in which c_1 is unity (e.g., 1 atm or 1 M), it will merely cancel the units of c_2. We can, therefore, write the entropy of an arbitrary molecule A as :

$S(A) = S^0(A) - k \ln[A],$

where S^0 is the entropy at standard conditions and $[A]$ denotes the concentration of A. The change in entropy for a reaction :

$aA + bB \rightarrow yY + zZ$

is then given by :

$$\Delta S_{rxn} = [yS(Y) + zS(Z)] - [aS(A) + bS(B)] = \Delta S^0_{rxn} - k \ln \frac{[Y]^y[Z]^z}{[A]^a[B]^b}.$$

We define the ratio in the last term as the reaction quotient :

$$Q = \frac{\prod_j a_j^{vj}}{\prod_i a_i^{vi}} \approx \frac{[Z]^z[Y]^y}{[A]^a[B]^b}.$$

where the numerator is a product of reaction product activities, $a_{j'}$ each raised to the power of a stoichiometric coefficient, $v_{j'}$ and the denominator is a similar product of reactant activities. All activities refer to a time t. Under certain circumstances each activity term such as a_j^{vj} may be replaced by a concentration term, $[A]$. In an electro-chemical cell, the cell potential E is the chemical potential available from redox reactions ($E = \mu_c/e$). E is related to the Gibbs energy change ΔG only by a constant: $\Delta G = -nFE$, where n is the number of electrons transferred and F is the Faraday constant. There is a negative sign because a spontaneous reaction has a negative free energy ΔG and a positive potential E. The Gibbs energy is related to the entropy by $G = H - TS$, where H is the enthalpy and T is the temperature of the system. Using these relations, we can now write the change in Gibbs energy,

$$\Delta G = \Delta H - TDS = \Delta G^0 + kT \ln Q,$$

and the cell potential,

$$E = E^0 - \frac{kT}{ne} \ln Q.$$

This is the more general form of the Nernst equation. For the redox reaction:

$$Ox + ne^- \rightarrow \text{Red}, \quad Q = \frac{[\text{Red}]}{[\text{Ox}]}, \text{ and we have :}$$

$$E = E^0 - \frac{kT}{ne} \ln \frac{[\text{Red}]}{[\text{Ox}]} = E^0 - \frac{RT}{nF} \ln \frac{[\text{Red}]}{[\text{Ox}]} = E^0 - \frac{RT}{nF} \ln Q.$$

The cell potential at standard conditions E^0 is often replaced by the formal potential $E^{0'}$, which includes some small corrections to the logarithm and is the potential that is actually measured in an electro-chemical cell.

Relation to Equilibrium

At equilibrium, $E = 0$ and $Q = K$. Therefore :

$$0 = E^0 - \frac{RT}{nF} \ln K$$

$$\ln K = \frac{nFE^0}{RT}$$

Or at standard temperature,

$$\log_{10} K = \frac{nE^0}{59.2 mV} \quad \text{at } T = 298 \text{ K.}$$

We have thus related the standard electrode potential and the equilibrium constant of a redox reaction.

Limitations

In dilute solutions, the Nernst equation can be expressed directly in terms of concentrations (since activity coefficients are close to unity). But at higher concentrations, the true activities of the ions must be used. This complicates the use of the Nernst equation, since estimation of non–ideal activities of ions generally requires experimental measurements.

The Nernst equation also only applies when there is no net current flow through the electrode. The activity of ions at the electrode surface changes when there is current flow, and there are additional overpotential and resistive loss terms which contribute to the measured potential.

At very low concentrations of the potential–determining ions, the potential predicted by Nernst equation approaches toward $\pm\infty$. This is physically meaningless because, under such conditions, the exchange current density becomes very low, and there is no thermodynamic equilibrium necessary for Nernst equation to hold. The electrode is called to be unpoised in such case. Other effects tend to take control of the electro-chemical behaviour of the system.

Significance to Related Scientific Domains

The equation has been involved in the scientific controversy involving cold fusion. The discoverers of cold fusion, Fleischmann and Pons, calculated that a palladium cathode immersed in a heavy water electrolysis cell could achieve up to 10^{27} atmospheres of pressure on the surface of the cathode, enough pressure to cause spontaneous nuclear fusion. In reality, only 10,000–20,000 atmospheres were achieved. John R. Huizenga claimed their original calculation was affected by a misinterpretation of Nernst equation. He cited a paper about Pd–Zr alloys.

CONCENTRATION CELL

A **concentration cell** is a limited form of a galvanic cell that has two equivalent half–cells of the same material differing only in concentrations. One can calculate the potential developed by such a cell using the Nernst Equation. A concentration cell produces a small voltage as it attempts to reach equilibrium. This equilibrium occurs when the concentration of reactant in both cells are equal. Because an order of magnitude concentration difference produces less than 30 millivolts at room temperature, concentration cells are not typically used for energy storage.

Concentration cells generate electricity from the thermodynamic free energy that can be extracted from the difference in chemical concentrations of reactants, in the same reaction. This energy is generated from thermal energy that the cells absorb as heat, as the electricity flows. This generation of electricity from ambient thermal energy, without a temperature gradient, is possible because the convergence of chemical concentrations in the two cells increases entropy, and this increase more than compensates for the entropy decreased when heat is converted into electrical energy.

Concentration cell methods of chemical analysis compare a solution of known concentration with an unknown, determining the concentration of the unknown *via* the Nernst Equation or comparison tables against a group of standards.

Concentration cell corrosion occurs when two or more areas of a metal surface are in contact with different concentrations of the same solution. There are three general types of concentration cells.

Metal Ion Concentration Cells

In the presence of water, a high concentration of metal ions will exist under faying surfaces and a low concentration of metal ions will exist adjacent to the crevice created by the faying surfaces. An electrical potential will exist between the two points. The area of the metal in contact with the high concentration of metal ions will be cathodic and will be protected, and the area of metal in contact with the low metal ion concentration will be anodic and corroded.

Oxygen Concentration Cells

Water in contact with the metal surface will normally contain dissolved oxygen. An oxygen cell can develop at any point where the oxygen in the air is not allowed to diffuse uniformly into the solution, thereby creating a difference in oxygen concentration between two points. Corrosion will occur at the area of low–oxygen concentration which are anodic.

Active–passive Cells

For metals that depend on a tightly adhering passive film (usually an oxide) for corrosion protection, salt that deposits on the metal surface in the presence of water, in areas where the passive film is broken, the active metal beneath the film will be exposed to corrosive attack. An electrical potential will develop between the large area of the cathode (passive film) and the small area of the anode (active metal). Rapid pitting of the active metal will result.

Chapter 4

ELECTROLYSIS

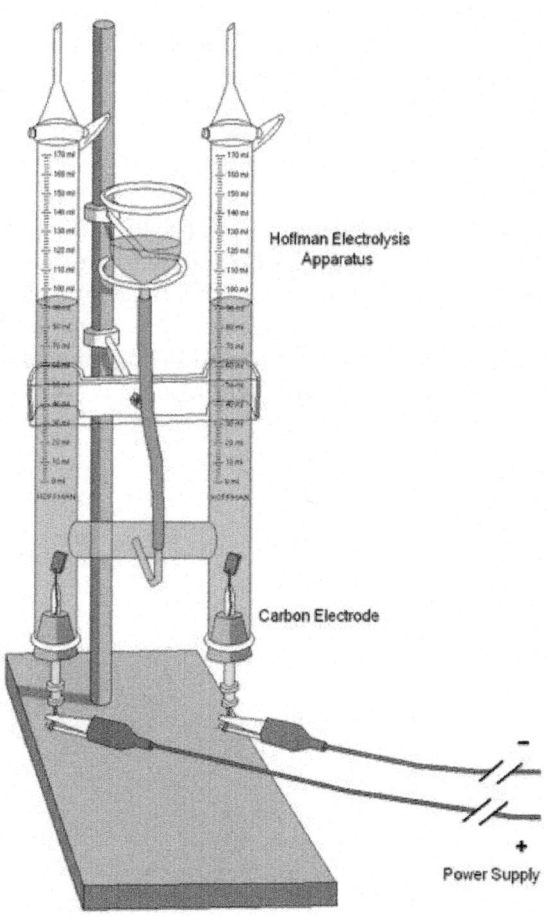

Hoffman Electrolysis
Apparatus

Carbon Electrode

Power Supply

Fig. : Illustration of an electrolysis apparatus used in a school laboratory.

In chemistry and manufacturing, **electrolysis** is a method of using a direct electric current (DC) to drive an otherwise non-spontaneous chemical reaction. Electrolysis is commercially highly important as a stage in the separation of elements from naturally occurring sources such as ores using an electrolytic cell.

HISTORY

The word electrolysis comes from the Greek ἤλεκτρον [ě lektron] "amber" and λύσις [lýsis] "dissolution".

- 1785–Martinus van Marum's electrostatic generator was used to reduce tin, zinc, and antimony from their salts using electrolysis.
- 1800–William Nicholson and Anthony Carlisle (view also Johann Ritter), decomposed water into hydrogen and oxygen.
- 1807–Potassium, sodium, barium, calcium and magnesium were discovered by Sir Humphry Davy using electrolysis.
- 1875–Paul Émile Lecoq de Boisbaudran discovered gallium using electrolysis.
- 1886–Fluorine was discovered by Henri Moissan using electrolysis.
- 1886–Hall–Héroult process developed for making aluminium
- 1890–Castner–Kellner process developed for making sodium hydroxide

OVERVIEW

Electrolysis is the passage of a direct electric current through an ionic substance that is either molten or dissolved in a suitable solvent, resulting in chemical reactions at the electrodes and separation of materials.

The main components required to achieve electrolysis are :

- *An electrolyte :* A substance containing free ions which are the carriers of electric current in the electrolyte. If the ions are not mobile, as in a solid salt then electrolysis cannot occur.
- *A direct current (DC) supply :* Provides the energy necessary to create or discharge the ions in the electrolyte. Electric current is carried by electrons in the external circuit.
- *Two electrodes :* An electrical conductor which provides the physical interface between the electrical circuit providing the energy and the electrolyte.

Electrodes of metal, graphite and semi-conductor material are widely used. Choice of suitable electrode depends on chemical reactivity between the electrode and electrolyte and the cost of manufacture.

PROCESS OF ELECTROLYSIS

The key process of electrolysis is the interchange of atoms and ions by the removal or addition of electrons from the external circuit. The desired products of electrolysis are often in a different physical state from the electrolyte and can be removed by some physical processes. For example, in the electrolysis of brine to produce hydrogen and chlorine, the products are gaseous. These gaseous products bubble from the electrolyte and are collected.

$$2\,NaCl + 2\,H_2O \rightarrow 2\,NaOH + H_2 + Cl_2$$

A liquid containing mobile ions (electrolyte) is produced by :

- Solvation or reaction of an ionic compound with a solvent (such as water) to produce mobile ions
- An ionic compound is melted (*fused*) by heating.

An electrical potential is applied across a pair of electrodes immersed in the electrolyte.

Each electrode attracts ions that are of the opposite charge. Positively charged ions (cations) move towards the electron–providing (negative) cathode, whereas negatively charged ions (anions) move towards the positive anode.

At the electrodes, electrons are absorbed or released by the atoms and ions. Those atoms that gain or lose electrons to become charged ions pass into the electrolyte. Those ions that gain or lose electrons to become uncharged atoms separate from the electrolyte. The formation of uncharged atoms from ions is called discharging.

The energy required to cause the ions to migrate to the electrodes, and the energy to cause the change in ionic state, is provided by the external source of electrical potential.

Oxidation and Reduction at the Electrodes

Oxidation of ions or neutral molecules occurs at the anode, and the reduction of ions or neutral molecules occurs at the cathode. For example, it is possible to oxidize ferrous ions to ferric ions at the anode :

Fe2+

aq $\rightarrow$ Fe3+

aq + e⁻

It is also possible to reduce ferricyanide ions to ferrocyanide ions at the cathode :

Fe(CN)3–

6 + e⁻$\rightarrow$ Fe(CN)4–6

Neutral molecules can also react at either of the electrodes. For example : p–Benzoquinone can be reduced to hydroquinone at the cathode :

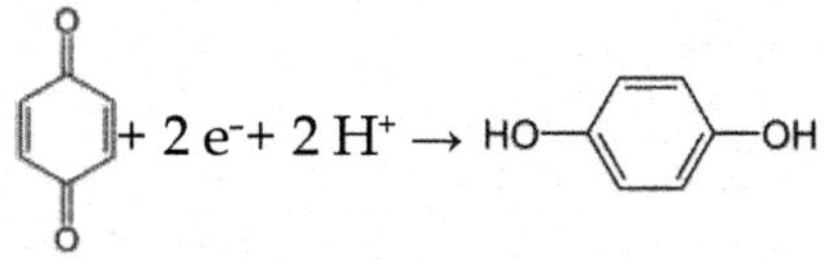

In the last example, H⁺ ions (hydrogen ions) also take part in the reaction, and are provided by an acid in the solution, or the solvent itself (water, methanol etc.). Electrolysis reactions involving H^+ ions are fairly common in acidic solutions. In alkaline water solutions, reactions involving OH^-(hydroxide ions) are common.

The substances oxidised or reduced can also be the solvent (usually water) or the electrodes. It is possible to have electrolysis involving gases. (Such as when using a Gas diffusion electrode)

Energy Changes During Electrolysis

The amount of electrical energy that must be added equals the change in Gibbs free energy of the reaction plus the losses in the system. The losses can (in theory) be arbitrarily close to zero, so the maximum thermodynamic efficiency equals the enthalpy change divided by the free energy change of the reaction. In most cases, the electric input is larger than the enthalpy change of the reaction, so some energy is released in the form of heat. In some cases, for instance, in the electrolysis of steam into hydrogen and oxygen at high temperature, the opposite is true. Heat is absorbed from the surroundings, and the heating value of the produced hydrogen is higher than the electric input.

Related Techniques

The following techniques are related to electrolysis :

* Electro-chemical cells, including the hydrogen fuel cell, utilise differences in Standard electrode potential in order to generate an electrical potential from which useful power can be extracted. Although related *via* the interaction of ions and electrodes, electrolysis and the operation of Electro-chemical cells are quite distinct. A chemical cell should *not* be thought of as performing "electrolysis in reverse".

FARADAY'S LAWS OF ELECTROLYSIS

Faraday's laws of electrolysis are quantitative relationships based on the Electro-chemical researches published by Michael Faraday in 1834.

Statements of the Laws

Several versions of the laws can be found in textbooks and the scientific literature. The most common statements resemble the following :

* *Faraday's 1st Law of Electrolysis* : The mass of a substance altered at an electrode during electrolysis is directly proportional to the quantity of electricity transferred at that electrode. Quantity of electricity refers to the quantity of electrical charge, typically measured in coulomb.
* *Faraday's 2nd Law of Electrolysis* : For a given quantity of D.C. electricity (electric charge), the mass of an elemental material altered at an electrode is directly

proportional to the element's equivalent weight. The equivalent weight of a substance will be explained in the next paragraph.

For an element the equivalent weight is the quantity that combines with or replaces 1.00797 grams (g) of hydrogen or 7.9997 g of oxygen; or, the weight of an element that is liberated in an electrolysis (chemical reaction caused by an electric current) by the passage of $9.64853399(24) \times 10^4$ coulombs of electricity. The equivalent weight of an element is its gram atomic weight divided by its valence (combining power). Some equivalent weights are : silver (Ag), 107.868 g; magnesium (Mg), 24.312/2 g; aluminum (Al), 26.9815/3 g; sulfur (S, in forming a sulfide), 32.064/2 g. For compounds that function as oxidizing or reducing agents (compounds that act as acceptors or donors of electrons), the equivalent weight is the gram molecular weight divided by the number of electrons lost or gained by each molecule; $e.g.$, potassium permanganate ($KMnO_4$) in acid solution, 158.038/5 g; potassium dichromate (K2Cr2O7), 294.192/6 g; and sodium thiosulfate (Na2S2O3 · 5H2O), 248.1828/1 g. For all oxidizing and reducing agents (elements or compounds) the equivalent weight is the weight of the substance that is associated with the loss or gain of 6.023×10^{23} electrons. The equivalent weight of an acid or base for neutralization reactions or of any other compound that acts by double decomposition is the quantity of the compound that will furnish or react with or be equivalent to 1.00797 g of hydrogen ion or 17.0074 g of hydroxide ion; $e.g.$, hydrochloric acid (HCl), 36.461 g; sulfuric acid (H2SO4), 98.078/2 g; sodium hydroxide (NaOH), 40 g; sodium carbonate (Na2CO3), 105.9892/2 g. The equivalent weight of a substance may vary with the type of reaction it undergoes. Thus, potassium permanganate reacting by double decomposition has an equivalent weight equal to its gram molecular weight, 158.038/1 g; as an oxidizing agent under different circumstances it may be reduced to the manganate ion ($MnO42-$), to manganese dioxide ($MnO2$), or to the manganous ion ($Mn2+$), with the equivalent weights of 158.038/1 g, 158.038/3 g, and 158.038/5 g, respectively. The number of equivalent weights of any substance dissolved in one litre of solution is called the normality of that solution.

Mathematical Form

Faraday's laws can be summarized by :

$$m = \left(\frac{Q}{F}\right)\left(\frac{M}{z}\right)$$

where :

- m is the mass of the substance liberated at an electrode in grams
- Q is the total electric charge passed through the substance
- $F = 96485$ C mol^{-1} is the Faraday constant
- M is the molar mass of the substance
- z is the valency number of ions of the substance (electrons transferred per ion).

Note that M/z is the same as the equivalent weight of the substance altered.

For Faraday's first law, M, F, and z are constants, so that the larger the value of Q the larger m will be.

For Faraday's second law, Q, F, and z are constants, so that the larger the value of M/z (equivalent weight) the larger m will be.

In the simple case of constant–current electrolysis, $Q = It$ leading to

$$m = \left(\frac{It}{F}\right)\left(\frac{M}{z}\right)$$

and then to

$$n = \left(\frac{It}{F}\right)\left(\frac{1}{z}\right)$$

where :

- n is the amount of substance ("number of moles") liberated : $n = m/M$
- t is the total time the constant current was applied.

In the more complicated case of a variable electrical current, the total charge Q is the electric current $I(\tau)$ integrated over time τ :

$$Q = \int_0^t I(\tau)\, d\tau$$

Here t is the *total* electrolysis time.

COMPETING HALF–REACTIONS IN SOLUTION ELECTROLYSIS

Using a cell containing inert platinum electrodes, electrolysis of aqueous solutions of some salts leads to reduction of the cations (*e.g.*, metal deposition with, *e.g.*, zinc salts) and oxidation of the anions (*e.g.* evolution of bromine with bromides). However, with salts of some metals (*e.g.* sodium) hydrogen is evolved at the cathode, and for salts containing some anions (*e.g.* sulfate SO_4^{2-}) oxygen is evolved at the anode. In both cases this is due to water being reduced to form hydrogen or oxidised to form oxygen. In principle the voltage required to electrolyze a salt solution can be derived from the standard electrode potential for the reactions at the anode and cathode. The standard electrode potential is directly related to the Gibbs free energy, ΔG, for the reactions at each electrode and refers to an electrode with no current flowing. An extract from the table of standard electrode potentials is shown below.

Half–reaction	E° (V)	Ref.
$Na^+ + e^- \rightleftharpoons Na(s)$	−2.71	
$Zn + 2e^- \rightleftharpoons Zn(s)$	−0.7618	
$2H^+ + 2e^- \rightleftharpoons H_2(g)$	$\equiv 0$	
$Br_2(aq) + 2e^- \rightleftharpoons 2Br^-$	+1.0873	

(Contd...)

(*Contd...*)

Half–reaction	$E°$ (V)	Ref.
$O_2(g) + 4H^+ + 4e^- \rightleftharpoons 2H_2O$	+1.23	
$Cl_2(g) + 2e^- \rightleftharpoons 2\,Cl^-$	+1.36	
$S2O^{2-}\,8 + 2e^- \rightleftharpoons 2SO2{-}4$	+2.07	

In terms of electrolysis, this table should be interpreted as follows :

- Oxidized species (often a cation) with a more negative cell potential are more difficult to reduce than oxidized species with a more positive cell potential. For example it is more difficult to reduce a sodium ion to a sodium metal than it is to reduce a zinc ion to a zinc metal.

- Reduced species (often an anion) with a more positive cell potential are more difficult to oxidize than reduced species with a more negative cell potential. For example it is more difficult to oxidize sulfate anions than it is to oxidize bromide anions.

Using the Nernst equation the electrode potential can be calculated for a specific concentration of ions, temperature and the number of electrons involved. For pure water (pH 7) :

- the electrode potential for the reduction producing hydrogen is −0.41 V
- the electrode potential for the oxidation producing oxygen is +0.82 V.

Comparable figures calculated in a similar way, for 1M zinc bromide, $ZnBr_2$, are −0.76 V for the reduction to Zn metal and +1.10 V for the oxidation producing bromine. The conclusion from these figures is that hydrogen should be produced at the cathode and oxygen at the anode from the electrolysis of water which is at variance with the experimental observation that zinc metal is deposited and bromine is produced. The explanation is that these calculated potentials only indicate the thermodynamically preferred reaction. In practice many other factors have to be taken into account such as the kinetics of some of the reaction steps involved. These factors together mean that a higher potential is required for the reduction and oxidation of water than predicted, and these are termed overpotentials. Experimentally it is known that overpotentials depend on the design of the cell and the nature of the electrodes.

For the electrolysis of a neutral (pH 7) sodium chloride solution, the reduction of sodium ion is thermodynamically very difficult and water is reduced evolving hydrogen leaving hydroxide ions in solution. At the anode the oxidation of chlorine is observed rather than the oxidation of water since the overpotential for the oxidation of chloride to chlorine is lower than the overpotential for the oxidation of water to oxygen. The hydroxide ions and dissolved chlorine gas react further to form hypochlorous acid. The aqueous solutions resulting from this process is called electrolyzed water and is used as a disinfectant and cleaning agent.

ELECTROLYSIS OF WATER

Electrolysis of water is the decomposition of water (H_2O) into oxygen (O_2) and hydrogen gas (H_2) due to an electric current being passed through the water.

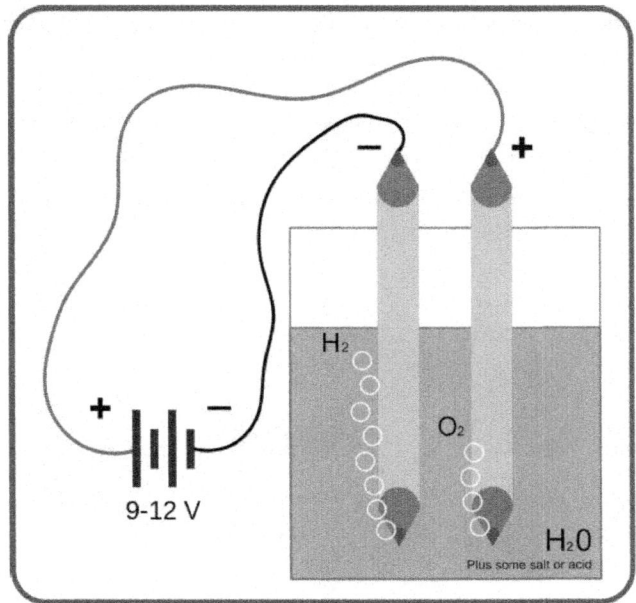

Fig. : Simple setup for demonstration of electrolysis of water at home.

History

Jan Rudolph Deiman and Adriaan Paets van Troostwijk used in 1789 an electrostatic machine to produce electricity which was discharged on gold electrodes in a Leyden jar with water. In 1800 Alessandro Volta invented the voltaic pile, and a few weeks later William Nicholson and Anthony Carlisle used it for the electrolysis of water. When Zénobe Gramme invented the Gramme machine in 1869 electrolysis of water became a cheap method for the production of hydrogen. A method of industrial synthesis of hydrogen and oxygen through electrolysis was developed by Dmitry Lachinov in 1888.

Principle

An electrical power source is connected to two electrodes, or two plates (typically made from some inert metal such as platinum, stainless steel or iridium) which are placed in the water. Hydrogen will appear at the cathode (the negatively charged electrode, where electrons enter the water), and oxygen will appear at the anode (the positively charged electrode). Assuming ideal faradaic efficiency, the amount of hydrogen generated is twice the number of moles of oxygen, and both are proportional to the total electrical charge conducted by the solution. However, in many cells competing side reactions dominate, resulting in different products and less than ideal faradaic efficiency.

Electrolysis of *pure* water requires excess energy in the form of overpotential to overcome various activation barriers. Without the excess energy the electrolysis of *pure* water occurs very slowly or not at all. This is in part due to the limited self–ionization of water. Pure water has an electrical conductivity about one millionth that of seawater. Many electrolytic cells may also lack the requisite electro-catalysts. The efficiency of electrolysis is increased through the addition of an electrolyte (such as a salt, an acid or a base) and the use of electro-catalysts.

Currently the electrolytic process is rarely used in industrial applications since hydrogen can currently be produced more affordably from fossil fuels.

Equations

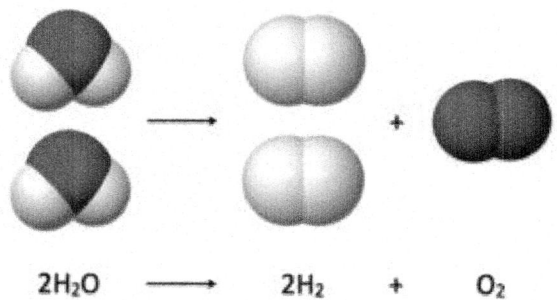

$$2H_2O \longrightarrow 2H_2 + O_2$$

Fig. : Diagram showing the overall chemical equation.

In pure water at the negatively charged cathode, a reduction reaction takes place, with electrons (e^-) from the cathode being given to hydrogen cations to form hydrogen gas (the half reaction balanced with acid) :

Reduction at cathode : $2\,H^+(aq) + 2e^- \rightarrow H_2(g)$

At the positively charged anode, an oxidation reaction occurs, generating oxygen gas and giving electrons to the anode to complete the circuit :

Oxidation at anode : $2\,H_2O(l) \rightarrow O_2(g) + 4\,H^+(aq) + 4e^-$

The same half reactions can also be balanced with base as listed below. Not all half reactions must be balanced with acid or base. Many do, like the oxidation or reduction of water listed here. To add half reactions they must both be balanced with either acid or base.

Cathode (reduction) : $2\,H_2O(l) + 2e^- \rightarrow H_2(g) + 2\,OH^-(aq)$

Anode (oxidation) : $4\,OH^-(aq) \rightarrow O_2(g) + 2\,H_2O(l) + 4e^-$

Combining either half reaction pair yields the same overall decomposition of water into oxygen and hydrogen :

Overall reaction : $2\,H_2O(l) \rightarrow 2\,H_2(g) + O_2(g)$

The number of hydrogen molecules produced is thus twice the number of oxygen molecules. Assuming equal temperature and pressure for both gases, the produced hydrogen gas has therefore twice the volume of the produced oxygen gas. The number of electrons pushed through the water is twice the number of generated hydrogen molecules and four times the number of generated oxygen molecules.

Thermodynamics of the Process

Decomposition of pure water into hydrogen and oxygen at standard temperature and pressure is not favourable in thermodynamic terms.

Anode (oxidation) : $2 H_2O(l) \rightarrow O_2(g) + 4 H^+(aq) + 4e^-$ Eo

ox =-1.23 V (Eo

red = 1.23))

Cathode (reduction) : $2 H^+(aq) + 2e^- \rightarrow H_2(g)$ Eo

red = 0.00 V

Thus, the standard potential of the water electrolysis cell is–1.23 V at 25°C at pH 0 ($H^+ = 1.0$ M). At 25°C with pH 7 ($H^+ = 1.0 \times 10^{-7}$ M), the potential is unchanged based on the Nernst equation. However, electrolysis will not generally proceed at these voltages, as the electrical input must provide the full amount of enthalpy of the H_2-O_2 products (286 kJ per mol). This takes the theoretical and real observed threshold of electrolysis to (–)1.48 V. This is a standard value derived from basic energy conservation for H_2 with a known molar enthalpy value of 286 kJ, (diatomic H_2 having 2 Faraday units of charge per mol), therefore the ideal voltage becomes 286,000/(2*96485) = 1.48 V.

The negative voltage indicates the Gibbs free energy for electrolysis of water is greater than zero for these reactions. This can be found using the $G = -nFE$ equation from chemical kinetics, where n is the moles of electrons and F is the Faraday constant. The reaction cannot occur without adding necessary energy, usually supplied by an external electrical power source.

Electrolyte Selection

If the above described processes occur in pure water, H^+ cations will accumulate at the anode and OH^- anions will accumulate at the cathode. This can be verified by adding a pH indicator to the water : the water near the anode is acidic while the water near the cathode is basic. The negative hydroxyl ions that approach the anode mostly combine with the positive hydronium ions (H_3O^+) to form water. The positive hydronium ions that approach the negative cathode mostly combine with negative hydroxyl ions to form water. Relatively few hydronium (hydroxyl) ions reach the cathode (anode). This can cause a concentration overpotential at both electrodes.

Pure water is a fairly good insulator since it has a low autoionization, $K_w = 1.0 \times 10^{-14}$ at room temperature and thus pure water conducts current poorly,

$0.055 \mu S \cdot cm^{-1}$. Unless a very large potential is applied to cause an increase in the autoionization of water the electrolysis of pure water proceeds very slowly limited by the overall conductivity.

If a water–soluble electrolyte is added, the conductivity of the water rises considerably. The electrolyte disassociates into cations and anions; the anions rush towards the anode and neutralize the buildup of positively charged H^+ there; similarly, the cations rush towards the cathode and neutralize the buildup of negatively charged OH^- there. This allows the continued flow of electricity.

Care must be taken in choosing an electrolyte, since an anion from the electrolyte is in competition with the hydroxide ions to give up an electron. An electrolyte anion with less standard electrode potential than hydroxide will be oxidized instead of the hydroxide, and no oxygen gas will be produced. A cation with a greater standard electrode potential than a hydrogen ion will be reduced in its stead, and no hydrogen gas will be produced.

The following cations have lower electrode potential than H^+ and are therefore suitable for use as electrolyte cations : Li^+, Rb^+, K^+, Cs^+, Ba^{2+}, Sr^{2+}, Ca^{2+}, Na^+, and Mg^{2+}. Sodium and lithium are frequently used, as they form inexpensive, soluble salts.

If an acid is used as the electrolyte, the cation is H^+, and there is no competitor for the H^+ created by disassociating water. The most commonly used anion is sulfate ($SO2-4$), as it is very difficult to oxidize, with the standard potential for oxidation of this ion to the peroxodisulfate ion being -2.05 volts.

Strong acids such as sulfuric acid (H_2SO_4), and strong bases such as potassium hydroxide (KOH), and sodium hydroxide ($NaOH$) are frequently used as electrolytes due to their strong conducting abilities.

A solid polymer electrolyte can also be used such as Nafion and when applied with a special catalyst on each side of the membrane can efficiently split the water molecule with as little as 1.5 Volts. There are also a number of other solid electrolyte systems that have been trialled and developed with a number of electrolysis systems now available commercially that use solid electrolytes.

Applications

About five per cent of hydrogen gas produced worldwide is created by electrolysis. The majority of this hydrogen produced through electrolysis is a side product in the production of chlorine. This is a prime example of a competing side reaction.

$$2NaCl + 2H_2O \rightarrow Cl_2 + H_2 + 2NaOH$$

The electrolysis of brine (saltwater), a water sodium chloride mixture, is only half the electrolysis of water since the chloride ions are oxidized to chlorine rather than water being oxidized to oxygen. The hydrogen produced from this process is either burned (converting it back to water), used for the production of specialty chemicals, or various other small scale applications.

Water electrolysis is also used to generate oxygen for the International Space Station.

Hydrogen may later be used in a fuel cell as a storage method of energy and water.

Efficiency

Thermodynamics

The electrolysis of water in standard conditions requires a theoretical minimum of 237 kJ of electrical energy input to dissociate each mole of water, which is the standard Gibbs free energy of formation of water. Very often, although incorrectly, the standard enthalpy of formation of liquid water is used as a reference (286 kJ/mol), or possibly that of water vapour (242 kJ/mol). These are also often indicated as the higher heating value (HHV) and lower heating value (LHV) of hydrogen.

Since each mole of water requires two moles of electrons, and given that the Faraday constant F represents the charge of a mole of electrons (96485 C/mol), it follows that the minimum voltage necessary for electrolysis is about 1.23 V. If the HHV is used as a reference instead, the obtained voltage is 1.48 V, the *thermoneutral* voltage.

In the case of water electrolysis, Gibbs free energy represents the minimum *work* necessary for the reaction to proceed, and the reaction enthalpy is the amount of energy (both work and heat) that has to be provided so the reaction products are at the same temperature as the reactants (*i.e.* standard temperature for the values given above). This implies that an ideal, 100% efficient electrolyser would produce hydrogen and oxygen at a lower temperature than the provided water. An electrolyser operating at 1.48 V would be only 83% efficient, and would produce hydrogen and oxygen at the same temperature as provided water.

Overpotential

Real water electrolysers require higher voltages for the reaction to proceed. The part that exceeds 1.23 V is called overpotential or overvoltage, and represents any kind of loss and non-ideality in the Electro-chemical process.

For a well designed cell the largest overpotential is the reaction overpotential for the four–electron oxidation of water to oxygen at the anode; electro-catalysts can facilitate this reaction, and platinum alloys are the state of the art for this oxidation. Developing a cheap, effective electro-catalyst for this reaction would be a great advance, and is a topic of current research. The simpler two–electron reaction to produce hydrogen at the cathode can be electro-catalyzed with almost no overpotential by platinum, or in theory a hydrogenase enzyme. If other, less effective, materials are used for the cathode (*e.g.* graphite), large overpotentials will appear.

Industrial State of the Art

Efficiency of modern hydrogen generators is measured by **Power consumed per standard volume of hydrogen** (MJ/m^3 or kWh/m^3), assuming standard temperature and pressure of the H_2. A 100%–efficient electrolyser would consume then 10.6 MJ/m^3 (2.94 kWh/m^3); the lower the actual, the higher efficiency.

Very often electrolyser vendors provide efficiencies based on enthalpy (LHV or HHV). These, being larger than the corresponding Gibbs free energy, make the calculated efficiency values appear significantly better than the actual values. To assess the claimed efficiency of an electrolyser it is therefore important to establish how it was defined by the vendor.

There are two main technologies available on the market, *alkaline* and *PEM* electrolysers. Alkaline electrolysers are cheaper in terms of investment, but less efficient; PEM electrolysers, conversely, are more expensive but also more efficient, and can be cheaper if the hydrogen production is large enough. Reported efficiencies are in the range 50–80%.

WATER SPLITTING

Water splitting is the general term for a chemical reaction in which water is separated into oxygen and hydrogen. Efficient and economical water splitting would be a key technology component of a hydrogen economy. Various techniques for water splitting have been issued in water splitting patents in the United States. In photosynthesis, water splitting donates electrons to power the electron transport chain in photo-system II.

High–pressure Electrolysis

High–pressure electrolysis (HPE) is the electrolysis of water by decomposition of water (H_2O) into oxygen (O_2) and hydrogen gas (H_2) due to the passing of an electric current through the water. The difference with a standard proton exchange membrane electrolyzer is the compressed hydrogen output around 12–20 megapascals (120–200 bar) at 70°C. By pressurising the hydrogen in the electrolyser the need for an external hydrogen compressor is eliminated, the average energy consumption for internal differential pressure compression is around 3%.

Approaches

As the required compression power for water is less than that for hydrogen–gas the water is pumped up to a high–pressure, in the other approach differential pressure is used. There is also an **importance** for the electrolyser stacks to be able to accept a fluctuating electrical input, such as that found with renewable energy. This then enables the ability to help with grid balancing and energy storage.

Ultra-high–pressure Electrolysis

Ultra-high–pressure electrolysis is high–pressure electrolysis operating at 34–69 megapascals (5,000–10,000 psi). At ultra–high pressures the water solubility and cross–permeation across the membrane of H_2 and O_2 is affecting hydrogen purity, modified PEMs are used to reduce cross–permeation in combination with catalytic H_2/O_2 recombiners to maintain H_2 levels in O_2 and O_2 levels in H_2 at values compatible with hydrogen safety requirements.

Research

The US DOE believes that high–pressure electrolysis, supported by ongoing research and development, will contribute to the enabling and acceptance of technologies where hydrogen is the energy carrier between renewable energy resources and clean energy consumers.

High–pressure electrolysis is being investigated by the DOE for efficient production of hydrogen from water. The target total in 2005 is $4.75 per gge H_2 at an efficiency of 64%. The total goal for the DOE in 2010 is $2.85 per gge H_2 at an efficiency of 75%. As of 2005 the DOE provided a total of $1,563,882 worth of funding for research.

Mitsubishi is pursuing such technology with its High–pressure hydrogen energy generator (HHEG) project.

The Forschungszentrum Jülich, in Jülich Germany is currently researching the cost reduction of components used in high–pressure PEM electrolysis in the EKOLYSER project. The primary goal of this research is to improve performance and gas purity, reduce cost and volume of expensive materials and reach the alternative energy targets set forth by the German government for 2050 in the Energy Concept published in 2010.

HIGH–TEMPERATURE ELECTROLYSIS

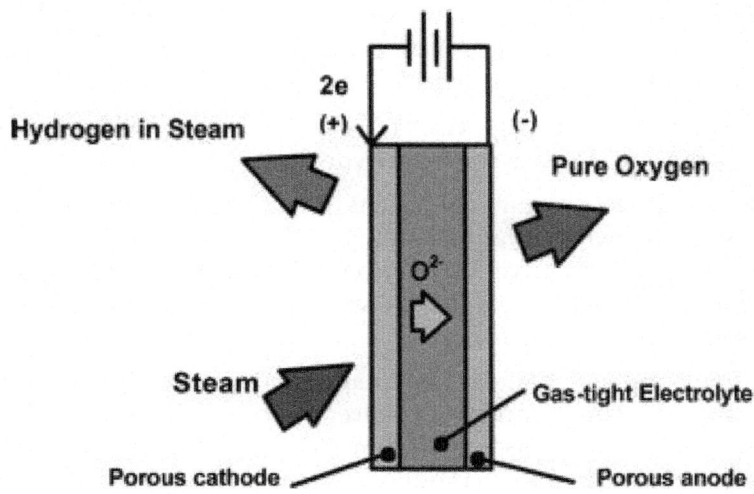

Fig. : High–temperature electrolysis schema.

High-temperature electrolysis (also **HTE** or **steam electrolysis**) is a method being investigated for the production of hydrogen from water with oxygen as a by-product.

Efficiency

High temperature electrolysis is more efficient economically than traditional room-temperature electrolysis because some of the energy is supplied as heat, which is cheaper than electricity, and because the electrolysis reaction is more efficient at higher temperatures. In fact, at 2500°C, electrical input is unnecessary because water breaks down to hydrogen and oxygen through thermolysis. Such temperatures are impractical; proposed HTE systems operate between 100°C and 850°C.

The efficiency improvement of high-temperature electrolysis is best appreciated by assuming the electricity used comes from a heat engine, and then considering the amount of heat energy necessary to produce one kg hydrogen (141.86 megajoules), both in the HTE process itself and also in producing the electricity used. At 100°C, 350 megajoules of thermal energy are required (41% efficient). At 850°C, 225 megajoules are required (64% efficient).

Materials

The selection of the materials for the electrodes and electrolyte in a solid oxide electrolyser cell is essential. One option being investigated for the process used yttria-stabilized zirconia (YSZ) electrolytes, nickel-cermet steam/hydrogen electrodes, and mixed oxide of lanthanum, strontium and cobalt oxygen electrodes.

Economic Potential

Even with HTE, electrolysis is a fairly inefficient way to store energy. Significant conversion losses of energy occur both in the electrolysis process, and in the conversion of the resulting hydrogen back into power.

At current hydrocarbon prices, HTE can not compete with pyrolysis of hydrocarbons as an economical source of hydrogen.

HTE is of interest as a more efficient route to the production of hydrogen, to be used as a carbon neutral fuel and general energy storage. It may become economical if cheap non-fossil fuel sources of heat (concentrating solar, nuclear, geothermal) can be used in conjunction with non-fossil fuel sources of electricity (such as solar, wind, ocean, nuclear).

Possible supplies of cheap high-temperature heat for HTE are all non-chemical, including nuclear reactors, concentrating solar thermal collectors, and geothermal sources. HTE has been demonstrated in a laboratory at 108 kilojoules (thermal) per gram of hydrogen produced, but not at a commercial scale. The first commercial generation IV reactors are expected around 2030.

The Market for Hydrogen Production

Given a cheap, high–temperature heat source, other hydrogen production methods are possible. In particular. Thermo-chemical production might reach higher efficiencies than HTE because no heat engine is required. However, large–scale thermo-chemical production will require significant advances in materials that can withstand high–temperature, high–pressure, highly corrosive environments.

The market for hydrogen is large (50 million metric tons/year in 2004, worth about $135 billion/year) and growing at about 10% per year. This market is met by pyrolysis of hydrocarbons to produce the hydrogen, which results in CO_2 emissions. The two major consumers are oil refineries and fertilizer plants (each consume about half of all production). Should hydrogen–powered cars become widespread, their consumption would greatly increase the demand for hydrogen.

Electrolysis and Thermodynamics

During electrolysis, the amount of electrical energy that must be added equals the change in Gibbs free energy of the reaction plus the losses in the system. The losses can (theoretically) be arbitrarily close to zero, so the maximum thermodynamic efficiency of any Electro-chemical process equals 100%. In practice, the efficiency is given by electrical work achieved divided by the Gibbs free energy change of the reaction.

In most cases, such as room temperature water electrolysis, the electric input is larger than the enthalpy change of the reaction, so some energy is released as waste heat. In the case of electrolysis of steam into hydrogen and oxygen at high temperature, the opposite is true. Heat is absorbed from the surroundings, and the heating value of the produced hydrogen is higher than the electric input. In this case the efficiency relative to electric energy input can be said to be greater than 100%. The maximum theoretical efficiency of a fuel cell is the inverse of that of electrolysis. It is thus impossible to create a perpetual motion machine by combining the two processes.

IN-SITU RESOURCE UTILIZATION

In space exploration, *in-situ* resource utilization (ISRU) describes the proposed use of resources found or manufactured on other astronomical objects (the Moon, Mars, Asteroids, etc.) to further the goals of a space mission.

According to NASA, "In-situ resource utilization will enable the affordable establishment of extra-terrestrial exploration and operations by minimizing the materials carried from Earth."

ISRU can provide materials for life support, propellants, construction materials, and energy to a science payload or a crew deployed on a planet, moon, or asteroid.

Fig. : ISRU Reverse Water Gas Shift Testbed (NASA KSC).

It is now very common for spacecraft to harness the solar radiation found *in–situ*, and it is likely missions to planetary surfaces will also use solar power. Beyond that, ISRU has not yet received any practical application, but it is seen by exploration proponents as a way to drastically reduce the amount of payload that must be launched from Earth in order to explore a given planetary body.

Proposals have been made for "mining" atmospheric gases for rocket propulsion, using what is called a Propulsive Fluid Accumulator.

Uses

Solar Cell Production

It has long been suggested that solar cells could be produced from the materials present on the lunar surface. In its original form, known as the solar power satellite, the proposal was intended as an alternate power source for Earth. Solar cells would be shipped to Earth orbit and assembled, the power being transmitted to Earth *via* microwave beams. Despite much work on the cost of such a venture, the uncertainty lay in the cost and complexity of fabrication procedures on the lunar surface. A more modest reincarnation of this dream is for it to create solar cells to power future lunar bases. One particular proposal is to simplify the process by using fluorine brought from Earth as potassium fluoride to separate the raw materials from the lunar rocks.

Rocket Propellant

Rocket propellant from water ice has also been proposed for the Moon, mainly from ice that has been found at the poles. The likely difficulties include working at extremely low temperatures and extraction from the regolith. Most schemes electrolyse the water and form hydrogen and oxygen and liquify and cryogenically store them. This requires large amounts of equipment and power to achieve. Alternatively it is possible to simply heat the water in a nuclear or solar thermal rocket, which seems to give very much more mass delivered to low Earth orbit (LEO) in spite of the much lower specific impulse, for a given amount of equipment.

The monopropellant hydrogen peroxide (H_2O_2) can be made from water on Mars and the Moon.

Aluminium as well as other metals have been proposed for use as rocket propellant made using lunar resources, and proposals include reacting the aluminium with water.

The spacecraft could use the propellant itself or supply a propellant depot.

Oxygen to Breathe and Water to Drink

Water ice could replenish a space ship's water tanks. Water is needed for hygiene and obviously to drink, but may also be used for radiation protection in deep space (living quarters inside a double-walled cylindrical water tank). Splitting water allows the creation of rocket propellant, but can also liberate oxygen that could be used to replenish the atmosphere in a closed-loop recycling system.

Metals for Construction or Return to Earth

Asteroid mining could also involve extraction of metals for construction material in space, which may be more cost-effective than bringing such material up out of Earth's deep gravity well, or that of any other large body like the Moon or Mars. Metallic asteroids contain huge amounts of siderophilic metals, including precious metals.

Locations

Mars

ISRU research for Mars is focused primarily on providing rocket propellant for a return trip to Earth — either for a manned or a sample return mission — or for use as fuel on Mars. Many of the proposed techniques utilize the well-characterised atmosphere of Mars as feedstock. Since this can be easily simulated on Earth, these proposals are relatively simple to implement, though it is by no means certain that NASA or the ESA will favour this approach over a more conventional direct mission.

A typical proposal for ISRU is the use of a Sabatier reaction, $CO_2 + 4H_2 \rightarrow CH_4 + 2H_2O$, in order to produce methane on the Martian surface, to be used as a propellant. Oxygen is liberated from the water by electrolysis, and the hydrogen recycled back into the Sabatier reaction. The usefulness of this reaction is that only the hydrogen (which is light) need be brought from Earth.

A similar reaction proposed for Mars is the reverse water gas shift reaction, $CO_2 + H_2 \rightarrow CO + H_2O$. This reaction takes place rapidly in the presence of an iron–chrome catalyst at 400 Celsius, and has been implemented in an Earth–based testbed by NASA. Again, oxygen is recycled from the water by electrolysis, and only a small amount of hydrogen is needed from Earth. The net result of this reaction is the production of oxygen, to be used as the oxidizer component of rocket fuel.

Another reaction proposed for production of oxygen is electrolysis of the atmosphere, $2CO_2$ (+ energy) $\rightarrow 2CO + O_2$.

Mars Surveyor 2001 Lander MIP (Mars ISPP Precursor) was to demonstrate manufacture of oxygen from the atmosphere of Mars, and test solar cell technologies and methods of mitigating the effect of Martian dust on the power systems. The proposed Mars 2020 rover mission might include ISRU technology demonstrator that would extract CO_2 from the atmosphere and produce O_2 for rocket fuel.

The Moon

On the moon, the lunar highland material anorthite is similar to the earth mineral bauxite, which is an aluminium ore. Smelters can produce pure aluminum, calcium metal, oxygen and silica glass from anorthite. Raw anorthite is also good for making fiberglass and other glass and ceramic products.

Over twenty different methods have been proposed for oxygen extraction on the moon. Oxygen is often found in iron rich lunar minerals and glasses as iron oxide. The oxygen can be extracted by heating the material to temperatures above 900°C and exposing it to hydrogen gas. The basic equation is : $FeO + H_2 \rightarrow Fe + H_2O$. This process has recently been made much more practical by the discovery of significant amounts of hydrogen–containing regolith near the moon's poles by the Clementine spacecraft.

Lunar materials may also be valuable for other uses. It has also been proposed to use lunar regolith as a general construction material, through processing techniques such as sintering, hot–pressing, liquification, and the cast basalt method. Cast basalt is used on Earth for construction of, for example, pipes where a high resistance to abrasion is required. Cast basalt has a very high hardness of 8 Mohs (diamond is 10 Mohs) but is also susceptible to mechanical impact and thermal shock which could be a problem on the moon.

Glass and glass fibre are straightforward to process on the moon and Mars, and it has been argued that the glass is optically superior to that made on the Earth because it can be made anhydrous. Successful tests have been performed on earth using two lunar regolith simulants MLS–1 and MLS–2. Basalt fibre has also been made from lunar regolith simulators.

In August 2005, NASA contracted for the production of 16 tonnes of simulated lunar soil, or "Lunar Regolith Simulant Material." This material, called JSC–1a, is now commercially available for research on how lunar soil could be utilized *in–situ*.

Martian Moons, Ceres, Asteroids

Other proposals are based on Phobos and Deimos. These moons are in reasonably high orbits above Mars, have very low escape velocities, and unlike Mars have return delta-v's from their surfaces to LEO which are less than the return from the Moon.

Ceres is further out than Mars, with a higher delta-v, but launch windows and travel times are better, and the surface gravity is just 0.028 g, with a very low escape velocity of 510 m/s. Researchers have speculated that the interior configuration of Ceres includes a water–ice–rich mantle over a rocky core.

Near Earth Asteroids and bodies in the asteroid belt could also be sources of raw materials for ISRU.

Low Orbit

Gases like oxygen and argon could be extracted from the atmosphere of planets like the Earth and Mars by Propulsive Fluid Accumulator satellites in low orbit.

ISRU Classification

In October 2004, NASA's Advanced Planning and Integration Office commissioned an ISRU capability roadmap team. The team's report, along with those of 14 other capability roadmap teams, were published May 22, 2005. The report identifies seven ISRU capabilities : (i) resource extraction, (ii) material handling and transport, (iii) resource processing, (iv) surface manufacturing with *in–situ* resources, (v) surface construction, (vi) surface ISRU product and consumable storage and distribution, and (vii) ISRU unique development and certification capabilities.

ARTIFICIAL PHOTOSYNTHESIS

A sample of a photoelectric cell in a lab environment. Catalysts are added to the cell, which is submerged in water and illuminated by simulated sunlight. The bubbles seen are oxygen (forming on the front of the cell) and hydrogen (forming on the back of the cell).

Artificial photosynthesis is a chemical process that replicates the natural process of photosynthesis, a process that converts sunlight, water, and carbon dioxide into carbohydrates and oxygen. The term is commonly used to refer to any scheme for capturing and storing the energy from sunlight in the chemical bonds of a fuel (a solar fuel). Photo-catalytic water splitting converts water into protons (and eventually hydrogen) and oxygen, and is a main research area in artificial photosynthesis. Light–driven carbon dioxide reduction is another studied process, replicating natural carbon fixation.

Research developed in this field encompasses design and assembly of devices (and their components) for the direct production of solar fuels, photo-electrochemistry and its application in fuel cells, and engineering of enzymes and photo-autotrophic micro-organisms for microbial bio-fuel and biohydrogen production from sunlight. Many, if not most, of the artificial approaches are bio–inspired, *i.e.*, they rely on biomimetics.

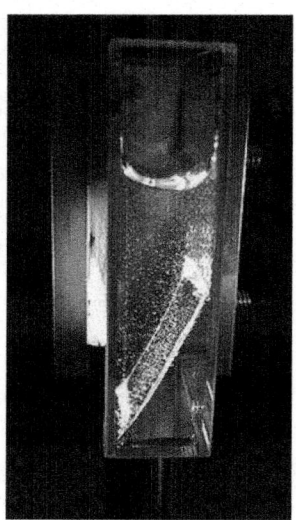

Overview

The photosynthetic reaction can be divided into two half–reactions (oxidation and reduction), both of which are essential to producing fuel. In plant photosynthesis, water molecules are photo–oxidized to release oxygen and protons. The second stage of plant photosynthesis (also known as the Calvin–Benson cycle) is a light-independent reaction that converts carbon dioxide into glucose. Researchers of artificial photosynthesis are developing photo-catalysts to perform both of these reactions separately. Furthermore, the protons resulting from water splitting can be used for hydrogen production. These catalysts must be able to react quickly and absorb a large percentage of solar photons.

Whereas photo-voltaics can provide direct electrical current from sunlight, the inefficiency of fuel production from photo-voltaic electricity (indirect process) and the fact sunshine is not constant throughout time sets a limit to its use. A way of using natural photosynthesis is via the production of bio-fuel through biomass, also an indirect process that suffers from low energy conversion efficiency (due to photosynthesis' own low efficiency in converting sunlight to biomass), and clashes with the increasing need of land mass for human food production. Artificial photosynthesis aims then to produce a fuel from sunlight that can be stored and used when sunlight is not available, by using direct processes, that is, to produce a solar fuel. With the development of catalysts able to reproduce the key steps of photosynthesis, water and sunlight would ultimately be the only needed sources for clean energy production. The only by–product would be oxygen, and production of a solar fuel has the potential to be cheaper than gasoline.

One process for the creation of a clean and affordable energy supply is the development of Photo-catalytic water splitting under solar light. This method of sustainable hydrogen production is a key objective in the development of alternative energy systems of the future. It is also predicted to be one of the more, if not the most, efficient ways of obtaining hydrogen from water. The conversion of solar energy into hydrogen via a water–splitting process assisted by photo-semi-conductor catalysts is one of the most promising technologies in development. This process has the potential for large quantities of hydrogen to be generated in an ecologically sound method. The conversion of solar energy into a clean fuel (H_2) under ambient conditions is one of the greatest challenges facing scientists in the twenty–first century.

Two approaches are generally recognized in the construction of solar fuel cells for hydrogen production :

- A homogeneous system is one where catalysts are not compartmentalized, that is, components are present in the same compartment. This means that hydrogen and oxygen are produced in the same location. This can be a drawback, since they compose an explosive mixture, demanding further gas purification. Also, all components must be active in approximately the same conditions (*e.g.*, pH).

- A heterogeneous system has two separate electrodes, an anode and a cathode, making possible the separation of oxygen and hydrogen production. Furthermore, different components do not necessarily need to work in the same conditions. However, the increased complexity of these systems makes them harder to develop and more expensive.

Another area of research within artificial photosynthesis is the selection and manipulation of photosynthetic micro-organisms, namely green microalgae and cyanobacteria, for the production of solar fuels. Many strains are able to produce hydrogen naturally, and scientists are working to improve them. Algae bio-fuels such as butanol and methanol are produced both at laboratory and commercial scales. This approach has benefited with the development of synthetic biology, which is also being explored by the J. Craig Venter Institute to produce a synthetic organism capable of bio-fuel production.

History

In the late 60s, Akira Fujishima discovered the Photo-catalytic properties of titanium dioxide, the so–called Honda–Fujishima effect, which could be used for hydrolysis.

The Swedish Consortium for Artificial Photosynthesis, the first of its kind, was established in 1994 as a collaboration between groups of three different universities, Lund, Uppsala and Stockholm, being presently active around Lund and the Ångström Laboratories in Uppsala. The consortium was built with a multidisciplinary approach to focus on learning from natural photosynthesis and applying this knowledge in biomimetic systems. Research in artificial photosynthesis is undergoing a boom at the beginning of the 21st century. In 2000, Commonwealth Scientific and Industrial Research Organisation (CSIRO) researchers publicize their intent to focus on carbon dioxide capture and conversion to hydrocarbons. In 2003, the Brookhaven National Laboratory announced the discovery of an important intermediate step in the reduction of CO_2 to CO (the simplest possible carbon dioxide reduction reaction), which could lead to better catalyst designing.

One of the drawbacks of artificial systems for water–splitting catalysts is their general reliance on scarce, expensive elements, such as ruthenium or rhenium. With the funding of the United States Air Force Office of Scientific Research, in 2008, MIT chemist and head of the Solar Revolution Project Daniel G. Nocera and post-doctoral fellow Matthew Kanan attempted to circumvent this issue by using a catalyst containing the cheaper and more abundant elements cobalt and phosphate. The catalyst was able to split water into oxygen and protons using sunlight, and could potentially be coupled to a hydrogen–producing catalyst such as platinum. Furthermore, while the catalyst broke down during catalysis, it could self–repair. This experimental catalyst design was considered a major breakthrough in the field by many researchers.

Whereas CO is the prime reduction product of CO_2, more complex carbon compounds are usually desired. In 2008, Princeton chemistry professor Andrew B. Bocarsly reported the direct conversion of carbon dioxide and water to methanol using solar energy in a highly efficient photo-chemical cell.

While Nocera and co-workers had accomplished water splitting to oxygen and protons, a light–driven process to produce hydrogen from protons still needed to be developed. In 2009, the Leibniz Institute for Catalysis reported inexpensive iron carbonyl complexes able to do just this. In the same year, researchers at the University of East Anglia also used iron carbonyl compounds to achieve photo-electro-chemical hydrogen production with 60% efficiency, this time using a gold electrode covered with layers of indium phosphide to which the iron complexes were linked. Both these processes used a molecular approach, where discrete nanoparticles are responsible for catalysis.

Visible light water splitting with a one piece multi-junction cell was first demonstrated and patented by William Ayers at Energy Conversion Devices in 1983. This group demonstrated water photolysis into hydrogen and oxygen, now referred to as an "artifical leaf" or "wireless solar water splitting" with a low cost,thin film amorphous silicon multi-junction cell directly immersed in water. Hydrogen evolved on the front amorphous silicon surface decorated with various catalysts while oxygen evolved off the back metal substrate which also eliminated the problem of mixed hydrogen/oxygen gas evolution. A Nafion membrane above the immersed cell provided a path for proton transport. The higher photo-voltage available from the multi-junction thin film cell with visible light was a major advance over previous photolysis attempts with UV sensitive single junction cells. The group's patent also lists several other semi-conductor multi-junction compositions in addition to amorphous silicon.

In 2009, F. del Valle and K. Domen showed the impact of the thermal treatment in a closed atmosphere using Cd1–xZnxS photo-catalysts. Cd1–xZnxS solid solution reports high activity in hydrogen production from water splitting under sunlight irradiation. A mixed heterogeneous/molecular approach by researchers at the University of California, Santa Cruz, in 2010, using both nitrogen–doped and cadmium selenide quantum dots–sensitized titanium dioxide nanoparticles and nanowires, also yielded photoproduced hydrogen.

Artificial photosynthesis remained an academic field for many years. However, in the beginning of 2009, Mitsubishi Chemical Holdings was reported to be developing its own artificial photosynthesis research by using sunlight, water and carbon dioxide to "create the carbon building blocks from which resins, plastics and fibers can be synthesized." This was confirmed with the establishment of the KAITEKI Institute later that year, with carbon dioxide reduction through artificial photosynthesis as one of the main goals.

In 2010, the DOE established, as one of its Energy Innovation Hubs, the Joint Center for Artificial Photosynthesis. The mission of JCAP is to find a cost–effective method to produce fuels using only sunlight, water, and carbon–dioxide as inputs. JCAP is led by a team from Caltech, led by Professor Nathan Lewis and brings together more than 120 scientists and engineers from Caltech and its lead partner, Lawrence Berkeley National Laboratory. JCAP also draws on the expertise and capabilities of key partners from Stanford University, the University of California at Berkeley, UCSB, UCI, and UCSD, and the Stanford Linear Accelerator. In addition, JCAP serves as a central hub for other solar fuels research teams across the United States, including 20 DOE Energy Frontier Research Center. The program has a budget of $122M over five years, subject to Congressional appropriation.

Also in 2010, a team led by professor David Wendell at the University of Cincinnati successfully demonstrated photosynthesis in an artificial construct consisting of enzymes suspended in frog foam.

In 2011, Daniel Nocera and his research team announced the creation of the first practical artificial leaf. In a speech at the 241st National Meeting of the American Chemical Society, Nocera described an advanced solar cell the size of a poker card capable of splitting water into oxygen and hydrogen, approximately ten times more efficient than natural photosynthesis. The cell is mostly made of inexpensive materials that are widely available, works under simple conditions, and shows increased stability over previous catalysts : in laboratory studies, the authors demonstrated that an artificial leaf prototype could operate continuously for at least forty–five hours without a drop in activity. In May 2012, Sun Catalytix, the startup based on Nocera's research, stated that it will not be scaling up the prototype as the device offers few savings over other ways to make hydrogen from sunlight.

Current Research

In energy terms, natural photosynthesis can be divided in three steps :

- Light–harvesting complexes in bacteria and plants capture photons and transduce them into electrons, injecting them into the photosynthetic chain.

- Proton–coupled electron transfer along several co-factors of the photosynthetic chain, causing local, spatial charge separation.
- Redox catalysis, which uses the aforementioned transferred electrons to oxidize water to dioxygen and protons; these protons can in some species be utilized for dihydrogen production.

Using biomimetic approaches, artificial photosynthesis tries to construct systems doing the same type of processes. Ideally, a triad assembly could oxidize water with one catalyst, reduce protons with another and have a photo-sensitizer molecule to power the whole system. One of the simplest designs is where the photo-sensitizer is linked in tandem between a water oxidation catalyst and a hydrogen evolving catalyst :

- The photo-sensitizer transfers electrons to the hydrogen catalyst when hit by light, becoming oxidized in the process.
- This drives the water splitting catalyst to donate electrons to the photo-sensitizer. In a triad assembly, such a catalyst is often referred to as a donor. The oxidized donor is able to perform water oxidation.

The state of the triad with one catalyst oxidized on one end and the second one reduced on the other end of the triad is referred to as a charge separation, and is a driving force for further electron transfer, and consequently catalysis, to occur. The different components may be assembled in diverse ways, such as supra-molecular complexes, compartmentalized cells, or linearly, covalently linked molecules.

Research into finding catalysts that can convert water, carbon dioxide, and sunlight to carbohydrates or hydrogen is a current, active field. By studying the natural oxygen–evolving complex, researchers have developed catalysts such as the "blue dimer" to mimic its function. Photo-electro-chemical cells that reduce carbon dioxide into carbon monoxide (CO), formic acid (HCOOH) and methanol (CH_3OH) are under development. However, these catalysts are still very inefficient.

Hydrogen Catalysts

Hydrogen is the simplest solar fuel to synthesize, since it involves only the transference of two electrons to two protons. It must, however, be done step-wise, with formation of an intermediate hydride anion :

$$2 e^- + 2 H^+ \leftrightarrow H^+ + H^- \leftrightarrow H_2$$

The proton–to–hydrogen converting catalysts present in nature are hydrogenases. These are enzymes that can either reduce protons to molecular hydrogen or oxidize hydrogen to protons and electrons. Spectroscopic and crystallographic studies spanning several decades have resulted in a good understanding of both the structure and mechanism of hydrogenase catalysis. Using this information, several molecules mimicking the structure of the active site of both nickel–iron and iron–iron hydrogenases have been synthesized. Other catalysts are not structural mimics of hydrogenase but rather functional ones. Synthesized catalysts include structural H–cluster models, a dirhodium photo-catalyst, and cobalt catalysts.

Water–oxidizing Catalysts

Water oxidation is a more complex chemical reaction than proton reduction. In nature, the oxygen–evolving complex performs this reaction by accumulating reducing equivalents (electrons) in a manganese–calcium cluster within photo-system II (PS II), then delivering them to water molecules, with the resulting production of molecular oxygen and protons :

$$2\,H_2O \rightarrow O_2 + 4\,H^+ + 4e^-$$

Without a catalyst (natural or artificial), this reaction is very endothermic, requiring high temperatures (at least 2500 K).

The exact structure of the oxygen–evolving complex has been hard to determine experimentally. As of 2011, the most detailed model was from a 1.9 Å resolution crystal structure of photo-system II. The complex is a cluster containing four manganese and one calcium ions, but the exact location and mechanism of water oxidation within the cluster is unknown. Nevertheless, bio–inspired manganese and manganese–calcium complexes have been synthesized, such as $[Mn_4O_4]$ cubanes, some with catalytic activity.

Some ruthenium complexes, such as the dinuclear μ–oxo–bridged "blue dimer" (the first of its kind to be synthesized), are capable of light–driven water oxidation, thanks to being able to form high valence states. In this case, the ruthenium complex acts as both photo-sensitizer and catalyst.

Many metal oxides have been found to have water oxidation catalytic activity, including ruthenium (IV) oxide (RuO_2), iridium (IV) oxide (IrO_2), cobalt oxides (including nickel–doped Co_3O_4), manganese oxide (including MnO_2 (birnessite), Mn_2O_3), and a mix of Mn_2O_3 with $CaMn_2O_4$. Oxides are easier to obtain than molecular catalysts, especially those from relatively abundant transition metals (cobalt and manganese), but suffer from low turnover frequency and slow electron transfer properties, and their mechanism of action is hard to decipher and, therefore, to adjust.

Recently Metal–Organic Framework (MOF)–based materials have been shown to be a highly promising candidate for water oxidation with first row transition metals. The stability and tunability of this system is projected to be highly beneficial for future development.

Photo-sensitizers

Nature uses pigments, mainly chlorophylls, to absorb a broad part of the visible spectrum. Artificial systems can use either one type of pigment with a broad absorption range or combine several pigments for the same purpose.

Ruthenium polypyridine complexes, in particular tris(bipyridine)ruthenium (II) and its derivatives, have been extensively used in hydrogen photo-production due to their efficient visible light absorption and long–lived consequent metal–to–ligand charge transfer excited state, which makes the complexes strong reducing agents. Other noble metal–containing complexes used include ones with platinum, rhodium and iridium.

Fig. : Structure of $[Ru(bipy)_3]^{2+}$, a broadly used photo-sensitizer.

Metal–free organic complexes have also been successfully employed as photo-sensitizers. Examples include eosin Y and rose bengal. Pyrrole rings such as porphyrins have also been used in coating nanomaterials or semi-conductors for both homogeneous and heterogeneous catalysis.

Carbon Dioxide Reduction Catalysts

In nature, carbon fixation is done by green plants using the enzyme RuBisCO as a part of the Calvin cycle. RuBisCO is a rather slow catalyst compared to the vast majority of other enzymes, incorporating only a few molecules of carbon dioxide into ribulose–1,5–bisphosphate per minute, but does so at atmospheric pressure and in mild, biological conditions. The resulting product is further reduced and eventually used in the synthesis of glucose, which in turn is a precursor to more complex carbohydrates, such as cellulose and starch. The process consumes energy in the form of ATP and NADPH.

Artificial CO_2 reduction for fuel production aims mostly at producing reduced carbon compounds from atmospheric CO_2. Some transition metal polyphosphine complexes have been developed for this end; however, they usually require previous concentration of CO_2 before use, and carriers (molecules that would fixate CO_2) that are both stable in aerobic conditions and able to concentrate CO_2 at atmospheric concentrations haven't been yet developed. The simplest product from CO_2 reduction is carbon monoxide (CO), but for fuel development, further reduction is needed, and a key step also needing further development is the transfer of hydride anions to CO.

Other Materials and Components

Charge separation is a key property of dyad and triad assemblies. Some nanomaterials employed are fullerenes (such as carbon nanotubes), a strategy that explores the pi–bonding properties of these materials. Diverse modifications (covalent and non–covalent) of carbon nanotubes have been attempted to increase the efficiency of charge separation, including the addition of ferrocene and pyrrole–like molecules such as porphyrins and phthalocyanines.

Since photodamage is usually a consequence in many of the tested systems after a period of exposure to light, bio–inspired photo-protectants have been tested, such as carotenoids (which are used in photosynthesis as natural protectants).

Light–driven Methodologies Under Development

Photo-electro-chemical cells

Photo-electro-chemical cells are a heterogeneous system that use light to produce either electricity or hydrogen. The vast majority of photo-electro-chemical cells use semi-conductors as catalysts. There have been attempts to use synthetic manganese complex–impregnated Nafion as a working electrode, but it has been since shown that the catalytically active species is actually the broken–down complex.

A promising, emerging type of solar cell is the dye–sensitized solar cell. This type of cell still depends on a semi-conductor (such as TiO_2) for current conduction on one electrode, but with a coating of an organic or inorganic dye that acts as a photo-sensitizer; the counter electrode is a platinum catalyst for H_2 production. These cells have a self–repair mechanism and solar–to–electricity conversion efficiencies rivaling those of solid–state semi-conductor ones.

Photo-catalytic Water Splitting in Homogeneous Systems

Direct water oxidation by photo-catalysts is a more efficient usage of solar energy than photo-electro-chemical water splitting because it avoids an intermediate thermal or electrical energy conversion step.

Bio–inspired manganese clusters have been shown to possess water oxidation activity when adsorbed on clays together with ruthenium photo-sensitizers, although with low turnover numbers.

As mentioned above, some ruthenium complexes are able to oxidize water under solar light irradiation. Although their photostability is still an issue, many can be reactivated by a simple adjustment of the conditions they work in. Improvement of catalyst stability has been tried resorting to polyoxometalates, in particular ruthenium–based ones.

Whereas a fully functional artificial system is usually envisioned when constructing a water splitting device, some mixed approaches have been tried. One of these involve the use of a gold electrode to which photo-system II is linked; an electrical current is detected upon illumination.

Hydrogen–producing Artificial Systems

A H–cluster FeFe hydrogenase model compound covalently linked to a ruthenium photo-sensitizer. The ruthenium complex absorbs light and transduces its energy to the iron compound, which can then reduce protons to H_2.

The simplest Photo-catalytic hydrogen production unit consists of a hydrogen–evolving catalyst linked to a photo-sensitizer. In this dyad assembly, a so–called sacrificial donor for the photo-sensitizer is needed, that is, one that is externally supplied and replenished; the photo-sensitizer donates the necessary reducing equivalents to the hydrogen–evolving catalyst, which uses protons from a solution where it is immersed or dissolved in. Cobalt compounds such as cobaloximes are some of the best hydrogen catalysts, having been coupled to both metal–containing and metal–free photo-sensitizers. The first H–cluster models linked to photo-sensitizers (mostly ruthenium photo-sensitizers, but also porphyrin–derived ones) were prepared in the early 2000s. Both types of assembly are under development to improve their stability and increase their turnover numbers, both necessary for constructing a sturdy, long–lived solar fuel cell.

As with water oxidation catalysis, not only fully artificial systems have been idealized : hydrogenase enzymes themselves have been engineered for photo-production of hydrogen, by coupling the enzyme to an artificial photo-sensitizer, such as $[Ru(bipy)_3]^{2+}$ or even photo-system I.

NADP+/NADPH Coenzyme–inspired Catalyst

In natural photosynthesis, the $NADP^+$ coenzyme is reducible to NADPH through binding of a proton and two electrons. This reduced form can then deliver the proton and electrons, potentially as a hydride, to reactions that culminate in the production of carbohydrates (the Calvin cycle). The coenzyme is recyclable in a natural photosynthetic cycle, but this process is yet to be artificially replicated.

A current goal is to obtain an NADPH–inspired catalyst capable of recreating the natural cyclic process. Utilizing light, hydride donors would be regenerated and produced where the molecules are continuously used in a closed cycle. Brookhaven chemists are now using a ruthenium–based complex to serve as the acting model. The complex is proven to perform correspondingly with NADP+/NADPH, behaving as the foundation for the proton and two electrons needed to convert acetone to isopropanol.

Currently, Brookhaven researchers are aiming to find ways for light to generate the hydride donors. The general idea is to use this process to produce fuels from carbon dioxide.

Photo-biological Production of Fuels

Some photo-autotrophic micro-organisms can, under certain conditions, produce hydrogen. Nitrogen–fixing micro-organisms, such as filamentous cyanobacteria, possess the enzyme nitrogenase, responsible for conversion of atmospheric N_2 into ammonia; molecular hydrogen is a by-product of this reaction, and is many times not released by the micro-organism, but rather taken up by a hydrogen-oxidizing (uptake) hydrogenase. One way of forcing these organisms to produce hydrogen is then to annihilate uptake hydrogenase activity. This has been done on a strain of *Nostoc punctiforme* : one of the structural genes of the NiFe uptake hydrogenase was inactivated by insertional mutagenesis, and the mutant strain showed hydrogen evolution under illumination.

Many of these photo-autotrophs also have bidirectional hydrogenases, which can produce hydrogen under certain conditions. However, other energy–demanding metabolic pathways can compete with the necessary electrons for proton reduction, decreasing the efficiency of the overall process; also, these hydrogenases are very sensitive to oxygen.

Several carbon–based bio-fuels have also been produced using cyanobacteria, such as 1–butanol.

Synthetic biology techniques are predicted to be useful in this field. Microbiological and enzymatic engineering have the potential of improving enzyme efficiency and robustness, as well as constructing new bio-fuel–producing metabolic pathways in photo-autotrophs that previously lack them, or improving on the existing ones. Another field under development is the optimization of photo-bioreactors for commercial application.

Employed Research Techniques

Research in artificial photosynthesis is necessarily a multi-disciplinary field, requiring a multitude of different expertise. Some techniques employed in making and investigating catalysts and solar cells include :

- Organic and inorganic chemical synthesis.
- Electro-chemistry methods, such as photo-electro-chemistry, cyclic voltammetry, and bulk electrolysis.
- Spectroscopic methods :
  - Fast techniques, such as time–resolved spectroscopy and ultra-fast laser spectroscopy;
  - Magnetic resonance spectroscopies, such as nuclear magnetic resonance, electron paramagnetic resonance;
  - X–ray absorption methods, such as EXAFS.

- Crystallography.
- Molecular biology, microbiology and synthetic biology methodologies.

Advantages, Disadvantages, and Efficiency

Advantages of Solar Fuel Production through Artificial Photosynthesis Include

- The solar energy can be immediately converted and stored. In photo-voltaic cells, sunlight is converted into electricity and then converted again into chemical energy for storage, with some necessary loss of energy associated with the second conversion.
- The by-products of these reactions are environmentally friendly. Artificially photosynthesized fuel would be a carbon–neutral source of energy, which could be used for transportation or homes.

Disadvantages Include

- Materials used for artificial photosynthesis often corrode in water, so they may be less stable than photo-voltaics over long periods of time. Most hydrogen catalysts are very sensitive to oxygen, being inactivated or degraded in its presence; also, photo-damage may occur over time.
- The overall cost is not yet advantageous enough to compete with fossil fuels as a commercially viable source of energy.

A concern usually addressed in catalyst design is efficiency, in particular how much of the incident light can be used in a system in practice. This is comparable with photosynthetic efficiency, where light–to–chemical–energy conversion is measured. Photosynthetic organisms are able to collect about 50% of incident solar radiation, but photo-chemical cells could use materials absorbing a wider range of solar radiation. It is however not straightforward to compare overall fuel production between natural and artificial systems : for example, plants have a theoretical threshold of 12% efficiency of glucose formation from photosynthesis, while a carbon reducing catalyst may go beyond this value. However, plants are efficient in using CO_2 at atmospheric concentrations, something that artificial catalysts still cannot perform.

Potential Global Impact

Being a renewable and carbon–neutral source of solar fuels, producing either hydrogen (which when burnt produces energy and fresh water) or carbohydrates, artificial photosynthesis is set apart from other popular renewable energy sources, specifically hydro-electric, solar photo-voltaic, geothermal, and wind-which produce electricity directly and centrally with no easily stored and transportable fuel intermediate. As such, artificial photosynthesis may become a very important source of fuel for transportation; unlike biomass energy, it does not require arable land and, consequently, will not compete with the food supply. One vision for globalizing artificial photosynthesis involves large light capture facilities linked to coastal metropolitan industrial plants where sea water is split to produce hydrogen and oxygen; the other involves all human structures covering the earth's surface (*i.e.*, roads, vehicles and buildings) doing photosynthesis more efficiently than plants.

At the fifteenth meeting of the International Congress of Photosynthesis Research (ISPR) in Beijing 27 August, 2010, a proposal was made for a "macroscience" Global Artificial Photosynthesis (GAP) Project, with seven models being presented for evaluation. An international conference on the subject took place between the fourteenth and eighteenth of August 2011 at Lord Howe Island under the auspices of the UNESCO Natural Sciences Sector. The meeting featured presentations from both scientists and non–scientific members of the society from across the globe, such as Peidong Yang, Dan Nocera and Michael Kirby and the papers presented have now been edited for a special open–source edition of the *Australian Journal of Chemistry*. Joint articles in Energy and Environmental Science by major researchers have supported a Global Artificial Photosynthesis (GAP) project and increasing the public policy profile of the field. It has been argued that photosynthesis in its natural and artificial forms should be declared common heritage of humanity under international law and that global artificial photosynthesis should be considered the moral culmination of nanotechnology. Its global use has been linked to the governance and policy concept of a socially and ecologically harmonious Sustainocene. In the Sustainocene, "instead of the cargo–cult ideology of perpetual economic growth through corporate pillage of nature, globalised artificial photosynthesis will facilitate a steady state economy and further technological revolutions" so that humans "no longer feel economically threatened, but rather proud, that their moral growth has allowed them to uphold Rights of Nature." Whether the world needs a Global Artificial Photosynthesis Project is the theme of a major international conference involving cutting edge science and policy to be held in July 2014 at Chicheley Hall, one of the finest country houses in Buckinghamshire owned by the Royal Society in association with the Kavli Foundation.

PHOTO-CATALYTIC WATER SPLITTING

Photo-catalytic water splitting is a general term used for the dissociation of water into its constituent parts, hydrogen (H_2) and oxygen (O_2), using either artificial or natural light. Hydrogen fuel production has gained increased attention as oil and other non-renewable fuels become increasingly depleted and expensive. Methods such as Photo-catalytic water splitting are being investigated to produce hydrogen fuel, which burns cleanly and can be used in a hydrogen fuel cell. Water splitting holds particular interest since it utilizes water, an inexpensive renewable resource. Photo-catalytic water splitting has the simplicity of using a powder in solution and sunlight to produce H_2 and O_2 from water and can provide a clean, renewable energy, without producing greenhouse gases or having many adverse effects on the atmosphere. Theoretically, only solar energy (photons), water, and a catalyst (photo-catalyst) are needed.

Concepts

When H

2O is split into O

2 and H

2, the stoichiometric ratio of its products is 2 : 1 :

$$2\,H_2O \underset{}{\overset{\text{photon energy} > 1.23eV}{\rightleftharpoons}} 2H_2 + O_2$$

The process of water–splitting is a highly endothermic process ($\Delta H > 0$). Water splitting occurs naturally in photosynthesis when photon energy is absorbed and converted into the chemical energy through a complex biological pathway. However, production of hydrogen from water requires large amounts of input energy, making it incompatible with existing energy generation. For this reason, most commercially produced hydrogen gas is produced from natural gas.

There are several strict requirements for a photo-catalyst to be useful for water splitting. The minimum potential difference (voltage) needed to split water is 1.23V at 0 pH. Since the minimum band gap for successful water splitting at pH=0 is 1.23 eV, corresponding to light of 1008 nm, the Electro-chemical requirements can theoretically reach down into infrared light, albeit with negligible catalytic activity. These values are true only for a completely reversible reaction at standard temperature and pressure (1 bar and 25°C).

Theoretically, infrared light has enough energy to split water into hydrogen and oxygen; however, this reaction is kinetically very slow because the wavelength is greater than 380 nm. The potential must be less than 3.0V to make efficient use of the energy present across the full spectrum of sunlight. Water splitting can transfer charges, but not be able to avoid corrosion for long term stability. Defects within crystalline photo-catalysts can act as recombination sites, ultimately lowering efficiency.

Materials used in Photo-catalytic water splitting fulfill the band requirements outlined previously and typically have dopants and/or co-catalysts added to optimize their performance. A sample semi-conductor with the proper band structure is TiO

2. However, due to the relatively positive conduction band of TiO

2, there is little driving force for H

2 production, so TiO

2 is typically used with a co–catalyst such as Pt to increase the rate of H

2 production. It is routine to add co–catalysts to spur H.

2 evolution in most photo-catalysts due to the conduction band placement. Most semi-conductors with suitable band structures to split water absorb mostly UV light; in order to absorb visible light, it is necessary to narrow the band gap. Since the conduction band is fairly close to the reference potential for H

2 formation, it is preferable to alter the valence band to move it closer to the potential for O

2 formation, since there is a greater natural overpotential.

Photo-catalysts can suffer from catalyst decay and recombination under operating conditions. Catalyst decay becomes a problem when using a sulfide-based photo-catalyst such as CdS, as the sulfide in the catalyst is oxidized to elemental sulfur at the same potentials used to split water. Thus, sulfide–based photo-catalysts are not viable without sacrificial reagents such as sodium sulfide to replenish any sulfur lost, which effectively changes the main reaction to one of hydrogen evolution as opposed to water splitting. Recombination of the electron–hole pairs needed for photo-catalysis can occur with any catalyst and is dependent on the defects and surface area of the catalyst; thus, a high degree of crystallinity is required to avoid recombination at the defects.

The conversion of solar energy to hydrogen by means of photo-catalysis is one of the most interesting ways to achieve clean and renewable energy systems. However if this process is assisted by photo-catalysts suspended directly in water instead of using a photo-voltaic and electrolytic system the reaction is in just one step, and can therefore be more efficient.

Method of Evaluation

Photo-catalysts must conform to several key principles in order to be considered effective at water splitting. A key principle is that H

2 and O

2 evolution should occur in a stoichiometric 2 : 1 ratio; significant deviation could be due to a flaw in the experimental setup and/or a side reaction, both of which do not indicate a reliable photo-catalyst for water splitting. The prime measure of photo-catalyst effectiveness is quantum yield (QY), which is :

QY (%) = (Photo-chemical reaction rate)/(Photon absorption rate) × 100%

This quantity is a reliable determination of how effective a photo-catalyst is; however, it can be misleading due to varying experimental conditions. To assist in comparison, the rate of gas evolution can also be used; this method is more problematic on its own because it is not normalized, but it can be useful for a rough comparison and is consistently reported in the literature. Overall, the best photo-catalyst has a high quantum yield and gives a high rate of gas evolution.

The other important factor for a photo-catalyst is the range of light absorbed; though UV–based photo-catalysts will perform better per photon than visible light–based photo-catalysts due to the higher photon energy, far more visible light reaches the Earth's surface than UV light. Thus, a less efficient photo-catalyst that absorbs visible light may ultimately be more useful than a more efficient photo-catalyst absorbing solely light with smaller wavelengths.

Photo-catalyst Systems

NaTaO3 : La

NaTaO3 : La yields the highest water splitting rate of photo-catalysts without using sacrificial reagents. This UV–based photo-catalyst was shown to be highly effective with water splitting rates of 9.7 mmol/h and a quantum yield of 56%. The nanostep structure of the material promotes water splitting as edges functioned as H_2 production sites and the grooves functioned as O_2 production sites. Addition of NiO particles as cocatalysts assisted in H_2 production; this step was done by using an impregnation method with an aqueous solution of $Ni(NO20{-}3) \bullet 6H_2O$ and evaporating the solution in the presence of the photo-catalyst. $NaTaO_3$ has a conduction band higher than that of NiO, so photogenerated electrons are more easily transferred to the conduction band of NiO for H_2 evolution.

K3Ta3B2O12

K3Ta3B2O12, another catalyst activated by solely UV light and above, does not have the performance or quantum yield of NaTaO3 :La. However, it does have the ability to split water without the assistance of cocatalysts and gives a quantum yield of 6.5% along with a water splitting rate of 1.21 mmol/h. This ability is due to the pillared structure of the photo-catalyst, which involves TaO_6 pillars connected by BO_3 triangle units. Loading with NiO did not assist the photo-catalyst due to the highly active H_2 evolution sites.

(Ga.82Zn.18)(N.82O.18)

(Ga.82Zn.18)(N.82O.18) has the highest quantum yield in visible light for visible light–based photo-catalysts that do not utilize sacrificial reagents as of October 2008. The photo-catalyst gives a quantum yield of 5.9% along with a water splitting rate of 0.4 mmol/h. Tuning the catalyst was done by increasing calcination temperatures for the final step in synthesizing the catalyst. Temperatures up to 600°C helped to reduce the number of defects, though temperatures above 700°C destroyed the local structure around zinc atoms and was thus undesirable. The treatment ultimately reduced the amount of surface Zn and O defects, which normally function as recombination sites, thus limiting Photo-catalytic activity. The catalyst was then loaded with Rh

2–yCryO3 at a rate of 2.5 wt% Rh and 2 wt% Cr to yield the best performance.

Pt/TiO2

TiO2 is the most effective photo-catalyst, as it yields both a high quantum number and a high rate of H_2 gas evolution. For example, Pt/TiO2 (anatase phase) is a catalyst used in water splitting. These photo-catalysts combine with a thin NaOH aqueous layer to make a solution that can split water into H_2 and O_2. TiO_2 absorbs only ultra-violet light due to its large band gap(>3.0ev), but outperforms most visible light photo-catalysts because it does not photo-corrode as easily.

Most ceramic materials have large band gaps and thus have stronger covalent bonds than other semi-conductors with lower band gaps.

Cobalt Based Systems

Photo-catalysts based on cobalt have been reported. Members are tris (bipyridine) cobalt (II), compounds of cobalt ligated to certain cyclic polyamines, and certain cobaloximes.

In 2014 researchers announced an approach that connected a chromophore to part of a larger organic ring that surrounded a cobalt atom. The process is less efficient than using a platinum catalyst, cobalt is less expensive, potentially reducing total costs. The process uses one of two supra-molecular assemblies based on Co (II)–templated coordination of Ru(bpy)32+ (bpy = 2,2'–bipyridyl) analogues as photo-sensitizers and electron donors to a cobaloxime macrocycle. The Co (II) centres of both assemblies are high spin, in contrast to most previously described cobaloximes. Transient absorption optical spectroscopies include that charge recombination occurs through multiple ligand states present within the photo-sensitizer modules.

Bismuth

Bismuth based systems have been demonstrated to have an efficiency of 5% with the advantage of a very simple and cheap catalyst.

BIOLOGICAL HYDROGEN PRODUCTION (ALGAE)

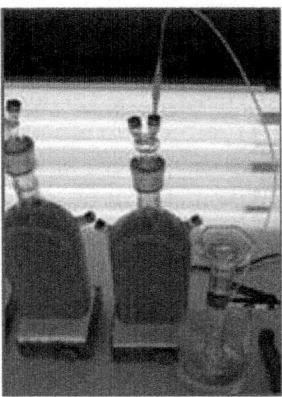

Fig. : An algae bioreactor for hydrogen production.

The **biological hydrogen production** with algae is a method of photo-biological water splitting which is done in a closed photo-bioreactor based on the production of hydrogen by algae. Algae produce hydrogen under certain conditions. In 2000 it was discovered that if *C. reinhardtii* algae are deprived of sulfur they will switch from the production of oxygen, as in normal photosynthesis, to the production of hydrogen.

History

In 1939 a German researcher named Hans Gaffron, while working at the University of Chicago, observed that the algae he was studying, *Chlamydomonas reinhardtii* (a green-algae), would sometimes switch from the production of oxygen to the production of hydrogen. He never discovered the cause for this change and for many years other scientists failed in their attempts at its discovery. In the late 1990s, professor Anastasios Melis a researcher at the University of California at Berkeley discovered that if the algae culture medium is deprived of sulfur it will switch from the production of oxygen (normal photosynthesis), to the production of hydrogen. He found that the enzyme responsible for this reaction is hydrogenase, but that the hydrogenase lost this function in the presence of oxygen. Melis found that depleting the amount of sulfur available to the algae interrupted its internal oxygen flow, allowing the hydrogenase an environment in which it can react, causing the algae to produce hydrogen. *Chlamydomonas moewusii* is also a good strain for the production of hydrogen. Scientists at the U.S. Department of Energy's Argonne National Laboratory are currently trying to find a way to take the part of the hydrogenase enzyme that creates the hydrogen gas and introduce it into the photosynthesis process. The result would be a large amount of hydrogen gas, possibly on par with the amount of oxygen created.

Milestones

1997 Professor Anastasios Melis discovered, after following Hans Gaffron's work, that the deprivation of sulfur will cause the algae to switch from producing oxygen to producing hydrogen. The enzyme, hydrogenase, he found was responsible for the reaction.

2006-Researchers from the University of Bielefeld and the University of Queensland have genetically changed the single-cell green alga *Chlamydomonas reinhardtii* in such a way that it produces an especially large amount of hydrogen. The Stm6 can, in the long run, produce five times the volume made by the wild form of alga and up to 1.6-2.0 per cent energy efficiency.

2007-It was discovered that if copper is added to block oxygen generation algae will switch from the production of oxygen to hydrogen

2007-Anastasios Melis studying solar-to-chemical energy conversion efficiency in *tlaX* mutants of *Chlamydomonas reinhardtii*, achieved 15% efficiency, demonstrating that truncated Chl antenna size would minimize wasteful dissipation of sunlight by individual cells This solar-to-chemical energy conversion process could be coupled to the production of a variety of bio-fuels including hydrogen.

2008-Anastasios Melis studying solar-to-chemical energy conversion efficiency in *tlaR* mutants of *Chlamydomonas reinhardtii*, achieved 25% efficiency out of a theoretical maximum of 30%.

2009-A team from the University of Tennessee, Knoxville and Oak Ridge National Laboratory stated that the process was more than 10 times more efficient as the temperature increased.

2011–Adding a bioengineered enzyme increases the rate of algal hydrogen production by about 400 per cent.

2011–A team at Argonne's Photosynthesis Group demonstrated how platinum nanoparticles can be linked to key proteins in algae to produce hydrogen fuel five times more efficiently.

Research

As of 2009, HydroMicPro is testing plate reactors.

As of 2013, Grow Energy has developed novel system for the large-scale production of hydrogen from structural bio-reactors.

Economics

It would take about 25,000 square kilometres to be sufficient to displace gasoline use in the US. To put this in perspective, this area represents approximately 10% of the area devoted to growing soya in the US.

In 2004, the US Department of Energy issued a selling price of $2.60 per kilogram ($1.18/lb) as a goal for making renewable hydrogen economically viable. 1 kg is approximately the energy equivalent to a gallon of gasoline. To achieve this, the efficiency of light–to–hydrogen conversion must reach 10% while 2004 achieved–efficiency is only 1% and the 2004 actual selling price is estimated at $13.53 per kilogram ($6.14/lb)

According to a 2004 DOE cost estimate, for a refueling station to supply 100 cars per day, it would need 300 kg. With current technology, a 300 kg per day stand–alone system will require 110,000 m^2 of pond area, 0.2 g/l cell concentration, a truncated antennae mutant and 10 cm pond depth.

Areas of research to increase efficiency include developing oxygen–tolerant FeFe–hydrogenases and increased hydrogen production rates through improved electron transfer.

Truncated Antenna

The chlorophyll (Chl) antenna size in green algae is minimized, or truncated, to maximize photo-biological solar conversion efficiency and H_2 production. The truncated Chl antenna size minimizes absorption and wasteful dissipation of sunlight by individual cells, resulting in better light utilization efficiency and greater photosynthetic productivity by the green alga mass culture.

Bioreactor Design Issues

- Restriction of photosynthetic hydrogen production by accumulation of a proton gradient.
- Competitive inhibition of photosynthetic hydrogen production by carbon dioxide.

- Requirement for bicarbonate binding at photo-system II (PSII) for efficient photosynthetic activity.
- Competitive drainage of electrons by oxygen in algal hydrogen production.
- Economics must reach competitive price to other sources of energy and the economics are dependent on several parameters.
- A major technical obstacle is the efficiency in converting solar energy into chemical energy stored in molecular hydrogen.

Attempts are in progress to solve these problems via bio-engineering.

THERMO-CHEMICAL CYCLE

Thermo-chemical cycles combine solely heat sources (*thermo*) with *chemical* reactions to split water into its hydrogen and oxygen components. The term *cycle* is used because aside of water, hydrogen and oxygen, the chemical compounds used in these processes are continuously recycled.

If work is partially used as an input, the resulting **thermo-chemical cycle** is defined as a hybrid one.

History

This concept was first postulated by Funk and Reinstrom (1966) as a reflexion about the most efficient way to produce fuels (*e.g.* hydrogen, ammonia) from stable and abundant species (*e.g.* water, nitrogen) and heat sources. Although fuel availability was scarcely considered before the oil crisis era, these researches were justified by niche markets. As an example, in the military logistics field, providing fuels for vehicles in remote battlefields is a key task. Hence, a mobile production system based on a portable heat source (a specific nuclear reactor was strikingly considered) was being investigated with the uttermost interest. Following the crisis, many programs (Europe, Japan, USA) were set up to design, test and qualify such processes for more peaceful purposes such as energy independence. High temperature (1000K) nuclear reactors were still considered as the heat sources. However, the optimistic expectations of the first thermodynamics studies were quickly moderated by more pragmatic analysis based on fair comparisons with standard technologies (thermodynamic cycles for electricity generation, coupled with the electrolysis of water) and by numerous practical issues (not high enough temperatures with nuclear reactors, slow reactivities, reactor corrosion, significant losses of intermediate compounds with time...). Hence, the interest for this technology was fading away during the next decades, or at least some tradeoffs (hybrid versions) were being considered with the use of electricity as a fractional energy input instead of only heat for the reactions (*e.g.* Hybrid sulfur cycle). A rebirth in the year 2000 can be explained by both new energy crisis and the rapid pace of development of concentrated solar power technologies whose potentially very high temperatures are ideal for thermo-chemical processes, while the environmentally friendly side of these researches attracts funding in a period with the peak oil shadow.

Principles

Water–splitting via a Single Reaction

Let us consider a system composed of chemical species (*e.g.* water–splitting) in thermodynamic equilibrium at constant pressure and thermodynamic temperature T :

$$H_2O(l) \rightleftharpoons H_2(g) + 1/2 \, O_2(g) \tag{1}$$

Equilibrium is displaced to the right only if energy (enthalpy change ΔH for water–splitting) is provided to the system under strict conditions imposed by Thermodynamics :

- One fraction must be provided as work, namely the Gibbs free energy change ΔG of the reaction : it consists of "noble" energy, *i.e.* under an organized state where matter can be controlled, such as electricity in the case of the electrolysis of water. Indeed, the generated electron flow can reduce protons (H^+)) at the cathode and oxidize anions (O^{2-}) at the anode (the ions exist because of the chemical polarity of water), yielding the desired species.

- The other one must be supplied as heat, *i.e.* by increasing the thermal agitation of the species, and is equal by definition of the entropy to the absolute temperature T times the entropy change ΔS of the reaction.

$$\Delta H = \Delta G + T\Delta S \tag{2}$$

Hence, for an ambient temperature T° of 298K (kelvin) and a pressure of 1 atm (atmosphere (unit)) ($\Delta G°$ and $\Delta S°$ are respectively equal to 237 kJ/mol and 163 J/mol/K, relative to the initial amount of water), more than 80% of the required energy ΔH must be provided as work in order for water–splitting to proceed.

If phase transitions are neglected for simplicity's sake (*e.g.* water electrolysis under pressure to keep water in its liquid state), one can assume that ΔH et ΔS do not vary significantly for a given temperature change. These parameters are thus taken equal to their standard values $\Delta H°$ et $\Delta S°$ at temperature T°. Consequently, the work required at temperature T is,

$$\Delta G = \Delta G^0 - (T - T^0)\Delta S^0 \tag{3}$$

As $\Delta S°$ is positive, a temperature increase leads to a reduction of the required work. This is the basis of high–temperature electrolysis. This can also be intuitively explained graphically. Chemical species can have various excitation levels depending on the absolute temperature T, which is a measure of the thermal agitation. The latter causes shocks between atoms or molecules inside the closed system such that energy spreading among the excitation levels increases with time, and stop (equilibrium) only when most of the species have similar excitation levels (a molecule in a highly excited level will quickly return to a lower energy state by collisions) (Entropy (statistical thermodynamics)).

Relative to the absolute temperature scale, the excitation levels of the species are gathered based on standard enthalpy change of formation considerations; *i.e.* their stabilities. As this value is null for water but strictly positive for oxygen and hydrogen, most of the excitation levels of these last species are above the ones of water. Then, the density of the excitation levels for a given temperature range is monotonically increasing with the species entropy. A positive entropy change for water–splitting means far more excitation levels in the products. Consequently,

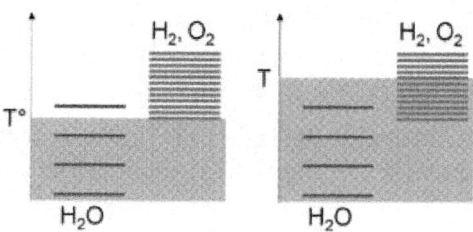

- A low temperature ($T°$), thermal agitation allow mostly the water molecules to be excited as hydrogen and oxygen levels required higher thermal agitation to be significantly populated (on the arbitrary diagram, 3 levels can be populated for water *vs* 1 for the oxygen/hydrogen sub-system),
- At high temperature (T), thermal agitation is sufficient for the oxygen/hydrogen sub-system excitation levels to be excited (on the arbitrary diagram, 4 levels can be populated for water *vs* 8 for the oxygen/hydrogen sub-system). According to the previous statements, the system will thus evolve toward the composition where most of its excitation levels are similar, *i.e.* a majority of oxygen and hydrogen species.

One can imagine that if T were high enough in Eq.(3), ΔG could be nullified, meaning that water–splitting would occur even without work (thermolysis of water). Though possible, this would required tremendously high temperatures : considering the same system naturally with steam instead of liquid water ($\Delta H°$ = 242 kJ/mol; $\Delta S°$ = 44 J/mol/K) would hence give required temperatures above 3000K, that make reactor design and operation extremely challenging.

Hence, a single reaction only offers one freedom degree (T) to produce hydrogen and oxygen only from heat (though using Le Chatelier's principle would also allow to slightly decrease the thermolysis temperature, work must be provided in this case for extracting the gas products from the system)

Water–splitting with Multiple Reactions

On the contrary, as shown by Funk and Reinstrom, multiple reactions (*e.g.* k steps) provide additional means to allow spontaneous water–splitting without work thanks to different entropy changes $\Delta S°_i$ for each reaction i. An extra-benefit compared with water thermolysis is that oxygen and hydrogen are separately produced, avoiding complex separations at high temperatures.

The first pre–requisites (Eqs.(4) and (5)) for multiple reactions i to be equivalent to water–splitting are trivial (cf. Hess's law) :

- $$\sum_i \Delta H_i^0 = \Delta H^0 \qquad\qquad\qquad (4)$$

- $$\sum_i \Delta S_i^0 = \Delta S^0 \tag{5}$$

Similarly, the work ΔG required by the process is the sum of each reaction work ΔG_i :

$$\Delta G = \sum_i \Delta G_i \tag{6}$$

As Eq.(3) is a general law, it can be used anew to develop each ΔG_i term. If the reactions with positive (p indice) and negative (n indice) entropy changes are expressed as separate summations, this gives,

$$\Delta G = \sum_p (\Delta G_i^0 - (T_i - T^0)\Delta S_i^0) + \sum_n (\Delta G_i^0 - (T_i - T^0)\Delta S_i^0) \tag{7}$$

Using Eq.(6) for standard conditions allows to factorize the ΔG_i° terms, yielding,

$$\Delta G = \Delta G^0 + \sum_p (T_i - T^0)(-\Delta S_i^0) + \sum_n (T_i - T^0)(-\Delta S_i^0) \tag{8}$$

Now let us consider the contribution of each summation in Eq.(8) : in order to minimize ΔG, they must be as negative as possible :

- $\sum_p (T_i - T^0)(-\Delta S_i^0)$: $-\Delta S^{\circ}_i$ are negative, so $(T - T^0)$ must be as high as possible

  : hence, one choose to operate at the maximum process temperature T_H

- $\sum_n (T_i - T^0)(-\Delta S_i^0)$: $-\Delta S^{\circ}_i$ are positive, $(T - T^0)$ should be ideally negative in

  order to decrease ΔG. Practically, one can only set T equals to T^0 as the minimum process temperature in order to get rid of this troublesome term (a process requiring a lower than standard temperature for energy production is a physical absurdity as it would required refrigerators and thus a higher work input than output). Consequently, Eq.(8) becomes,

$$\Delta G = \Delta G^0 - (T_H - T^0) \sum_p \Delta S_i^0 \tag{9}$$

Finally, one can deduce from this last equation the relationship required for a null work requirement ($\Delta G \leq 0$) :

$$\sum_p \Delta S_i^0 \geq \frac{\Delta G^0}{(T_H - T^0)} \tag{10}$$

Consequently, a thermo-chemical cycle with i steps can be defined as sequence of i reactions equivalent to water–splitting and satisfying equations (4), (5) and (10). The key point to remember in that case is that the process temperature T_H can theoretically be arbitrary chosen (1000K as a reference in most of the past studies, for high temperature nuclear reactors), far below the water thermolysis one.

This equation can alternatively (and naturally) be derived *via* the Carnot's theorem, that must be respected by the system composed of a thermo-chemical process coupled with a work producing unit (chemical species are thus in a closed loop) :

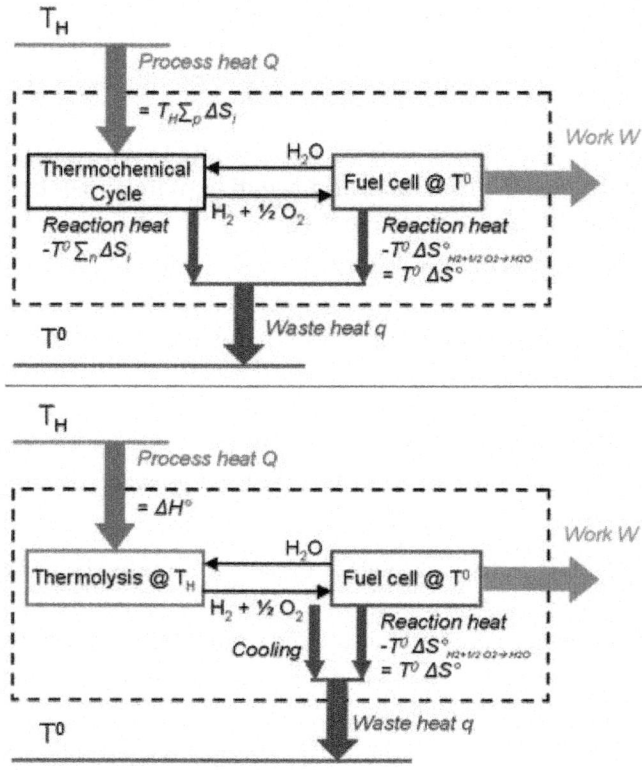

- At least two heat sources of different temperatures are required for cyclical operation, otherwise perpetual motion would be possible. This is trivial in the case of thermolysis, as the fuel is consumed *via* an inverse reaction. Consequently, if there is only one temperature (the thermolysis one), maximum work recovery in a fuel cell is equal to the opposite of the Gibbs free energy of the water–splitting reaction at the same temperature, *i.e.* null by definition of the thermolysis. Or differently said, a fuel is defined by its instability, so if the water/hydrogen/oxygen system only exists as hydrogen and oxygen (equilibrium state), combustion (engine) or use in a fuel cell would not be possible.

- Endothermic reactions are chosen with positive entropy changes in order to be favoured when the temperature increases, and the opposite for the exothermic reactions.

- Maximal heat–to–work efficiency is the one of a Carnot heat engine with the same process conditions, *i.e.* a hot heat source at T_H and a cold one at T^0,

$$\frac{W}{Q} \leq \frac{T_H - T^0}{T_H} \tag{11}$$

- The work output W is the "noble" energy stored in the hydrogen and oxygen products (*e.g.* released as electricity during fuel consumption in a fuel cell). It thus corresponds to the free Gibbs energy change of water–splitting ΔG, and is maximum according to Eq.(3) at the lowest temperature of the process (T^0) where it is equal to $\Delta G°$.

- The heat input Q is the heat provided by the hot source at temperature T_H to the i endothermic reactions of the thermo-chemical cycle (the fuel consumption sub-system is exothermic) :

$$Q = \sum_i q_i \tag{12}$$

Hence, each heat requirement at temperature T_H is,

$$q_i = T_H \Delta S_i \tag{13}$$

Replacing Eq.(13) in Eq.(12) yields :

$$Q = T_H \sum_p \Delta S_i \tag{14}$$

Consequently, replacing W ($\Delta G°$) and Q (Eq.(14)) in Eq.(11) gives after reorganization Eq.(10) (assuming that the ΔS_i do not change significantly with the temperature, *i.e.* are equal to $\Delta S°_i$).

Equation (10) has practical implications about the minimum number of reactions for such a process according to the maximum process temperature T_H. Indeed, a numerical application ($\Delta G°$ equals to 229 kJ/K for water considered as steam) in the case of the originally chosen conditions (high–temperature nuclear reactor with T_H and T° respectively equal to 1000K and 298K) gives a minimum value around *330 J/mol/K* for the summation of the positive entropy changes $\Delta S°_i$ of the process reactions.

This last value is very high as most of the reactions have entropy change values below 50 J/mol/K, and even an elevated one (e.g. water–splitting from liquid water : 163 J/mol/K) is twice lower. Consequently, thermo-chemical cycles composed of less than three steps are practically impossible with the originally planned heat sources (below 1000K), or require "hybrid" versions

Hybrid Thermo-chemical Cycles

In this case, an extra-freedom degree is added *via* a relatively small work input W_{add} (maximum work consumption, Eq.(9) with $\Delta G \le W_{add}$), and Eq.(10) becomes,

$$\sum_p \Delta S_i^0 \ge \frac{\Delta G^0 - W_{add}}{(T_H - T^0)} \tag{15}$$

If W_{add} is expressed as a fraction f of the process heat Q (Eq.(14)), Eq.(15) becomes after reorganization,

$$\sum_p \Delta S_i^0 \ge \frac{\Delta G^0}{((1+f)T_H - T^0)} \tag{16}$$

Using a work input equals to a fraction f of the heat input is equivalent relative to the choice of the reactions to operate a pure similar thermo-chemical cycle but with a hot source with a temperature increased by the same proportion f.

Naturally, this decreases the heat–to–work efficiency in the same proportion *f*. Consequently, if one want a process similar to a thermo-chemical cycle operating with a 2000K heat source (instead of 1000K), the maximum heat–to–work efficiency is twice lower. As real efficiencies are often significantly lower than ideal one, such a process is thus strongly limited.

Practically, use of work is restricted to key steps such as product separations, where techniques relying on work (*e.g.* electrolysis) might sometimes have fewer issues than those using only heat (*e.g.* distillations)

Particular case : Two–step thermo-chemical cycles

According to equation (10), the minimum required entropy change (right term) for the summation of the positive entropy changes decreases when T_H increases. As an example, performing the same numerical application but with T_H equals to 2000K would give a twice lower value (around 140 kJ/mol), which allows thermo-chemical cycles with only two reactions. Such processes can be realistically coupled with concentrated solar power technologies like solar towers. As an example in Europe, this is the goal of the Hydrosol–2 project (Greece, Germany (German Aerospace Center), Spain, Denmark, England) and of the researches of the solar department of the ETH Zurich and the Paul Scherrer Institute (Switzerland).

Examples of reactions satisfying high entropy changes are metal oxide dissociations, as the products have more excitation levels due to their gaseous state (metal vapours and oxygen) than the reactant (solid with crystalline structure, so symmetry dramatically reduces the number of different excitation levels). Consequently, these entropy changes can often be larger than the water–splitting one and thus a reaction with a negative entropy change is required in the thermo-chemical process so that Eq.(5) is satisfied. Furthermore, assuming similar stabilities of the reactant ($\Delta H°$) for both thermolysis and oxide dissociation, a larger entropy change in the second case explained again a lower reaction temperature (Eq.(3)).

Let us assume two reactions, with positive (*1* subscript, at T_H) and negative (2 subscript, at $T°$) entropy changes. An extra-property can be derived in order to have T_H strictly lower than the thermolysis temperature : *The standard thermodynamic values must be unevenly distributed among the reactions.*

Indeed, according to the general equations (2) (spontaneous reaction), (4) and (5), one must satisfy,

$$\frac{\Delta H_1^0}{\Delta S_1^0} < \frac{\Delta H_1^0 + \Delta H_2^0}{\Delta S_1^0 + \Delta S_2^0} \tag{17}$$

Hence, if $\Delta H°_1$ is proportional to $\Delta H°_2$ by a given factor, and if $\Delta S°_1$ and $\Delta S°_2$ follow a similar law (same proportionality factor), the inequality (17) is broken (equality instead, so T_H equals to the water thermolysis temperature).

Examples

Hundreds of such cycles have been proposed and investigated. This task has been eased by the availability of computers, allowing a systematic screening of chemical reactions sequences based on thermodynamic databases. Only the main "families" will be described in this article.

Cycles with More Than 3 Steps or Hybrid Ones

Cycles Based on the Sulfur Chemistry

An advantage of the sulfur chemical element is its high covalence. Indeed, it can form up to 6 chemical bonds with other elements such as oxygen (*e.g.* sulfates), *i.e.* a wide range of oxidation states. Hence, there exist several redox reactions involving such compounds. This freedom allows numerous chemical steps with different entropy changes, and thus offer more odds to meet the criteria required for a thermo-chemical cycle (cf. Principles). Most of the first studies were performed in the USA, as an example at the Kentucky University for sulfide–bases cycles. Sulfate–based cycles were studied in the same laboratory and also at Los Alamos National Laboratory and at General Atomics. Significant researches based on sulfates (*e.g.* $FeSO_4$ and $CuSO_4$) were also performed in Germany and in Japan. However, the cycle which has given rise to the highest interests is probably the (Sulfur–iodine cycle) one (acronym : *S–I*) discovered by General Atomics.

Cycles Based on The Reversed Deacon Process

Above 973K, the Deacon reaction is reversed, yielding hydrogen chloride and oxygen from water and chlorine :

$$H_2O + Cl_2 \rightarrow 2\,HCl + 1/2\,O_2$$

PHOTOHYDROGEN

Photohydrogen is hydrogen produced with the help of artificial or natural light This is how the leaf of a tree splits water molecules into protons (hydrogen ions), electrons (to make carbohydrates) and oxygen (released into the air as a waste product). Photohydrogen may also be produced by the photodissociation of water by ultra-violet light.

Photohydrogen is sometimes discussed in the context of obtaining renewable energy from sunlight, by using microscopic organisms such as bacteria or algae. These organisms create hydrogen with the help of hydrogenase enzymes which convert protons derived from the water splitting reaction into hydrogen gas which can then be collected and used as a bio-fuel.

WATER–GAS SHIFT REACTION

The **water–gas shift reaction** (WGSR) describes the reaction of carbon monoxide and water vapour to form carbon dioxide and hydrogen (the mixture of carbon monoxide and hydrogen is known as water gas) :

$$CO + H_2O \rightleftharpoons CO_2 + H_2$$

The water gas shift reaction was discovered by Italian physicist Felice Fontana in 1780. It was not until much later that the industrial value of this reaction was realized. Before the early 20th century, hydrogen was obtained by reacting steam under high pressure with iron to produce iron, iron oxide and hydrogen. With the development of industrial processes that required hydrogen, such as the Haber-Bosch ammonia synthesis, the demand for a cheaper and more efficient method of hydrogen production was needed. As a resolution to this problem, the WGSR was combined with the gassification of coal to produce a pure hydrogen product. As the ideal of Hydrogen economy gains popularity, the focus on hydrogen as a replacement fuel source for hydrocarbons is increasing.

Applications

The WGSR is an important industrial reaction that is used in the manufacture of ammonia, hydrocarbons, methanol, and hydrogen. It is also often used in conjunction with steam reformation of methane and other hydrocarbons. In the Fischer–Tropsch process, the WGSR is one of the most important reactions used to balance the H_2/CO ratio. It provides a source of hydrogen at the expense of carbon monoxide, which is important for the production of high purity hydrogen for use in ammonia synthesis.

The water–gas shift reaction may be an undesired side reaction in processes involving water and carbon monoxide, *e.g.* the rhodium–based Monsanto process. The iridium–based Cativa process uses less water, which suppresses this reaction.

The equilibrium of this reaction shows a significant temperature dependence and the equilibrium constant decreases with an increase in temperature, that is, higher carbon monoxide conversion is observed at lower temperatures. In order to take advantage of both the thermodynamics and kinetics of the reaction, the industrial scale water gas shift reaction is conducted in multiple adiabatic stages consisting of a high temperature shift (HTS) followed by a low temperature shift (LTS) with inter-system cooling. The initial HTS takes advantage of the high reaction rates, but is thermodynamically limited, which results in incomplete conversion of carbon monoxide and a 2–4% carbon monoxide exit composition. To shift the equilibrium toward hydrogen production, a subsequent low temperature shift reactor is employed to produce a carbon monoxide exit composition of less than 1%. The transition from the HTS to the LTS reactors necessitates inter-system colling. Due to the different reaction conditions, different catalysts must be employed at each stage to ensure optimal activity. The commercial HTS catalyst is the iron oxide–chromium oxide catalyst and the LTS catalyst is a copper–based catalyst. The order proceeds from high to low temperature due to the susceptibility of the copper catalyst to poisoning by sulfur that may remain after the steam reformation process. This necessitates the removal of the sulfur compounds prior to the LTS reactor by a guard bed in order to protect the copper catalyst. Conversely, the iron used in the HTS reaction is generally more robust and resistant toward poisoning by sulfur compounds. While both the HTS and LTS catalysts are commercially available, their specific composition varies based on vendor. An important limitation for the HTS is the H_2O/CO ratio where low ratios may lead to side reactions such as the formation of metallic iron, methanation, carbon deposition, and Fischer–Tropsch reaction.

Low Temperature Shift (LTS)

The typical composition of commercial LTS catalyst has been reported as 32–33% CuO, 34–53% ZnO, 15–33% Al_2O_3. The active catalytic species is CuO. The function of ZnO is to provide structural support as well as prevent the poisoning of copper by sulfur. The Al_2O_3 prevents dispersion and pellet shrinkage. The LTS shift reactor operates at a range of 200 °C to 250 °C. Low reaction temperatures must be maintained due to the susceptibility of copper to thermal sintering. These lower temperatures also reduce the occurrence of side reactions that are observed in the case of the HTS.

High Temperature Shift Catalysts

The typical composition of commercial HTS catalyst has been reported as 74.2% Fe_2O_3, 10.0% Cr_2O_3, 0.2% MgO (remaining percentage attributed to volatile components). The chromium acts to stabilize the iron oxide and prevents sintering. The operation of HTS catalysts occurs within the temperature range of 310°C to 450°C. The temperature increases along the length of the reactor due to the exothermic nature of the reaction. As such, the inlet temperature is maintained at 350°C to prevent the exit temperature from exceeding 550°C. Industrial reactors operate at a range from atmospheric pressure to 8375 kPa.

Fuel Cells

The WGSR can aid in the efficiency of fuel cells by increasing hydrogen production. The WGSR is considered a critical component in the reduction of carbon monoxide concentrations in cells that are susceptible to carbon monoxide poisoning such as the proton exchange membrane (PEM) fuel cell. The benefits of this application are two-fold : not only would the water gas shift reaction effectively reduce the concentration of carbon monoxide, but it would also increase the efficiency of the fuel cells by increasing hydrogen production. Unfortunately, current commercial catalysts that are used in industrial water gas shift processes are not compatible with fuel cell applications. With the high demand for clean fuel and the critical role of the water gas shift reaction ion hydrogen fuel cells, the development of water gas shift catalysts for the application in fuel cell technology is an area of current research interest.

Catalysts for fuel cell application would need to operate at low temperatures. Since the WGSR is slow at lower temperatures where equilibrium favours hydrogen production, WGS reactors require large amounts of catalysts, which increases their cost and size beyond practical application. The commercial LTS catalyst used in large scale industrial plants is also pyrophoric in its inactive state and therefore presents safety concerns for consumer applications. Developing a catalyst that can overcome these limitations is relevant to implementation of a hydrogen economy.

Reaction Conditions

While the WGSR has been extensively studied for over a hundred years, the mechanism remains under debate. A universal rate expression and mechanistic understanding have proven elusive, reflecting the many reaction variables, vagaries of the catalyst, and the proprietary nature of commercial processes.

Temperature Dependence

The water gas shift reaction is a moderately exothermic reversible reaction. Therefore with increasing temperature the reaction rate increases but the conversion of reactants to products becomes less favourable. Due to its exothermic nature, high carbon monoxide conversion is thermodynamically favoured at low temperatures. Despite the thermodynamic favourability at low temperatures, the reaction is kinetically favoured at high temperatures. The water–gas shift reaction is sensitive to temperature, with the tendency to shift towards reactants as temperature increases due to Le Chatelier's principle. Over the temperature range 600–2000 K, the logarithm of the equilibrium constant for the WGSR is given by the following equation :

$$\log K_{eq} = -2.4198 + 0.0003855T + \frac{2180.6}{T}$$

The value of K_{eq} approaches 1 at 1100 K. The following plot depicts the temperature dependence of K_{eq} as shown by this equation.

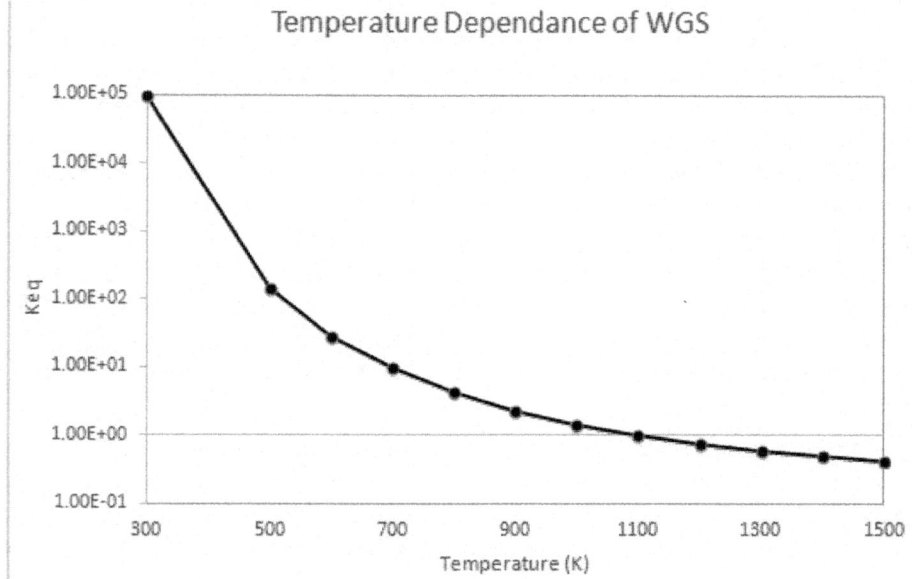

Mechanism

Two main mechanisms have been proposed : an associative 'Langmuir–Hinshel-wood' mechanism, and a regenerative 'redox' mechanism. While the regenerative mechanism is generally implemented to describe the WGS at higher temperatures, at low temperature both the redox and associative mechanisms are suitable explanations.

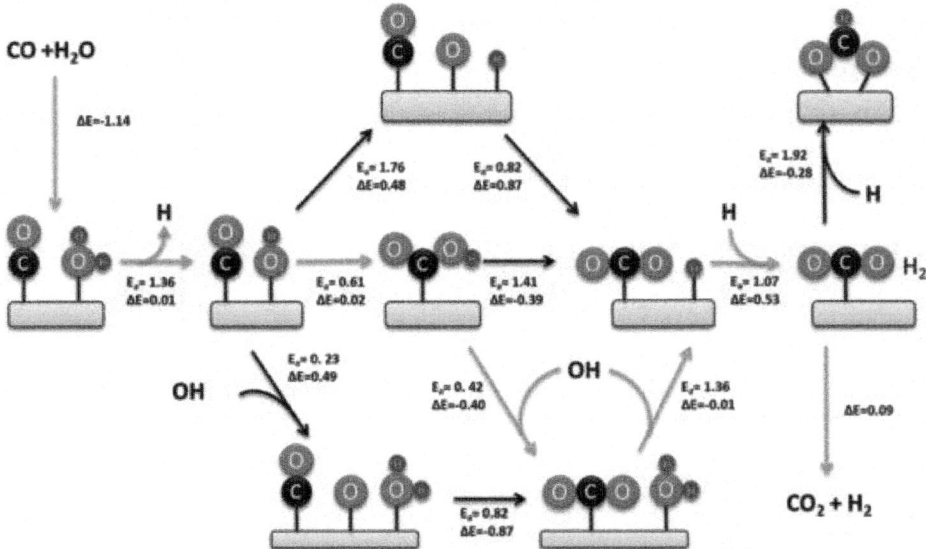

Fig. : Associative and Redox Mechanism of the Water Gas Shift.

Associative Mechanism

In 1920 Armstrong and Hilditch first proposed the associative mechanism. In this mechanism CO and H_2O are adsorbed onto the surface of the metal catalyst followed by the formation of an intermediate and the desorption of H_2 and CO_2. In the initial step, H_2O dissociates into a metal adsorbed OH and H. The hydroxide then reacts with CO to form a carboxyl or formate intermediate which subsequently decomposes into CO_2 and the metal adsorbed H, which ultimately yields H_2. While this mechanism may be valid under LTS conditions, the redox mechanism which does not involve any long lived surface intermediates is a more suitable explanation of the WGS mechanism at higher temperatures.

Redox Mechanism

The regenerative 'redox' mechanism is the most commonly accepted mechanism for the WGSR. It involves a regenerative change in the oxidation state of the catalytic metal. In this mechanism, H_2O is activated first by the abstraction of H from water followed by dissociation or disproportionation of the resulting OH to afford atomic O. The CO is then oxidized by the atomic O forming CO_2 which returns the catalytic surface back to its pre–reaction state. Alternatively, CO may be directly oxidized by the OH to form a carboxyl intermediate, followed by the dissociation or disproportionation of the carboxyl. Finally H is recombined to H_2 and CO_2 and H_2 are desorbed from the metal. The principal difference in these mechanisms is the formation of CO_2. The redox mechanism generates CO_2 by reaction with adsorbed oxygen, while the associative mechanism forms CO_2 via the dissociation of an intermediate. The mechanism of decarboxylation is debated; it may involve β–hydride elimination, or it may require the action of an external base.

Homogeneous Models

Useful mechanistic insights resulted from the discovery that metal carbonyls catalyze the WGSR. The basic reaction entails nucleophilic attack of water or hydroxide on a M–CO center, generating a metallacarboxylic acid.

Reverse Water Gas Shift

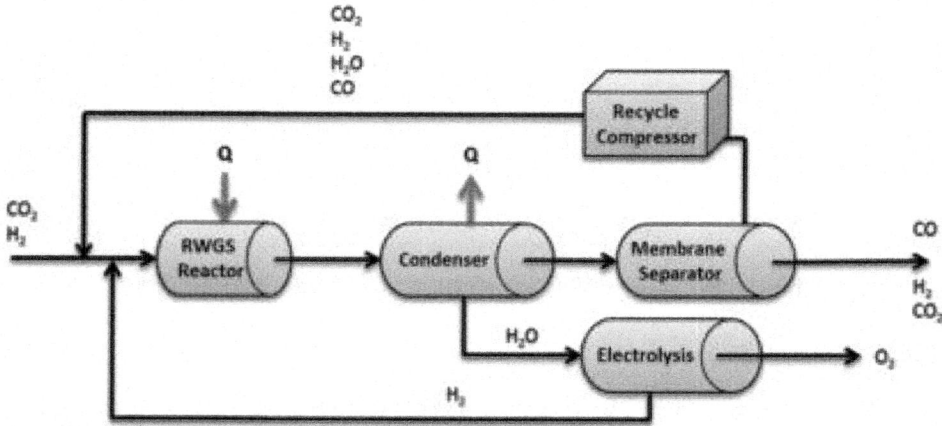

Fig. : Reverse Water Gas Shift Flow Cylcle.

Depending on the reaction conditions, the equilibrium for the water gas shift can be pushed in either the forward or reverse direction. The reversibility of the WGSR is important in the production of ammonium, methanol, and Fischer–Tropsch synthesis where the ratio of H_2/CO is critical. Many other industrial companies exploit the reverse water gas shift reaction (RWGS) reaction as a source of the synthetically valuable CO from cheap CO_2. Typically, It is done using a copper on aluminium catalyst. The RWGS reaction is also gaining interest in the context of the human missions to Mars primarily for its potential to produce water and oxygen. The Mars atmosphere is about 95% CO_2 which can be utilized by the RWGS reaction given a source of hydrogen. Coupling the RWGS with the water electrolysis process will yield methane and oxygen. Post electrolysis, the hydrogen produced can be recycled back into the RWGS reactor for the continued conversion of CO_2. Because this reaction is only mildly endothermic, the thermal power needed to drive this reaction can potentially be produced by a Sabatier reactor.

ACID NEUTRALIZING CAPACITY

Acid–neutralizing capacity or **ANC** in short is a measure for the overall buffering capacity against acidification for a solution, *e.g.* surface water or soil water.

ANC is defined as the difference between cations of strong bases and anions of strong acids, or dynamically as the amount of acid needed to change the pH value from the sample's value to a chosen different value. The concepts alkalinity are nowadays often used as a synonym to positive ANC and similarly acidity is often used to mean negative ANC. Alkalinity and acidity however also have definitions based on an experimental setup (titration).

ANC is often used in models to calculate acidification levels from acid rain pollution in different geographical areas, and as a basis for calculating critical loads for forest soils and surface waters.

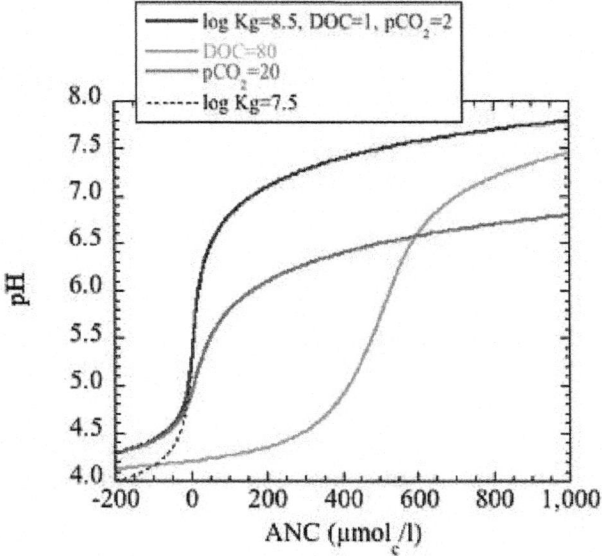

The relation between pH and ANC in natural waters depends on three conditions : Carbon dioxide, organic acids and aluminium solubility. The amount of dissolved carbon dioxide is usually higher than would be the case if there was an equilibrium with the carbon dioxide pressure in the atmosphere. This is due to biological activity : Decomposition of organic material releases carbon dioxide and thus increases the amount of dissolved carbon dioxide. An increase in carbon dioxide decreases pH but has no effect on ANC. Organic acids, often expressed as dissolved organic carbon (DOC), also decrease pH and have no effect on ANC. Soil water in the upper layers usually have higher organic content than the lower soil layers. Surface waters with high DOC are typically found in areas where there is a lot of peat and bogs in the catchment. Aluminium solubility is a bit tricky and there are several curve fit variants used in modelling, one of the more common being

$$[Al^{3+}] = k_G[H^+]^3$$

In the illustration to the right, the relation between pH and ANC is shown for four different solutions. In the blue line the solution has 1 mg/l DOC, a dissolved amount of carbon dioxide that is equivalent to a solution being in equilibrium with an atmosphere with twice the carbon dioxide pressure of our atmosphere. For the other lines, all three parameters except one is the same as for the blue line. Thus the orange line is a solution loaded with organic acids, having a DOC of 80 mg/l (typically very brown lake water or water in the top soil layer in a forest soil). The red line has a high amount of dissolved carbon dioxide (pCO_2=20 times ambient), a level that is not uncommon in ground water. Finally the black dotted line is a water with a lower aluminium solubility.

The reason why ANC is often defined as the difference between cations of strong bases and anions of strong acids is that ANC is derived from a charge balance : If we for simplicity consider a solution with only a few species and use the fact that a water solution is electrically neutral we get :

$$[H^+]+2[Ca^{2+}]+[Na^+]+3[Al^{3+}]+2[Al(OH)^{2+}]+[Al(OH)_2^+] = [OH^-]+[Cl^-]+2[CO_3^{2-}]+[HCO_3^-]+[R^-]$$

where R^-denote an anion of an organic acid. ANC is then defined by collecting all species controlled by equilibrium (*i.e.* species related to weak acids and weak bases) on one side and species not controlled by equilibrium (*i.e.* species related to strong acids and strong bases) on the other side. Thus, with the species above we get :

$$ANC = +2[Ca^{2+}] + [Na^+] - [Cl^-]$$

or

$$ANC = [OH^-]+ 2[CO_3^{2-}]+[HCO_3^-]+[R^-]-[H^+]-3[Al^{3+}]-2[Al(OH)^{2+}]-[Al(OH)_2^-]$$

Note :

1. That a change in DOC or CO_2 (or for that matter Aluminium solubility, but Aluminium solubility is not something that is easily controlled) does NOT have any effect on ANC.

2. That once a pH–ANC relation for has been established for a lake the pH–ANC relation can be used to easily calculate the amount of limestone needed to raise lake pH to *e.g.* 5.5

3. Not all acid lakes are acid due to human influence since high DOC gives low pH.

4. That the concentrations are multiplied with the charge of the species, hence the unit mol charge per liter.

POLYMER ELECTROLYTE MEMBRANE ELECTROLYSIS

Polymer electrolyte membrane (PEM) electrolysis is the electrolysis of water in a cell equipped with a solid polymer electrolyte (SPE) that is responsible for the conduction of protons, separation of product gases, and electrical insulation of the electrodes. The PEM electroyzer was introduced to overcome the issues of partial load, low current density, and low pressure operation currently plaguing the alkaline electrolyzer. Electrolysis is an important new technology for the production of hydrogen to be used as an energy carrier. With fast dynamic response times, large operational ranges, high efficiencies, and very high gas purities (99.999%), PEM electrolysis is a promising alternative for energy storage coupled with renewable energy sources.

Polymer electrolyte membrane electrolysis

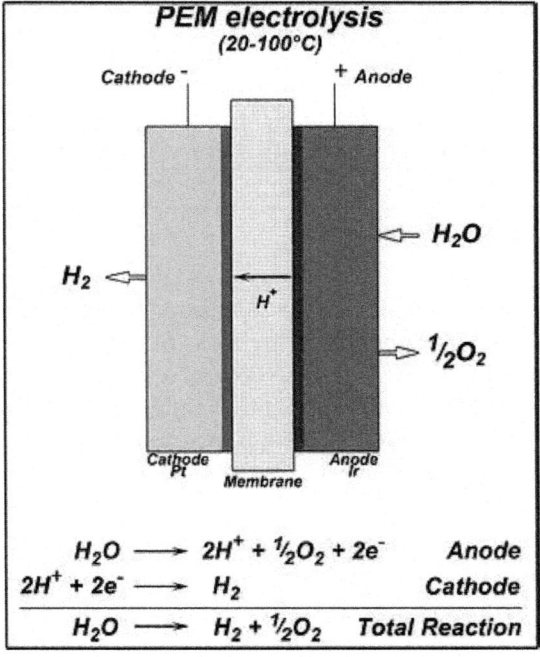

Fig. : Diagram of PEM electrolysis reactions.

Table : Typical Materials

Type of Electrolysis :	PEM Electrolysis
Style of membrane/diaphragm :	Solid polymer
Bipolar/separator plate material :	Titanium or gold and platinum coated titanium
Catalyst material on the anode :	Iridium
Catalyst material on the cathode :	Platinum
Anode PTL material :	Titanium
Cathode PTL material :	Carbon paper/carbon fleece
State–of–the–art Operating Ranges	
Cell temperature :	50–80C
Stack pressure :	<30 bar
Cell area :	<300 cm^2
Current density :	0.6–2.0 mA/cm^2

(Contd...)

(Contd...)

Cell voltage :	1.75–2.20 V
Power density :	to 4.4 mW/cm²
Part–load range :	0–10%
Spec. energy consumption stack :	4.2–5.6 kWh/Nm³
Spec. energy consumption system :	4.5–7.5 kWh/Nm³
Cell voltage efficiency :	57–69%
System hydrogen production rate :	30 Nm³/h
Lifetime stack :	<60,000 h
Acceptable degradation rate :	<14 µV/h
System Lifetime :	10–20 a

History

The use of a PEM for electrolysis was first introduced in the 1960s by General Electric, developed to overcome the drawbacks to the alkaline electrolyis technology. The initial performances yielded 1.88 V at 1.0 A/cm² which was, compared to the alkaline electrolysis technology of that time, very efficient. In the late 1970s the alkaline electrolyzers were reporting performances around 2.06 V at 0.215 A/cm², thus prompting a sudden interest in the late 1970s and early 1980s in polymer electrolytes for water electrolysis.

A thorough review of the historical performance from the early research to that of today can be found in chronological order with many of the operating conditions in the 2013 review by Carmo *et. al.*

Advantages of PEM Electrolysis

One of the largest advantages to PEM electrolysis is its ability to operate at high current densities. This can in result in reduced operational costs, especially for systems coupled with very dynamic energy sources such as wind and solar, where sudden spikes in energy input would otherwise result in uncaptured energy. The polymer electrolyte allows the PEM electrolyzer to operate with a very thin membrane (~100–200µm) while still allowing high pressures, resulting in low ohmic losses, primarily caused by the conduction of protons across the membrane (0.1 S/cm) and a compressed hydrogen output.

The polymer electrolyte membrane, due to its solid structure, exhibits a low gas crossover rate resulting in very high product gas purity. Maintaining a high gas purity is important for storage safety and for the direct usage in a fuel cell. The safety limits for H_2 in O_2 are at standard conditions 4 mol–% H_2 in O_2.

Science

An electrolyzer is an Electro-chemical device to convert electricity and water into hydrogen and oxygen, these gases can then be used as a means to store energy for later use. This use can range from electrical grid stabilization from dynamic electrical sources such as wind turbines and solar cells to localized hydrogen production as a fuel for fuel cell vehicles. The PEM electrolyzer utilizes a solid polymer electrolyte (SPE) to conduct protons from the anode to the cathode while insulating the electrodes electrically. Under standard conditions the enthalpy required for the formation of water is 285.9 kJ/mol. A portion of the required energy for a sustained electrolysis reaction is supplied by thermal energy and the remainder is supplied through electrical energy.

Reactions

The actual value for open circuit voltage of an operating electrolyzer will lie between the 1.23 V and 1.48 V depending how the cell/stack design utilizes the thermal energy inputs. This is however quite difficult to determine or measure because an operating electrolyzer also experiences other voltage losses from internal electrical resistances, proton conductivity, mass transport through the cell and catalyst utilization to name a few.

Anode Reaction

The half reaction taking place on the anode side of a PEM electrolyzer is commonly referred to as the Oxygen Evolution Reaction (OER). Here the liquid water reactant is supplied to catalyst where the supplied water is oxidized to oxygen, protons and electrons.

$$2H_2O(l) \rightarrow O_2(g) + 4H^+(aq) + 4e^-$$

Cathode Reaction

The half reaction taking place on the cathode side of a PEM electrolyzer is commonly referred to as the Hydrogen Evolution Reaction (HER). Here the supplied electrons and the protons that have conducted through the membrane are combined to create gaseous hydrogen.

$$2H^+(aq) + 2e^- \rightarrow H_2(g)$$

Second Law of Thermodynamics

As per the second law of thermodynamics the enthalpy of the reaction is :

$$\Delta H = \underbrace{\Delta G}_{\text{elec.}} + \underbrace{T \Delta S}_{\text{heat}}$$

Where ΔG is the Gibb's free energy of the reaction, T is the temperature of the reaction and ΔS is the change in entropy of the system.

$$H_2O(l) + \Delta H = H_2 + \tfrac{1}{2}O_2$$

The overall cell reaction with thermodynamic energy inputs then becomes :

$$H_2O_{(l)} + \underbrace{237.2 \text{ kJ / mol}}_{\text{electricity}} + \underbrace{48.6 \text{ kJ / mol}}_{\text{heat}} \rightarrow H_2 + \tfrac{1}{2}O_2$$

The thermal and electrical inputs shown above represent the minimum amount of energy that can be supplied by electricity in order to obtain an electrolysis reaction. Assuming that the maximum amount of heat energy (48.6 kJ/mol) is supplied to the reaction, the reversible cell voltage V^0_{rev} can be calculated.

Open Circuit Voltage (OCV)

$$V^0_{rev} = \frac{\Delta G^0}{n \cdot F} = \frac{237 \text{ kJ/mol}}{2 \times 96,485 \text{ C/mol}} = 1.23 \text{V}$$

where n is the number of electrons and F is Faraday's constant. The calculation of cell voltage assuming no irreversibilities exist and all of the thermal energy is utilized by the reaction is referred to as the lower heating value (LHV). The alternative formulation, using the higher heating value (HHV) is calculated assuming that all of the energy to drive the electrolysis reaction is supplied by the electrical component of the required energy which results in a higher reversible cell voltage. When using the HHV the voltage calculation is referred to as the thermoneutral voltage.

$$V^0_{th} = \frac{\Delta G^0}{n \cdot F} = \frac{285.9 \text{ kJ/mol}}{2 \times 96,485 \text{ C/mol}} = 1.48 \text{V}$$

Voltage Losses

The performance of electrolysis cells, like fuel cells, are typically compared by plotting their polarization curves, which is obtained by plotting the cell voltage against the current density. The primary sources of increased voltage in a PEM electrolyzer (the same also applies for PEM fuel cells) can be categorized into thee main areas, ohmic loss Ohmic losses, activation losses and mass transport losses. Due to the reversal of operation between a PEM fuel cell and a PEM electrolyzer, the degree of impact for these various losses is different between the two processes.

$$V_{cell} = E + V_{act} + V_{trans} + V_{ohm}$$

The performance of a PEM electrolysis system is typically compared by plotting the overpotential versus the cells current density. This essentially results in a curve that represents the power per square centimeter of cell area required to produce hydrogen and oxygen. Conversely to the PEM fuel cell, the better the PEM electrolyzer the lower the cell voltage at a given current density. The figure below is the result of a simulation from the Forschungszentrum Jülich of a 25 cm^2 single cell PEM electrolyzer under thermoneutral operation depicting the primary sources of voltage loss and their contributions for a range of current densities.

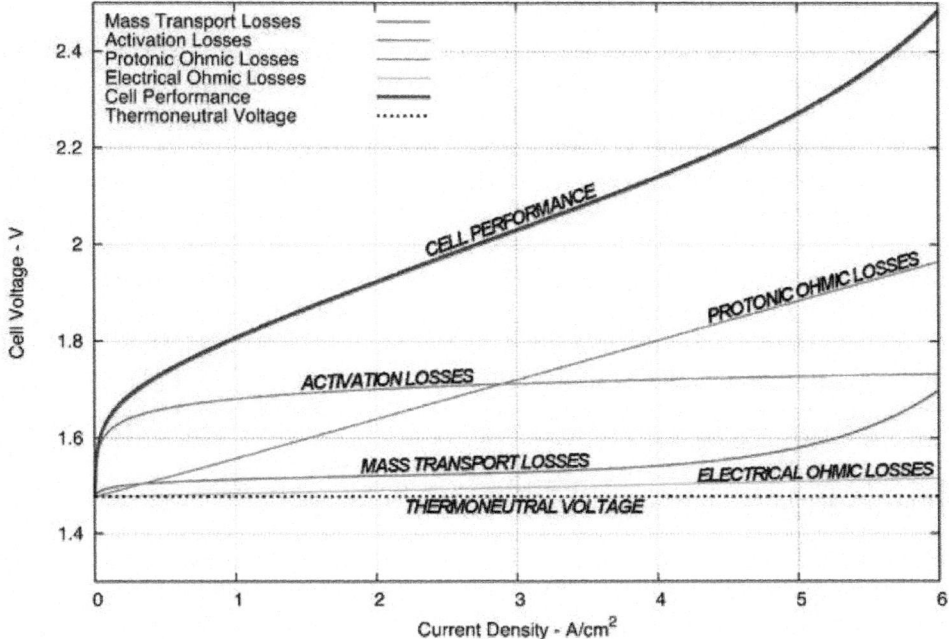

Ohmic Losses

Ohmic losses are an electrical overpotential introduced to the electrolysis process by the internal resistance of the cell components. This loss then requires an additional voltage to maintain the electrolysis reaction, the prediction of this loss follows Ohm's law and holds a linear relationship to the current density of the operating electrolyzer.

$$V = I \cdot R$$

The energy loss due to the electrical resistance is not entirely lost. The voltage drop due to resistivity is associated with the conversion the electrical energy to heat energy through a process known as Joule heating. Much of this heat energy is carried away with the reactant water supply and lost to the environment, however a small portion of this energy is then recaptured as heat energy in the electrolysis process. The amount of heat energy that can be recaptured is dependent on many aspects of system operation and cell design.

$$Q \, \alpha \, I^2 \cdot R$$

The Ohmic losses due to the conduction of protons contribute to the loss of efficiency which also follows Ohm's law, however without the Joule heating effect. The proton conductivity of the PEM is very dependent on the hydration, temperature, heat treatment, and ionic state of the membrane.

Faradaic Losses and Crossover

Faradaic losses describe the efficiency losses that are correlated to the current, that is supplied without leading to hydrogen at the cathodic gas outlet. The produced hydrogen and oxygen can permeate across the membrane, referred to as crossover. Mixtures of both gases at the electrodes result. At the cathode, oxygen can be catalytically reacted with hydrogen on the platinum surface of the cathodic catalyst. At the anode, hydrogen and oxygen do not react at the iridium oxide catalyst. Thus, safety hazards due to explosive anodic mixtures hydrogen in oxygen can result. The supplied energy for the hydrogen production is lost, when hydrogen is lost due to the reaction with oxygen at the cathode and permeation from the cathode across the membrane to the anode corresponds. Hence, the ratio of the amount of lost and produced hydrogen determines the faradaic losses. At pressurized operation of the electrolyzer the crossover and the correlated faradaic efficiency losses increase.

Hydrogen Compression During Water Electrolysis

Hydrogen evolution due to pressurized electrolysis is comparable to an isothermal compression process, which is in terms of efficiency preferable compared to mechanical isotropical compression. However, the contributions of the afore mentioned faradaic losses increase with operating pressures. Thus, in order to produce compressed hydrogen, the *in–situ* compression during electrolysis and subsequent compression of the gas have to be pondered under efficiency considerations.

PEM Electrolysis System Operation

The ability of the PEM electrolyzer to operate, not only under highly dynamic conditions, but also in part–load and overload conditions is one of the reasons for the recently renewed interest in this technology. The demands of an electrical grid are relatively stable and predictable, however when coupling these to energy sources such as wind and solar, the demand of the grid rarely matches the generation of the renewable energy. This means energy produced from renewable sources such as wind and solar must have a buffer, or a means of storing off–peak energy.

HOFMANN VOLTAMETER

A **Hofmann voltameter** is an apparatus for electrolyzing water, invented by August Wilhelm von Hofmann (1818–1892) in 1866. It consists of three joined upright cylinders, usually glass. The inner cylinder is open at the top to allow addition of water and an ionic compound to improve conductivity, such as a small amount of sulphuric acid. A platinum electrode is placed inside the bottom of each of the two side cylinders, connected to the positive and negative terminals of a source of electricity. When current is run through Hofmann's voltameter, gaseous oxygen forms at the anode and gaseous hydrogen at the cathode. Each gas displaces water and collects at the top of the two outer tubes.

Name

The name 'voltameter' was coined by Daniell who shortened Faraday's original name of "volta–electrometer". Hofmann voltameters are no longer used as electrical measuring devices. However, before the invention of the ammeter, voltameters were often used to measure direct current, since current through a voltameter with iron or copper electrodes electroplates the cathode with an amount of metal from the anode directly proportional to the total coulombs of charge transferred (Faraday's law of electrolysis). The modern name is "Electro-chemical coulometer". Although the correct spelling of Hofmann contains only one 'f', it is often incorrectly depicted as Hoffmann.

Uses

The amount of electricity that has passed through the system can then be determined by weighing the cathode. Thomas Edison used voltameters as electricity meters. (A Hofmann voltameter cannot be used to weigh electric current in this fashion, as the platinum electrodes are too inert for plating.)

A Hofmann voltameter is often used as a demonstration of stoichiometric principles, as the two–to–one ratio of the volumes of hydrogen and oxygen gas produced by the apparatus illustrates the chemical formula of water, H_2O. However, this is only true if oxygen and hydrogen gases are assumed to be diatomic. If hydrogen gas were monatomic and oxygen diatomic, the gas volume ratio would be 4 : 1. The volumetric composition of water is the ratio by volume of hydrogen to oxygen present This value is 2 : 1 experimentally,this value is determined using Hofmann's water voltameter.

HETEROGENEOUS WATER OXIDATION

Heterogeneous Water Oxidation

Water oxidation is one of the half reactions of water splitting :

$2H_2O \rightarrow O_2 + 4H^+ + 4e^-$ Oxidation (generation of dioxygen)

$4H^+ + 4e^- \rightarrow 2H_2$ Reduction (generation of dihydrogen)

$2H_2O \rightarrow 2H_2 + O_2$ Total Reaction.

Of the two half reactions, the oxidation step is the most demanding because it requires the coupling of 4 electron and proton transfers and the formation of an oxygen–oxygen bond. This process occurs naturally in plants photo-system II to provide protons and electrons for the photosynthesis process and release oxygen to the atmosphere. Since hydrogen can be used as an alternative clean burning fuel, there has been a need to split water efficiently. However, there are known materials that can mediate the reduction step efficiently therefore much of the current research is aimed at the oxidation half reaction also known as the Oxygen Evolution Reaction (OER). Current research focuses on understanding the mechanism of OER and development of new materials that catalyze the process.

Thermodynamics

Both the oxidation and reduction steps are pH dependent.

2 half reactions (at pH = 0)

Oxidation $2H_2O \rightarrow 4H^+ + 4e^- + O_2$, E° = 1.23 V *vs.* NHE

Reduction $4H^+ + 4e^- \rightarrow 2H_2$, E° = 0.00 V *vs.* NHE

Overall $2H_2O \rightarrow 2H_2 + O_2$ E°cell = 1.23 V; ΔG = 475 kJ/mol.

Water splitting can be done at higher pH values as well however the standard potentials will vary according to the Nernst equation and therefore shift by –59 mV for each pH unit increase. However, the total cell potential (difference between oxidation and reduction half cell potentials) will remain 1.23 V. This potential can be related to Gibbs free energy (ΔG) by :

$$\Delta G°cell = -nFE°cell$$

Where n is the number of electrons per mole products and F is the Faraday constant. Therefore, it takes 475 kJ of energy to make one mole of O_2 as calculated by thermodynamics. However, in reality no process can be this efficient. Systems always suffer from an overpotential that arise from activation barriers, concentration effects and voltage drops due to resistance. The activation barriers or activation energy is associated with high energy transition states that are reached during the Electro-chemical process of OER. The lowering of these barriers would allow for OER to occur at lower overpotentials and faster rates.

Mechanism

Heterogeneous OER is sensitive to the surface which the reaction takes place and is also affected by the pH of the solution. The general mechanism for acidic and alkaline solutions is shown below. Under acidic conditions water binds to the surface with the irreversible removal of one electron and one proton to form a platinum hydroxide. In an alkaline solution a reversible binding of hydroxide ion coupled to a one electron oxidation is thought to precede a turnover–limiting Electro-chemical step involving the removal of one proton and one electron to form a surface oxide species. The shift in mechanism between the pH extremes has been attributed to the kinetic facility of oxidizing hydroxide ion relative to water. Using the Tafel equation, one can obtain kinetic information about the kinetics of the electrode material such as the exchange current density and the Tafel slope. OER is presumed to not take place on clean metal surfaces such as Platinum, but instead an oxide surface is formed prior to oxygen evolution.

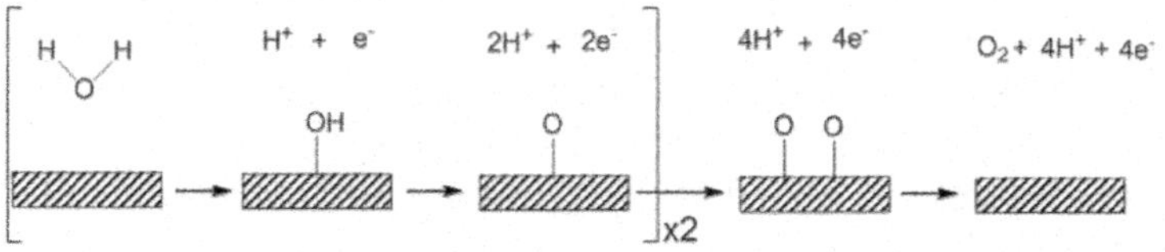

Fig. : OER under acidic conditions.

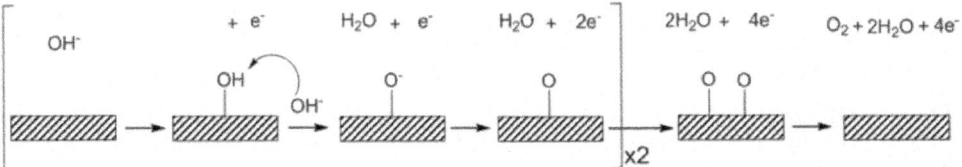

Fig. : OER under alkaline conditions.

Catalyst Materials

OER has been studied on a variety of materials including :

- Platinum surfaces
- Transition metal oxides
- First–row transition metal spinels and perovskites. Recently Metal–Organic Framework (MOF)–based materials have been shown to be a highly promising candidate for water oxidation with first row transition metals.;

Preparation of the surface and electrolysis conditions have a large effect on reactivity (defects, steps, kinks, low coordinate sites) therefore, it is difficult to predict an OER material's properties by its bulk structure. Surface effects have a large influence on the kinetics and thermodynamics of OER.

Platinum

Platinum has been a widely studied material for OER because it is the catalytically most active element for this reaction. It exhibits exchange current values on the order of 10^{-9} A/cm^2. Much of the mechanistic knowledge of OER was gathered from studies on platinum and its oxides. It was observed that there was a lag in the evolution of oxygen during electrolysis. Therefore an oxide film must first form at the surface before OER begins. The Tafel slope, which is related to the kinetics of the electro-catalytic reaction, was shown to be independent of the oxide layer thickness at low current densities but becomes dependant on oxide thickness at high current densities.

Ruthenium Oxide

Ruthenium oxide (RuO_2) shows some of the best performance as an OER material in acidic environments. It has been studied since the early 1970s as a water oxidation catalyst with one of the lowest reported overpotentials for OER at the time. It has since been investigated for OER in Ru(110) single crystal oxide surfaces, compact films, Titanium supported films. RuO_2 films can be prepared by thermal decomposition of ruthenium chloride on inert substrates.

PARTICLE AGGREGATION

Particle aggregation refers to formation of clusters in a colloidal suspension and represents the most frequent mechanism leading to destabilization of colloidal systems. During this process, which normally occurs within short periods of time (seconds to hours), particles dispersed in the liquid phase stick to each other, and spontaneously form irregular particle clusters, flocs, or aggregates. This phenomenon is also referred to as *coagulation* or *flocculation* and such a suspension is also called *unstable*. Particle aggregation can be induced by adding salts or an other chemical referred to as coagulant or flocculant. Some people refer to specifically to flocculation when aggregation is induced by addition of polymers or polyelectrolytes, while coagulation is a more widely used term.

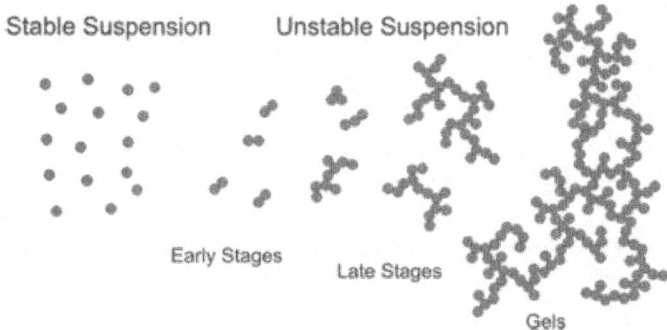

Scheme of particle aggregation. Particles are dispersed individually in a stable suspension, while they aggregate in an unstable suspension. As aggregation proceed from early to later states, the aggregates grow in size, and may eventually gel.

Particle aggregation is normally an irreversible process. Once particle aggregates have formed, they will not easily disrupt. In the course of aggregation, the aggregates will grow in size, and as a consequence they may settle to the bottom of the container, which is referred to as sedimentation. Alternatively, a colloidal gel may form in concentrated suspensions which changes its rheological properties. The reverse process whereby particle aggregates are disrupted and dispersed as individual particles, referred to as peptization, hardly occurs spontaneously, but may occur under stirring or shear.

Colloidal particles may also remain dispersed in liquids for long periods of time (days to years). This phenomenon is referred to as *colloidal stability* and such a suspension is said to be *stable*. Stable suspensions are often obtained at low salt concentrations or by addition of chemicals referred to as *stabilizers* or *stabilizing agents*.

Similar aggregation processes occur in other dispersed systems too. In emulsions, they may also be coupled to droplet coalescence, and not only lead to sedimentation but also to creaming. In aerosols, airborne particles may equally aggregate and form larger clusters (*e.g.*, soot).

Early Stages

A well dispersed colloidal suspension consists of individual, separated particles and is stabilized by repulsive inter–particle forces. When the repulsive forces weaken or become attractive through the addition of a coagulant, particles start to aggregate. Initially, particle doublets A_2 will form from singlets A_1 according to the scheme

$$A_1 + A_1 \rightarrow A_2$$

In the early stage of the aggregation process, the suspension mainly contains particle monomers and some dimers. The rate of this reaction is characterized by the aggregation rate coefficient k. Since doublet formation is a second order rate process, the units of this coefficients are m^3s^{-1} since particle concentrations are expressed as particle number per unit volume (m^{-3}). Since absolute aggregation rates are difficult to measure, one often refers to the dimensionless stability ratio $W = k_{fast}/k$ where k_{fast} is the aggregation rate coefficient in the fast regime, and k the coefficient at the conditions of interest. The stability ratio is close to unity in the fast regime, increases in the slow regime, and becomes very large when the suspension is stable.

When the interaction potential between the particles is purely attractive, the aggregation process is solely limited by mutual diffusion (or Brownian motion) of the particles, one refers to *fast, rapid* or *diffusion limited aggregation* (DLA). When the interaction potential shows an intermediate barrier, the aggregation is slowed down by the fact that numerous attempts will be necessary to overcome this barrier, and one refers to *slow* or *reaction limited aggregation* (RLA). The aggregation can be tuned from fast to slow by varying the concentration of salt, pH, or an other additive. Since the transition from fast to slow aggregation occurs in a narrow concentration range, and one refers to this range as the critical coagulation concentration (CCC).

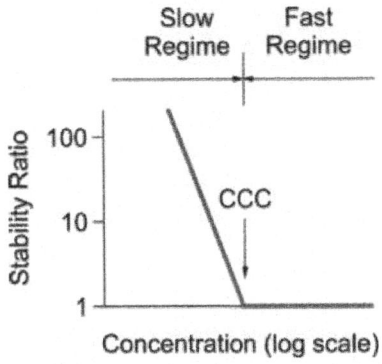

Fig. : Schematic stability plot of a colloidal suspension versus the salt concentration.

Often, colloidal particles are suspended in water. In this case, they accumulate a surface charge and an electrical double layer forms around each particle. The overlap between the diffuse layers of two approaching particles results in a repulsive double layer interaction potential, which leads to particle stabilization. When salt is added to the suspension, the electrical double layer repulsion is screened, and van der Waals attraction become dominant and induce fast aggregation.

The table below summarizes CCC ranges for different net charge of the counter ion. The charge is expressed in units of elementary charge. This dependence reflects the Schulze–Hardy rule, which states that the CCC varies as the inverse sixth power of the counter ion charge. The CCC also depends on the type of ion somewhat, even if they carry the same charge. This dependence may reflect different particle properties or different ion affinities to the particle surface. Since particles are frequently negatively charged, multivalent metal cations thus represent highly effective coagulants.

Charge	CCC ($\times\ 10^{-3}$ mol/L)
1	50–300
2	2–30
3	0.03–0.5

Adsorption of oppositely charged species (*e.g.*, protons, specifically adsorbing ions, surfactants, or polyelectrolytes) may destabilize a particle suspension by charge neutralization or stabilize it by buildup of charge, leading to a fast aggregation near the charge neutralization point, and slow aggregation away from it.

Quantitative interpretation of colloidal stability was first formulated within the DLVO theory. This theory confirms the existence slow and fast aggregation regimes, even though in the slow regime the dependence on the salt concentration is often predicted to be much stronger than observed experimentally. The Schulze–Hardy rule can be derived from DLVO theory as well.

Other mechanisms of colloid stabilization are equally possible, particularly, involving polymers. Adsorbed or grafted polymers may form a protective layer around the particles, induce steric repulsive forces, and lead to steric stabilization. When polymers chains adsorb to particles loosely, a polymer chain may bridge two particles, and induce bridging forces. This situation is referred to as bridging flocculation.

When particle aggregation is solely driven by diffusion, one refers to *perikinetic* aggregation. Aggregation can be enhanced through shear stress (*e.g.*, stirring). The latter case is called *orthokinetic* aggregation.

Later Stages

As the aggregation process continues, larger clusters form. The growth occurs mainly through encounters between different clusters, and therefore, one refers to cluster–cluster aggregation process. The resulting clusters are irregular, but statistically self-similar. They are examples of mass fractals, whereby their mass M grows with their typical size characterized by the radius of gyration R_g as a power–law :

$$M \propto R_g^{d}$$

where d is the mass fractal dimension. Depending whether the aggregation is fast or slow, one refers to diffusion limited cluster aggregation (DLCA) or reaction limited cluster aggregation (RLCA). The clusters have different characteristics in each regime. DLCA clusters are loose and ramified ($d \approx 1.8$), while the RLCA clusters are more compact ($d \approx 2.3$). The cluster size distribution is also different in these two regimes. DLCA clusters are relatively monodisperse, while the size distribution of RLCA clusters is very broad.

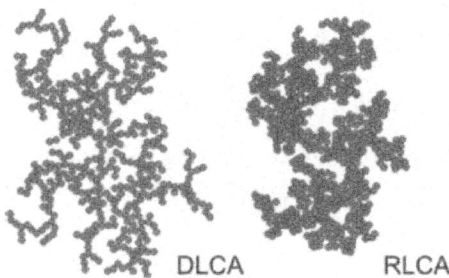

Structure of larger aggregates formed can be different. In the fast aggregation regime or DLCA regime, the aggregates are more ramified, while in the slow aggregation regime or RCLA regime, the aggregates are more compact.

The larger the cluster size, the faster their settling velocity. Therefore, aggregating particles sediment and this mechanism provides a way for separating them from suspension. At higher particle concentrations, the growing clusters may interlink, and form a particle gel. Such a gel is an elastic solid body, but differs from ordinary solids by having a very low elastic modulus.

Homoaggregation *versus* Heteroaggregation

When aggregation occurs in a suspension composed of similar monodisperse colloidal particles, the process is called *homoaggregation* (or *homocoagulation*). When aggregation occurs in a suspension composed of dissimilar colloidal particles, one refers to *heteroaggregation* (or *heterocoagulation*). The simplest heteroaggregation process occurs when two types of monodisperse colloidal particles are mixed. In the early stages, three types of doublets may form :

$A + A \rightarrow A_2$

$B + B \rightarrow B_2$

$A + B \rightarrow AB$

While the first two processes correspond to homoaggregation in pure suspensions containing particles A or B, the last reaction represents the actual heteroaggregation process. Each of these reactions is characterized by the respective aggregation coefficients k_{AA}, k_{BB}, and k_{AB}. For example, when particles A and B bear positive and negative charge, respectively, the homoaggregation rates may be slow, while the heteroaggregation rate is fast. In contrast to homoaggregation, the heteroaggregation rate accelerates with decreasing salt concentration. Clusters formed at later stages of such heteroaggregation processes are even more ramified that those obtained during DLCA ($d \approx 1.4$).

An important special case of a heteroaggregation process is the deposition of particles on a substrate. Early stages of the process correspond to the attachment of individual particles to the substrate, which can be pictures as another, much larger particle. Later stages may reflect blocking of the substrate through repulsive interactions between the particles, while attractive interactions may lead to multi-layer growth, and is also referred to as ripening. These phenomena are relevant in membrane or filter fouling.

Experimental Techniques

Numerous experimental techniques have been developed to study particle aggregation. Most frequently used are time–resolved optical techniques that are based on transmittance or scattering of light.

Light Transmission : The variation of transmitted light through an aggregating suspension can be studied with a regular spectro-photometer in the visible region. As aggregation proceeds, the medium becomes more turbid, and its absorbance increases. The increase of the absorbance can be related to the aggregation rate constant k and the stability ratio can be estimated from such measurements. The advantage of this techniques is its simplicity, but its disadvantage is that it can be only reliably used for larger particles or that detailed corrections due to the presence of larger clusters must be considered. Small particles aggregate rapidly, and in such systems it is normally difficult to extract the stability ratio from the transmittance quantitatively.

Light Scattering : These techniques are based on probing the scattered light from an aggregating suspension in a time–resolved fashion. Static light scattering yields the change in the scattering intensity, while dynamic light scattering the variation in the apparent hydrodynamic radius. At early–stages of aggregation, the variation of each of these quantities is directly proportional to the aggregation rate constant k. At later stages, one can obtain information on the clusters formed (*e.g.*, fractal dimension). Light scattering works well for a wide range of particle sizes. Multiple scattering effects may have to be considered, since scattering becomes increasingly important for larger particles or larger aggregates. Such effects can be neglected in weakly turbid suspensions. Aggregation processes in strongly scattering systems have been studied with back-scattering techniques or diffusing–wave spectroscopy.

Single Particle Counting : This technique offers excellent resolution, whereby clusters made out of tenths of particles can be resolved individually. The aggregating suspension is forced through a narrow capillary particle counter and the size of each aggregate is being analyzed by light scattering. From the scattering intensity, one can deduce the size of each aggregate, and construct a detailed aggregate size distribution. If the suspensions contain high amounts of salt, one could equally use a Coulter counter. As time proceeds, the size distribution shifts towards larger aggregates, and from this variation aggregation and breakup rates involving different clusters can be deduced. The disadvantage of the technique is that the aggregates are forced through a narrow capillary under high shear, and the aggregates may disrupt under these conditions.

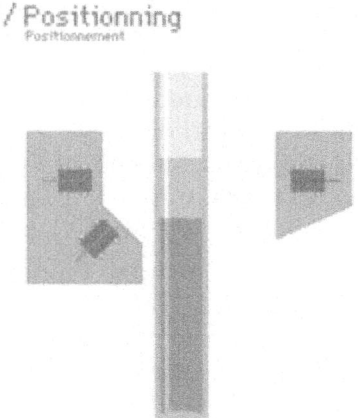

Fig. : Probing aggregation of a setting colloidal suspension with light scattering coupled with vertical scanning.

Indirect Techniques : As many properties of colloidal suspensions depend on the state of aggregation of the suspended particles, various indirect techniques have been used to monitor particle aggregation too. While it can be difficult to obtain quantitative information on aggregation rates or cluster properties from such experiments, they can be most valuable for practical applications. Among these techniques settling tests are most relevant. When one inspects a series of test tubes with suspensions prepared at different concentration of the flocculant, stable suspensions often remain dispersed, while the unstable ones settle. Automated instruments based on light scattering to monitor suspension settling have been developed, and they can be used to probe particle aggregation. One must realize, however, that these techniques may not always reflect the actual aggregation state of a suspension correctly. For example, larger primary particles may settle even in the absence of aggregation, or aggregates that have formed a colloidal gel will remain in suspension. Other indirect techniques capable to monitor the state of aggregation include, for example, filtration, rheology, absorption of ultra-sonic waves, or dielectric properties.

Relevance

Particle aggregation is a widespread phenomenon, which spontaneously occurs in nature but is also widely explored in manufacturing. Some examples include.

Formation of River Delta : When river water carrying suspended sediment particles reaches salty water, particle aggregation may be one of the factors responsible for river delta formation. Charged particles are stable in river's fresh water containing low levels of salt, but they become unstable in sea water containing high levels of salt. In the latter medium, the particles aggregate, the larger aggregates sediment, and thus create the river delta.

Paper-making : Retention aids are added to the pulp to accelerate paper formation. These aids are coagulating aids, which accelerate the aggregation between the cellulose fibers and filler particles. Frequently, cationic polyelectrolytes are being used for that purpose.

Water Treatment : Treatment of municipal waste water normally includes a phase where fine solid particles are removed. This separation is achieved by addition of a flocculating or coagulating agent, which induce the aggregation of the suspended solids. The aggregates are normally separated by sedimentation, leading to sewage sludge. Commonly used flocculating agents in water treatment include multivalent metal ions (*e.g.*, Fe^{3+} or Al^{3+}), polyelectrolytes, or both.

Cheese Making : The key step in cheese production is the separation of the milk into solid curds and liquid whey. This separation is achieved by inducing the aggregation processes between casein micelles by acidifying the milk or adding rennet. The acidification neutralizes the carboxylate groups on the micelles and induces the aggregation.

Chapter 5

ELECTRO-CHEMICAL REACTION

Electro-chemical reaction, any process either caused or accompanied by the passage of an electric current and involving in most cases the transfer of electrons between two substances — one a solid and the other a liquid.

Under ordinary conditions, the occurrence of a chemical reaction is accompanied by the liberation or absorption of heat and not of any other form of energy; but there are many chemical reactions that — when allowed to proceed in contact with two electronic conductors, separated by conducting wires — liberate what is called electrical energy, and an electric current is generated. Conversely, the energy of an electric current can be used to bring about many chemical reactions that do not occur spontaneously. A process involving the direct conversion of chemical energy when suitably organized constitutes an electrical cell. A process whereby electrical energy is converted directly into chemical energy is one of electrolysis; *i.e.,* an electrolytic process. By virtue of their combined chemical energy, the products of an electrolytic process have a tendency to react spontaneously with one another, reproducing the substances that were reactants and were therefore consumed during the electrolysis. If this reverse reaction is allowed to occur under proper conditions, a large proportion of the electrical energy used in the electrolysis may be regenerated. This possibility is made use of in accumulators or storage cells, sets of which are known as storage batteries. The charging of an accumulator is a process of electrolysis; a chemical change is produced by the electric current passing through it. In the discharge of the cell, the reverse chemical change occurs, the accumulator acting as a cell that produces an electric current.

Finally, the passage of electricity through gases generally causes chemical changes, and this kind of reaction forms a separate branch of electro-chemistry that will not be treated here.

GENERAL PRINCIPLES

Substances that are reasonably good conductors of electricity may be divided into two groups : the metallic, or electronic, conductors and the electrolytic conductors. The metals and many non-metallic substances such as graphite, manganese dioxide, and lead sulfide exhibit metallic conductivity; the passage of an electric current through them produces heating and magnetic effects but no chemical changes. Electrolytic conductors, or electrolytes, comprise most acids, bases, and salts, either in the molten condition or in solution in water or other solvents. Plates or rods composed of a suitable metallic conductor dipping into the fluid electrolyte are employed to conduct the current into and out of the liquid; *i.e.*, to act as electrodes. When a current is passed between electrodes through an electrolyte, not only are heating and magnetic effects produced but also definite chemical changes occur. At or in the neighbourhood of the negative electrode, called the cathode, the chemical change may be the deposition of a metal or the liberation of hydrogen and formation of a basic substance or some other chemical reduction process; at the positive electrode, or anode, it may be the dissolution of the anode itself, the liberation of a non-metal, the production of oxygen and an acidic substance, or some other chemical oxidation process.

An electrolyte, prepared either by the melting of a suitable substance or by the dissolving of it in water or other liquid, owes its characteristic properties to the presence in it of electrically charged atoms or groups of atoms produced by the spontaneous splitting up or dissociation of the molecules of the substance. In solutions of the so–called strong electrolytes, most of the original substance, or in some solutions perhaps all of it, has undergone this process of electrolytic dissociation into charged particles, or ions. When an electrical potential difference (*i.e.*, a difference in degree of electrification) is established between electrodes dipping into an electrolyte, positively charged ions move toward the cathode and ions bearing negative charges move toward the anode. The electric current is carried through the electrolyte by this migration of the ions. When an ion reaches the electrode of opposite polarity, its electrical charge is donated to the metal, or an electric charge is received from the metal. The ion is thereby converted into an ordinary neutral atom or group of atoms. It is this discharge of ions that gives rise to one of the types of chemical changes occurring at electrodes.

HISTORY

The study of electro-chemistry began in the 18th century, bloomed until the early 20th century, and then faded, owing to an excessive use of thermodynamic principles in analyzing the processes that take place at points in the system where the various parts form interfaces. Since about 1950 electro-chemistry has undergone a change. The study of processes in solutions has been less stressed, but the study of the transfer of electrons between metals and solution has increased explosively. With this new emphasis electro-chemistry is becoming a core science. It promises to be an important part of the foundation of the ecology–oriented society of the future, because electricity is not a pollutant. The pollution associated with some methods of generating electricity must, however, be reduced.

The first electro-chemical reactions studied, in 1796, were those in the cell of silver and zinc plates with blotting paper wetted by aqueous salt solution between them; these cells were constructed by the Italian scientist Alessandro Volta, for whom the term volt was named. This cell was the first primary battery used for the production of electricity.

Michael Faraday formulated the laws of electro-chemical stoichiometry, which deals with the application of laws of definite proportions and of the conservation of matter and energy to chemical activity. These state that a coulomb of electricity, a unit of charge, reacts with fixed quantities of a substance (*e.g.*, with 1.11800 milligrams of silver ions) or else that 1 gram equivalent of any substance reacts with 96,485 coulombs. This latter number represents a fundamental quantity known as one faraday of electricity. The relationship between the chemical affinity of the reactants in the cell and the voltage of the cell when it is operating was precisely defined by the U.S. chemist Josiah Willard Gibbs in 1875, while the relation of this affinity to the potential of the electro-chemical cell was initially formulated by the German physical chemist Walther Hermann Nernst in 1889.

The period 1910 to 1950 was one of decline in electro-chemistry, until it became limited mainly to the study of solutions. There was almost no progress in the understanding of electro-chemical reactions outside of equilibrium conditions and reversibility, and knowledge of these was applied invalidly to reactions occurring at a net rate — *i.e.*, reactions not in equilibrium and not totally reversible. From about 1950 the study of electrified interfaces, with special reference to the study of the transfer of electrons (called electrodics), gained in importance and became the main aspect of electro-chemistry. From about 1960, electrodics began to develop as an inter-disciplinary area in the search for solutions to problems such as the source of energy in space flights from fuel cells, the stability of metals in moist environments, the electro-chemical aspects of biological functions, extractions from mixtures, and the replacement of fossil fuels such as coal and petroleum and their by-products, by electricity produced or stored electro-chemically in transportation.

THE ELECTRO-CHEMICAL PROCESS

Interactions of matter associated with the passage of an electric current depend upon the characteristics of the negatively charged electron. As the basic particle of electricity, the electron has an affinity for positively charged particles of matter, protons, whether in atoms, groups of atoms, or molecules. This affinity is analogous to the chemical affinity that particles exhibit among themselves. In fact, all chemical reactions result from a shift in the electron structure of atoms, and free electrons can combine with particles of matter (reduction) or be released by them (oxidation). The quantitative relationship between the free electrons of an electric current and the particles of a substance in which they cause a reaction is defined by the laws of Faraday. The substances that take part in electro-chemical reactions, called electrolytes or ionic conductors, have been described above.

Electrons are available in large quantities in a relatively free (mobile) state only in substances called electronic conductors, among which metals are the most important. Thus, an electron conductor must be present as a basic component of any system in which electro-chemical reactions are to occur. Furthermore, the availability of electrons in a conductor is limited by energy distribution to such an extent that electro-chemical reactions take place only in the immediate vicinity of the electronic conductor's surface — i.e., a few angstroms from the conductor into the solution. These reactions are, therefore, normally considered as occurring at the interface, or common boundary, between an electronic conductor, such as an electrode, and an ionic conductor of electricity, such as an electrolytic solution. Electro-chemical reaction will take place, however, only to the extent that electricity can flow through such a system as a whole. To achieve this, it is necessary for the system to form a closed loop, electronically speaking.

To summarize, if at one metal–solution interface electrons are coming out of the metal, reducing a component of the solution, there must exist a second metal–solution interface where electrons are going into the metal in the process of oxidation.

The two electrodes and the ionic conductor in between (e.g., an aqueous solution of some ionized salt) represent an electro-chemical cell. The process occurring in the cell as a whole is a redox process with the reduction of one species spatially separated from the oxidation of another one. As a consequence of Faraday's law, the rates of electro-chemical reactions at electrodes (expressed in terms of gram moles per second per square centimetre of electrode surface) are directly proportional to the current density (expressed in amperes per square centimetre) — i.e., current flowing through the cell divided by the electrode surface area.

Sites of Electro-chemical Reactions

Electro-chemical reactions take place where the electron conductor meets the ionic conductor — i.e., at the electrode–electrolyte interface. Characteristic of this region, considered to be a surface phase, is the existence of a specific structure of particles and the presence of an electric field of considerable intensity (up to 10,000,000 volts per centimetre) across it; the field is caused by the separation of charges that are present between the two bulk phases in contact. For most purposes the surface phase can be considered as a parallel plate condenser, with one plate on the centre of the ions that have been brought to the electrode, at the distance of their closest approach to it, and with the second plate at the metal surface; between the two plates and acting as a dielectric (i.e., a non-conducting material) are oriented water molecules.

Thermal motion of the positive ions in the solution makes the condenser plate on the electrolyte side of the interface diffuse — i.e., the ions are distributed in a cloud-like way. This condition justifies the division of the potential change between the bulk of metal and the bulk of electrolyte into two parts : first, that between the metal surface and the first ionic layer at the distance of closest approach (called the outer Helmholtz plane, in which the ions are usually surrounded by solvent particles — i.e., are solvated); and second, that between the first ionic layer and the bulk of the solution, the diffuse part of the double layer. The picture is further complicated by the presence of ions in the electrode surface layer in addition to those that are present for electrostatic reasons – *i.e.*, by the force of attraction or repulsion between electric charges. Such electrode surface layer ions are said to be specifically adsorbed on the electrode surface. Since this species of ions is attracted by the surface to a distance closer than the "distance of the closest approach" of ions, further sub-division of the inner part of the electric double layer is justified. Hence, the inner Helmholtz plane is introduced as the plane formed by the centres of specifically adsorbed ions. Adsorption of neutral molecules on the surface can also change the properties of the electric double layer. This change occurs as a consequence of replacing the water molecules, and thus changes that part of the potential (electrical) difference across the double layer that is caused by the adsorbed dipoles (water molecules that have a polarity – *i.e.*, they behave like minute magnets – because of their hydrogen–oxygen structure, making one end of the molecule positive and the other end negative).

The absolute value of electrical potential difference, symbolized in calculation by the Greek letters delta and psi, $\Delta\psi$, between the bulk of a metal electrode and the bulk of an electrolyte cannot be measured. Instead, the voltage of a special cell, composed of the specific electrode being studied and of an arbitrarily selected reference electrode, is normally measured; the voltage is referred to as the relative electrode potential, E. Of special interest is that state of the electrode at which there is no net charge (in this case, no unbalanced, or extra-positive, charge) at the metal side of the double layer. The relative potential at which this state is achieved is characteristic of each metal. This point is termed the potential of zero charge. At that potential, the field across the double layer is due to orientation of water molecules and other dipoles at the surface only.

Most of the knowledge of the detailed structure of the interface between a metal and an electrolyte arises from experimentation with mercury, the only metal that is liquid at ordinary temperatures; the double layer structure turns out to have surface tensions that must be measured, and this measurement is difficult with solid metals. By 1970, however, it had been shown that it is possible to measure surface tension changes at the metal–solution interface. Thus, the way to the determination of the double layer structure involving solids was opened.

Substances that are semi-conductors can also be employed as electron carriers in electro-chemical reactions. Semi-conductors are substances which range between serving as insulators at low temperatures and as metallic–type conductors at high temperatures. In the case of Semi-conductors, however, the electric double layer has a more complex structure inasmuch as the condenser plate at the electrode side of the double layer also becomes diffuse. Thus, the overall potential difference between bulks of the phases in contact comprises also the potential difference between the bulk of the semi-conductor and its surface.

TYPES OF ELECTRO-CHEMICAL REACTIONS

There are several types of electro-chemical reactions.

Simple Redox Reactions

A simple redox reaction is one that involves a change in the electrical charge of a charge carrier, usually a simple or complex ion in the solution, by its taking away, an electron from the electrode (reduction), or its giving an electron to the electrode (oxidation). The same carrier may be present in solution in two states of charge. The higher, more positive charge is called the oxidized state, and the lower, less positive charge is called the reduced state. For example, when ferric and ferrous ions are both present in solution in significant quantity, and when electron exchange with the electrode is sufficiently fast, redox equilibrium is established at the electrode, giving it a well-defined potential, or reversible redox potential.

Reactions that Produce Gases

When hydrogen ions in solution react with electrons ejected from a metal, hydrogen atoms are formed at the surface, where they combine among themselves or with other hydrogen ions and electrons to give gaseous hydrogen molecules. If all the reactions are fast enough, an equilibrium is attained between hydrogen ions and gaseous hydrogen. A metal in contact with solution at which such a situation exists is called the reversible hydrogen electrode, and its electrical potential is arbitrarily taken to be zero; every other electrode can thus be compared with it as it represents the basis for constituting the hydrogen scale of relative electrode potentials. Similarly, negative hydroxyl ions in solution (OH^-) can be made to give up electrons to a metal and, in a series of reactions, the final one is the formation of gaseous oxygen. Chlorine is another gaseous product; it evolves upon electro-chemical oxidation of chloride ions in concentrated solutions of neutral and acid salts.

Reactions that Deposit And Dissolve Metals

When a metal ion is reduced and discharged as a neutral atom, or species, it tends to build into the metal lattice of the electrode. Thus, metals can be deposited at electrodes. Conversely, if electrons are taken away from the metal electrode by applying positive potentials to it, the metal ions thus formed can cross the double layer of electric charge at the interface, undergo hydration (combination with water), and enter the solution. The metal electrode thus dissolves. Many metals establish well-defined electric potentials when they are in contact with their own ions in solution.

Oxidation and Reduction of Organic Compounds

A reaction of the oxidation and reduction of organic compounds can also be done at electrodes. Such reactions, however, are mostly irreversible in the literal sense that they lead to products that cannot easily be converted back into the original substance. Exceptions are some oxygen-and nitrogen-containing compounds (quinones, amines, and nitrous compounds) that can give fairly well-defined reversible potentials.

Mechanism of Charge Transfer

The causes of the thermodynamically irreversible behaviour of electrode reactions are found in the nature of the elementary act of charge transfer. Like any chemical reaction, this act is inhibited by the existence of an energy barrier between the oxidized and the reduced state. This barrier implies that the reaction could take place only in the special circumstances when, during the course of numerous interactions with other species (atoms, ions, molecules, etc.) surrounding it, a molecule attains an excited state in which it has an abnormal energy content. In most chemical reactions, this energy content must be sufficient for the species to come into what is called the transition state; the transition state characterizes the top of the energy barrier just before a reaction begins. If such a model is applied to electron transfer at an interface, calculation shows that electron exchange reactions at electrodes would be prohibitively slow, a conclusion at variance with the observed phenomena; quantum mechanical laws, however, govern the motion of electrons, and their inclusion changes the calculations to fit reality. Quantum mechanics require that for fast electron exchange to take place, electrons in a particle outside the double layer (*e.g.*, a hydrated ion at the outer Helmholtz plane) must attain certain well-defined quantized energy levels equal to those in which free electrons exist in the metal. Since such states can be attained by the particle at a lower energy-content than that needed for its transfer over the top of the energy barrier, according to the classical view, this fast process of electron exchange between the electrode and a particle in solution is termed electron tunnelling through the energy barrier.

Whereas the rate of chemical processes, or what may loosely be termed the speed of reaction, can be influenced only by changing the concentrations of reactants or by changing the temperature or both, the rate of electro-chemical processes also can be manipulated by changing the electrode potential. Making the electrode more negative increases the number of electrons in the metal ready to tunnel to ions, and hence the rate of the reduction process increases. Conversely, making the potential more positive decreases this rate and increases the number of particles ready to give away electrons, thus increasing the rate of the oxidation process.

It can be deduced that there must exist a direct proportionality between the rate of reaction and the concentration of the reacting species and at the same time an exponential proportionality between the rate of reaction and the electrode potential.

At any electrode potential, both reduction of one species and oxidation of the product of reduction are taking place but at different rates; the rate of each reaction is determined by the respective concentration and by the corresponding effects of potential. The rate of an electro-chemical reaction can best be described as the electric current density — *i.e.*, a measure of the quantity of electrons moving in a certain volume of space during a specified unit of time.

The relationships can be represented quantitatively by an equation in which the net, or resulting, current (the difference of the rate of electron ejection across the interface to particles in solution, diminished by the rate at which particles in solution inject electrons into the metal) is equated to the difference of the rates of reduction and oxidation and the variables and constants that relate to these reactions. Equation (2), known as the Nernst equation, which can be derived from equation (1), gives the value of the electrode potential when the rate of oxidation exactly equals the rate of reduction. Using this value for the electrode potential, equation (3), called the Butler–Volmer equation, can be derived; it represents the most fundamental relationship in electrodic chemistry.

Electro-crystallization

Deposition of metals and other substances at electrodes as a consequence of an electrode process exhibits a number of specific features. The electrode process is followed by crystal building, and this results in a continuous change of the electrode surface. This change, in turn, affects the electro-chemical properties of the system — the double-layer capacity and the rate constants of the charge transfer processes. Hence, if electro-chemical properties are to be studied, transient methods should be employed that allow measurements to be made before major changes in surface morphology (structure) take place.

Since the two steps, the discharge of ions at the electrode and the incorporation of the discharged ions into the crystal lattice, are separated in time and space, an intermediate species exists at the surface, that of relatively loosely bound and freely moving atoms, called adatoms. Since the electrons tend to join the rest in the bulk of the metal, adatoms appear to have a partial charge, less than that of the elementary positive charge. The adatoms therefore attract solvent molecules, and the species is partially solvated. This reaction justifies considering an adatom as a kind of adsorbed ion, called an adion, which, however, has already undergone partial discharge.

Because such an intermediate state is possible, ions need not be reduced to the neutral atomic state at the point of incorporation into the crystal lattice. Indeed, energy considerations reveal this reduction to be improbable. Instead, discharge of ions is favoured on crystal planes. Their motion from the location of discharge to a site on the crystal occurs by surface diffusion. When such surface diffusion is inhibited, it can, in some cases, control the rate of reaction.

Problems of the crystal–building step in electro-crystallization are in many respects identical with those of crystal growth from the gas phase. Crystal building can occur in any one of three possible ways.

Kinks, naturally existing at any metal surface, form a suitable half–lattice position at which an atom is surrounded by one–half of the number of atoms that would surround it in the bulk of the metal; there, adatoms can be successively trapped and thus the crystal lattice is extended along a crystal edge and further on across the surface. The mechanism, however, has a limited capacity for crystal growth. A step can move as far as the edge of a crystal, and step growth would lead to smoothing of the surface to perfection, but then further growth would cease.

Mechanisms associated with screw dislocations, or twinning edges, can provide for a continuous growth of crystals. The screw dislocation mechanism, is made possible by a specific fault often found in the crystal lattice that may be called a dislocation originating from a shift of one atom in the lattice with respect to a perfect arrangement. This shift may then result in the formation of a mono-atomic edge projecting above the electrode surface at which new atoms can be stacked. The stacking produces a turning of the edge around the base atom as a centre, the process producing a spiral growth.

If growth sites are rare, or if the substrate at which deposition should take place is foreign to the depositing metal, the charge transfer results in an accumulation of adatoms to a concentration considerably larger than that which can exist there at equilibrium with the crystal lattice. In such a situation, termed supersaturation, agglomeration of adatoms to form crystal nuclei is favoured. Surface energy requirements show that, at any degree of supersaturation, nuclei of certain dimensions are stable and can represent sites for further growth.

The formation of deposits is controlled by the above mechanism as long as the discharge process supplies ample amounts of adatoms as building material over the entire surface. If the rate of deposition is increased, so as to produce near the surface considerable depletion of ions, uneven deposition will start. This is caused by the protrusions of a normally rough metal surface being closer to the bulk of solution than the recessed parts and, hence, getting a somewhat faster supply of the discharging species. Once such a situation is established, it tends to develop further. The faster growing points penetrate into ever richer layers of solution, resulting in ever faster growth. Thus, a natural consequence of deposition under transport–controlled rate is the amplification of the original surface roughness and the appearance of protruding spikes, called dendrites, as deposits.

A converse process is the smoothing of the original surface irregularities, which may occur when some foreign species, called an additive, is present in solution and adsorbs on the surface and inhibits the process of discharge. If those molecules are incorporated into a growing deposit, a situation may arise in which their supply to recessed parts of the surface becomes slower than to elevated parts. As a result, deposition becomes faster at recessed parts than at elevated ones and leveling of the surface occurs. This process has considerable technical application. Metal dissolution can sometimes be governed by a similar, transport–controlled mechanism, in which case a polishing effect is obtained. Electro-polishing has found wide application in practice.

Organic Electrode Reactions

A very large number of electro-chemical reactions involving organic molecules are known. An example is the oxidation of ethylene according to the equation :

$$\underset{\text{(ethylene)}}{C_2H_4} + \underset{\text{(water)}}{4H_2O} \rightarrow \underset{\text{(carbon dioxide)}}{2CO_2} + \underset{\text{(hydrogenion)}}{12H^+} + \underset{\text{(electrons)}}{12e^-}$$

Many chemical organic reactions can be made to function electro-chemically, a general advantage being that the rate of the reaction is easily controlled by controlling the potential. A change of potential may change the path of the reaction and hence a certain product may be tuned in. In particular, polymerization reactions at electrodes can be stimulated; such a method is used as a step in the production of nylon.

Multi-electrode Systems

So far, systems have been considered in which a single electrode process takes place. In principle, at any electrode potential all species present in the system fall into two categories : those that are stable, and those that undergo oxidation or reduction. The stable species are those that at the given electrode potential would not decrease their free energy by giving off or accepting electrons.

At any potential there should occur a codischarge of all unstable species. Thus, the system can be considered as a multi-electrode system, consisting of as many electrodes as there are redox couples present. The rate at which different processes occur, however, can be so widely different that usually a single process is by far the dominant one. Systems at which two electrode processes occur at comparable rates are of considerable importance. If two kinds of metal ions are discharged simultaneously, an alloy is formed upon crystallization. The properties of the alloy would in most cases be those determined by the phase diagram (a plot of the temperature of melting *versus* the composition of the mixed system) for the ratio of quantities of discharged metals given by the rates at which the discharges take place. In many cases, non-equilibrium metal phases are formed giving unusual properties to the alloy.

If a process of metal dissolution (an oxidation process) can occur at a rate comparable to that of some reduction process on the same metal, a corrosion couple is established. Thus, zinc immersed in acid solution tends to establish a potential sufficiently positive for the metal to eject metal ions into the solution (oxidation). At the same time, it is also sufficiently negative for the reduction of hydrogen ions, present in any aqueous solution, to hydrogen gas. Hence, a spontaneous process of hydrogen evolution and of dissolution of the metal will take place. The rate at which this corrosion process occurs is governed by rate laws of the type given by equation (3). The mixed potential, spontaneously established by the corroding metal, is obtained by equating expressions for the anodic and cathodic currents of the two processes in the corrosion couple.

Corrosion can be prevented in two ways. One is by using an external source to make the potential of the metal sufficiently negative to bring it into the potential region in which the metal is stable, called cathodic protection; and the other is by provoking by some means the formation of a film on the surface that would slow the process. Such films could consist of an oxide or a layer of organic molecules that prevents dissolution and hence is called an inhibitor.

APPLICATIONS

Electro-chemical processes are used in many ways and their use is likely to increase because they can replace polluting chemical situations with non-polluting electro-chemical ones. In many fields, however, applications have been profitable for some time. Major categories are listed below.

Metallurgy

All technologically important metals, except iron and steel, are either obtained or refined by electro-chemical processes; for example, aluminum, titanium, alkaline earth, and alkali metals are obtained by electro-deposition from molten salts, and copper is refined by electrolysis in aqueous copper sulfate solutions.

Electro-plating

One of the major ways of both decorating objects and improving their resistance to corrosion is by electro-plating them. All major metal–working industries, particularly the automobile industry, have large electro-plating plants.

Chemical Industry

Electrolysis of brine to obtain chlorine and caustic soda is an electro-chemical process that has become one of the largest volume productions in the chemical industry. Modern processes cover a wide field, from the production of a variety of inorganic compounds to the production of such synthetic fibres as nylon. Intensive research in organic electro-chemistry promises major developments in application, particularly with the prospect of greatly reduced electricity costs expected eventually to arise from the development of controlled fusion.

Batteries

Electro-chemical storage of electricity is effected in batteries. Such devices are electro-chemical cells and consist of two electrodes per unit. As the electricity to be stored is accepted on the plates of the cell, it converts substances on the plates to new substances having a higher energy than the old ones. When it is desired to make the electricity available again, the terminals of the battery are connected to the load and the substances on the battery plates retransform themselves to those originally present, giving off electricity as a product of their electro-chemical reactions. The steadily rising production of the lead–acid battery is largely the result of its use for starting the internal–combustion engine, which has had an equally steady rise. Other electro-chemical systems are also used as storers. The nickel–iron (Edison cell) and nickel–cadmium battery with alkaline electrolyte are both used in applications where longer lives than those of the lead–acid battery are needed; the silver–zinc battery is used to start airplane engines because of its high power per unit of weight. A variety of new systems is being investigated for covering other needs. One of the greatest challenges to electro-chemists and electro-chemical engineers is that of producing a battery with sufficient power and energy density to run an automobile the way gasoline (petrol) does. Even if the best hypothetical predictions for removal of polluting chemicals from automobile exhausts is realized, the cleanup will not be sufficient because the expected growth of the automobile population will continue to increase the pollutant rate.

Fuel Cells

The energy of chemical reactions is converted into electrical energy in fuel cells. In these, the fuel (*e.g.*, hydrogen, hydrazine) is fed continuously to one electrode, while oxygen from the air is reacting at the other one. The efficiency of energy conversion in fuel cells is more than twice that attainable by conventional means — for example, by means of internal combustion.

Analytical Chemistry

In analytical chemistry, most modern automated instrumental analysis is based on electrode processes — for example, potentiometry, used to measure ionization constant.

Biological Research

In biology the idea that many biological processes, from blood clotting to the transfer of nerve impulses, are electro-chemical in nature continues to spread. The biological conversion of the chemical energy of food to mechanical energy takes place at an efficiency so high that it is difficult to explain without electro-chemical mechanisms. Intensive research is developing in various directions in bio-electro-chemistry.

CALCULATIONS

The three equations referred to above are stated in this section, and other mathematical considerations are also included. The rate of an electro-chemical reaction in terms of oxidation and reduction reactions, the concentration of the reacting species, the electrode potentials and the current densities can all be related quantitatively according to equation (1) :

$$i = \vec{i} - \overleftarrow{i} = \vec{k} C_{red} \exp\left(\frac{\alpha_a F}{RT} E\right) - \overleftarrow{k} C_{ox} \exp\left(\frac{\alpha_c F}{RT} E\right)$$

in which i is the net current density, $i{\rightarrow}$ and $i{\leftarrow}$ are the partial current densities of the oxidation and reduction respectively, C_{red} and C_{ox} are the concentrations of the reducing and oxidizing agents, respectively, $k{\rightarrow}$ and $k{\leftarrow}$ are the corresponding rate constants, while α_a and α_c are the so-called transfer coefficients — that is, specific constants giving a proper influence factor to the exponential dependence of the rate on the potential, E. In the case of a simple one–electron transfer, these factors are termed symmetry factors, for they, in a way, reflect the symmetry of the energy barrier. It can be proved that $\alpha_c + \alpha_a = n$, the number of electrons exchanged in a single act of an electrode reaction.

THE NERNST EQUATION

For a particular value of E the two partial current densities must become equal. This value of potential is the reversible electrode potential. From equation (1) one can deduce equation (2) :

$$
E_{rev} = \frac{RT}{(\alpha_c + \alpha_a)F} \ln \frac{\vec{k}}{\overleftarrow{k}} + \frac{RT}{(\alpha_c + \alpha_a)F} \ln \frac{C_{ox}}{C_{red}}
$$

$$
= E^0 + \frac{RT}{nF} \ln \frac{C_{ox}}{C_{red}}.
$$

This equation is known as the Nernst equation; $E°$ is the standard electrode potential (at $C_{ox} = C_{red} = 1$) characteristic of the given redox couple.

The standard electrode potential on the hydrogen scale is related to the thermodynamics of the electrode process. It reflects the standard free energy change of the redox reaction between the electron and the given redox couple, relative to the free energy change that takes place in the hydrogen electrode process.

THE BUTLER–VOLMER EQUATION

The reversible electrode potential can be introduced into equation (1) and the potentials taken relative to its value. When so expressed, they are termed overpotentials and can be stated as $\eta = E - E_{rev}$; equation (1) then transforms to equation (3) :

$$
i = i_0 \left\{ \frac{(C_{red})_i}{(C_{red})_o} \exp\left(\frac{\alpha_a F\eta}{RT} \right) - \frac{(C_{ox})_i}{(C_{ox})_o} \exp\left(-\frac{\alpha_c F\eta}{RT} \right) \right\}
$$

in which i_0 represents the value of either of the (equal) electron–emitting and electron–accepting partial current densities at the reversible potential and is termed the exchange current density. Equation (3) is called the Butler–Volmer equation and represents one of the most fundamental relationships of electro-chemistry.

As overpotentials, either positive or negative, become larger than about 5×10^{-2} volts (V), the second or the first term of equation (3) becomes negligible, respectively. Hence, simple exponential relationships between current (*i.e.*, rate) and overpotential are obtained, or the overpotential can be considered as logarithmically dependent on the current density. This theoretical result is in agreement with the experimental findings of the German physical chemist Julius Tafel (1905), and the usual plots of overpotential *versus* log current density are known as Tafel lines. The slope of a Tafel plot reveals the value of the transfer coefficient α for the given direction of the electrode reaction.

DIFFICULTIES IN TRANSPORT OF REACTION SPECIES

The above conclusions about the overpotential–current density relationship are valid as long as the ratios of concentrations at the electrode surface of the species involved at current density i, C_i and, in the absence of current, C_o, stay close to unity. As the current density is increased, the concentration gradient needed to maintain a corresponding diffusion flux of the species concerned must begin to become appreciable. This condition is possible only if the concentration of the species at the surface starts to differ appreciably from the bulk value; *i.e.*, $(C_i)_i/(C_i)_o \neq 1$. The change in concentration of the discharging species at the electrode surface with time can, in principle, be obtained by using a second order partial differential equation (Fick's law), which, however, has explicit solutions only for a limited number of well-defined boundary conditions.

When significant concentration changes set in, no more exponential dependence of current density on potential can be obtained. It can be derived that, instead, a transition toward a limiting value takes place.

The important case is that in which the concentration of the discharging species at the electrode surface becomes equal to zero. The steady-state (*i.e.*, independent of time) current density obtained in such a case is the highest possible for the given set of conditions (diffusion limiting current density). The value of the concentration gradient in this case is directly proportional to the bulk concentration of the species involved and inversely proportional to the thickness of the diffusion layer (*i.e.*, the layer close to the electrode in which the concentration of the species differs from the species concentration in the bulk). This layer most often has a thickness fixed by hydrodynamic conditions in the solution surrounding the electrode. The definition used most often for the diffusion layer thickness is that of the German physical chemist Walther Hermann Nernst (1864–1941), according to whom this quantity is equal to the distance from the electrode at which the concentration would reach the bulk value if the concentration gradient were constant and equal to that at the electrode surface.

If a current larger than the limiting current is forced upon the electrode, the given electrode process will be able to sustain it only in the initial period in which the layer of solution close to the electrode is not completely exhausted of the discharging ions. As the concentration of ions tends toward zero, the electrode potential will change and another electrode process will start. The time at which the abrupt change of potential toward a new process takes place is termed the transition time. The relationship between transition time, current density, and concentration of the discharging species is given by Sand's equation :

$$\tau^{1/2} = \frac{\pi^{1/2} n F C_0 D^{1/2}}{2i} .$$

Since τ is a well-defined function of the concentration of the discharging species, the observation of the transition time can also be used as an analytical tool (chronopotentiometry).

Chapter 6

ELECTRO-SYNTHESIS

Electro-synthesis in organic chemistry is the synthesis of chemical compounds in an electrochemical cell. The main advantage of electro-synthesis over an ordinary redox reaction is avoidance of the potential wasteful other half–reaction and the ability to precisely tune the required potential. Electro-synthesis is actively studied as a science and also has many industrial applications.

EXPERIMENTAL SETUP

The basic setup in electro-synthesis is a galvanic cell, a potentiostat and two electrodes. Good electro-synthetic conditions use a solvent and electrolyte combination that minimizes electrical resistance. Protic conditions often use alcohol–water or dioxane–water solvent mixtures with an electrolyte such as a soluble salt, acid or base. Aprotic conditions often use an organic solvent such as acetonitrile or dichloromethane with electrolytes such as lithium perchlorate or tetrabutylammonium acetate. Electrodes are selected which provide favourable electron transfer properties towards the substrate while maximizing the activation energy for side reactions. This activation energy is often related to an overpotential of a competing reaction. For example, in aqueous conditions the competing reactions in the cell are the formation of oxygen at the anode and hydrogen at the cathode. In this case a graphite anode and lead cathode could be used effectively because of their high overpotentials for oxygen and hydrogen formation respectively. Many other materials can be used as electrodes. Other examples include platinum, magnesium, mercury (as a liquid pool in the reactor), stainless steel or reticulated vitreous carbon. Some reactions use a sacrificial electrode is used which is consumed during the reaction like zinc or lead. The two basic cell types are undivided cell or divided cell type. In divided cells the cathode and anode chambers are separated with a semi-porous membrane. Common membrane materials include sintered glass, porous porcelain, polytetrafloroethane or polypropylene. The purpose of the divided cell is to permit the diffusion of ions while restricting the flow of the products and reactants. This is important when unwanted side reactions are possible. An example of a reaction requiring a divided cell is the reduction of nitrobenzene to phenylhydroxylamine, where the latter chemical is susceptible to oxidation at the anode.

REACTIONS

Organic oxidations take place at the anode with initial formation of radical cations as reactive intermediates. Compounds are reduced at the cathode to radical anions. The initial reaction takes place at the surface of the electrode and then the intermediates diffuse into the solution where they participate in secondary reactions.

The yield of an electro-synthesis is expressed both in terms the chemical yield and current efficiency. Current efficiency is the ratio of Coulombs consumed in forming the products to the total number of Coulombs passed through the cell. Side reactions decrease the current efficiency.

The potential drop between the electrodes determines the rate constant of the reaction. Electro-synthesis is carried out with either constant potential or constant current. The reason one chooses one over the other is due to a trade off of ease of experimental conditions versus current efficiency. Constant potential uses current more efficiently because the current in the cell decreases with time due to the depletion of the substrate around the working electrode (stirring is usually necessary to decrease the diffusion layer around the electrode). This is not the case under constant current conditions however. Instead as the substrate's concentration decreases the potential across the cell increases in order to maintain the fixed reaction rate. This consumes current in side reactions produced outside the target voltage.

Anodic Oxidations

* The most well–known electro-synthesis is the Kolbe electrolysis, in which two carboxylic acids decarboxylate, and the remaining structures bond together :

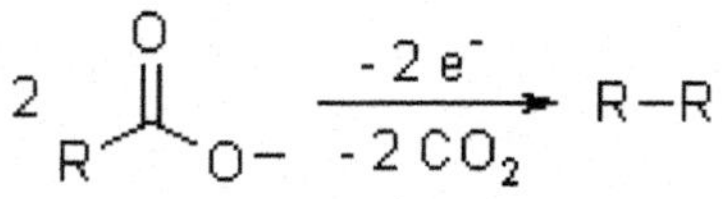

* A variation is called the **non–Kolbe reaction** when a heteroatom (nitrogen or oxygen) is present at the α–position. The intermediate oxonium ion is trapped by a nucleophile usually solvent.

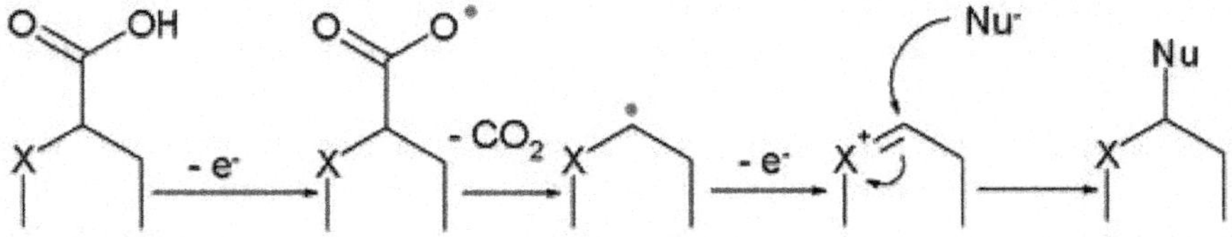

- Amides can be oxidized to *N*-acyliminium ions, which can be captured by various nucleophiles, for example :

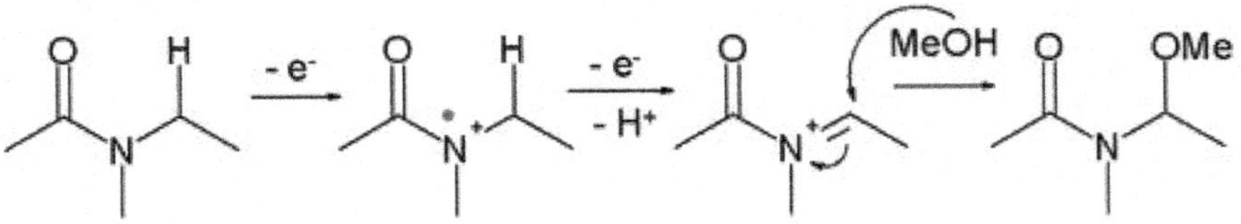

This reaction type is called a **Shono oxidation**. An example is the α–methoxylation of *N*–carbomethoxypyrrolidine

- Oxidation of a carbanion can lead to a coupling reaction for instance in the electro-synthesis of the tetramethyl ester of ethanetetracarboxylic acid from the corresponding malonate ester.

Cathodic Reductions

- In the Markó–Lam deoxygenation, an alcohol could be almost instantaneously deoxygenated by electro-reducing their toluate ester.

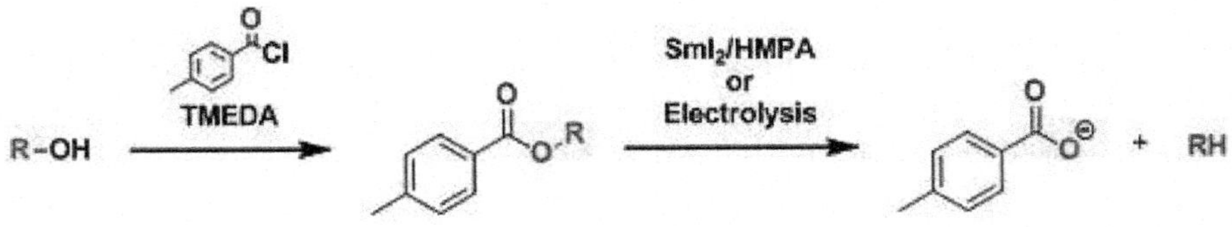

- The cathodic hydroisomerization of activated olefins is applied industrially in the synthesis of adiponitrile from 2 equivalents of acrylonitrile :

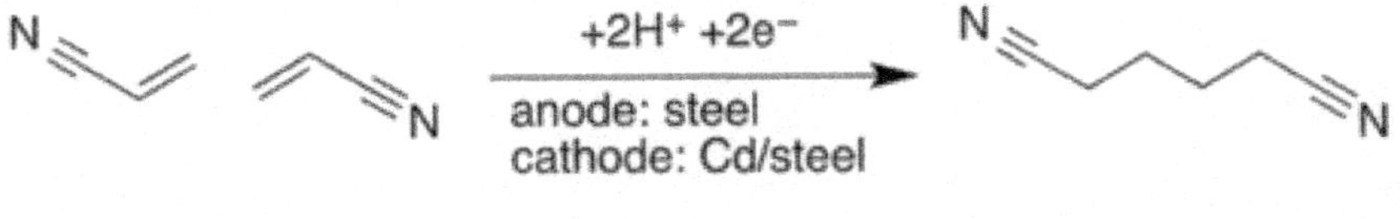

- The cathodic reduction of arene compounds to the 1,4–dihydro derivatives is similar to a Birch reduction. Examples from industry are the reduction of phthalic acid :

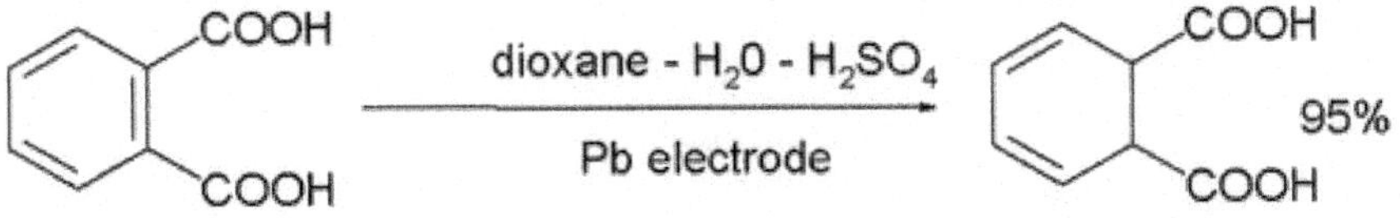

and the reduction of 2–methoxynaphthalene :

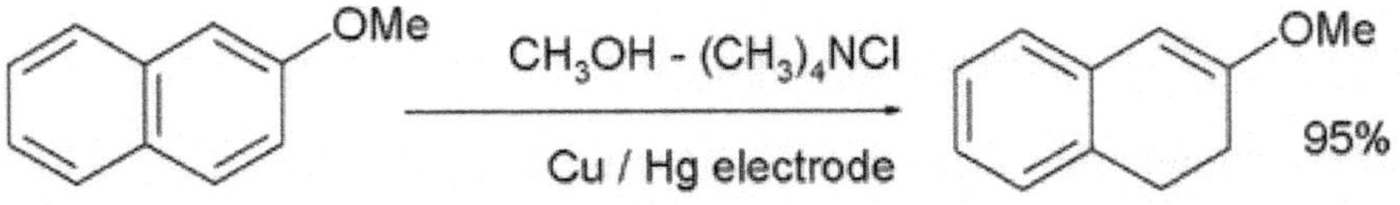

- The Tafel rearrangement, named for Julius Tafel, was at one time an important method for the synthesis of certain hydrocarbons from alkylated ethyl acetoacetate, a reaction accompanied by the re-arrangement reaction of the alkyl group :

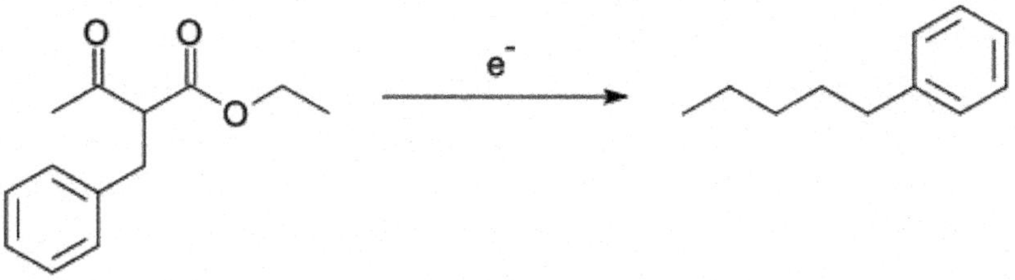

- The cathodic reduction of a nitrile to a primary amine in a divided cell :

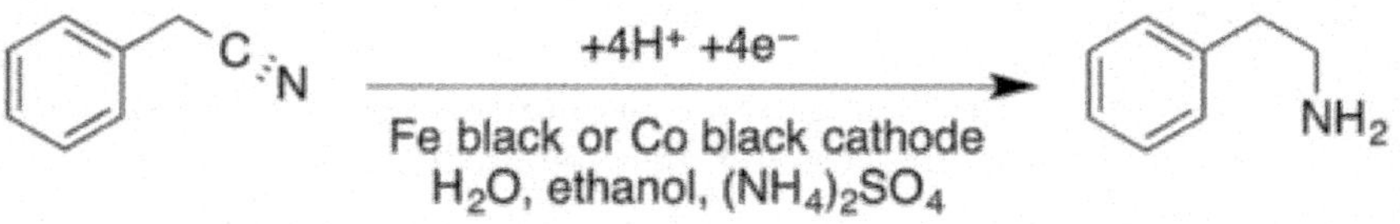

- Cathodic reduction of a nitroalkene can give the oxime in good yield. At higher negative reduction potentials, the nitroalkene can be reduced further, giving the primary amine but with lower yield.

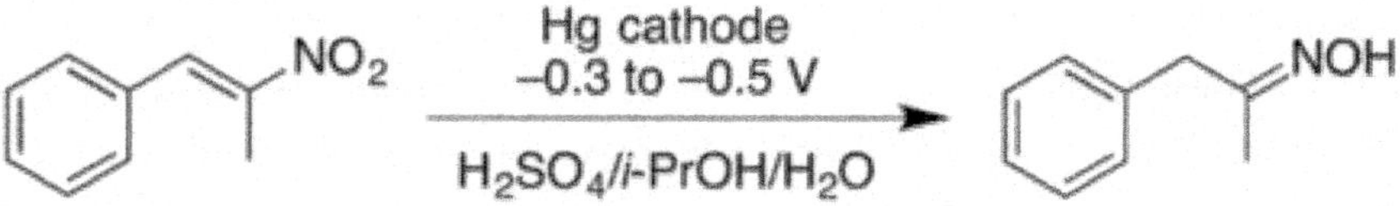

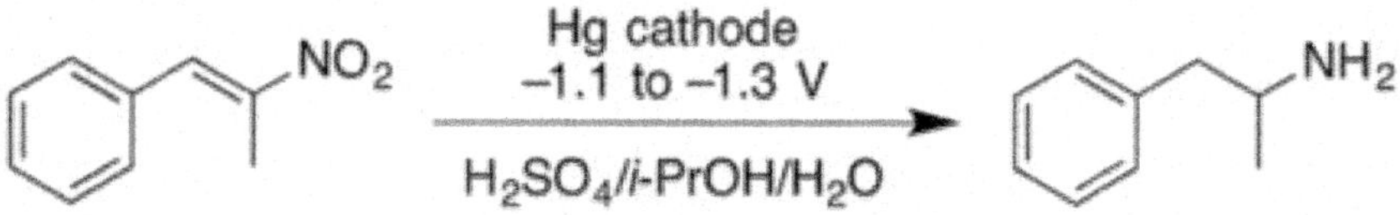

- Cathodic reduction of a carboxylic acid (oxalic acid) to an aldehyde (glyoxylic acid, shows as the rare aldehyde form) in a divided cell :

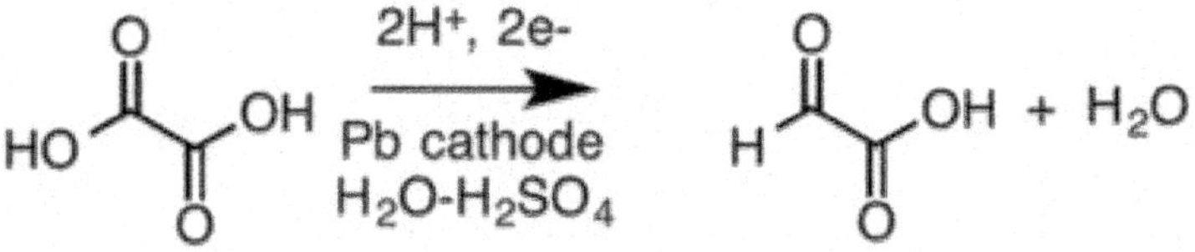

ELECTRO-CHEMICAL FLUORINATION

Electro-chemical fluorination (ECF), or **electro-fluorination**, is a foundational organo-fluorine chemistry method for the preparation of fluorocarbon–based organo-fluorine compounds. The general approach represents an application of electro-synthesis. The fluorinated chemical compounds produced by ECF are useful because of their distinctive solvation properties and the relative inertness of carbon–fluorine bonds. Two ECF synthesis routes are commercialized and commonly applied, the Simons Process and the Phillips Petroleum Process. Additionally, it is also possible to electro-fluorinate in various organic media. Prior to the development of these methods, fluorination with fluorine, a dangerous oxidant, was a dangerous and wasteful process. Also, ECF can be cost effective but it may also result in low yields.

Simons Process

The Simons Process entails electrolysis of a solution of an organic compound in a solution of hydrogen fluoride. An individual reaction can be described as :

$$R_3C-H + HF \rightarrow R_3C-F + H_2$$

In the course of a typical synthesis, this reaction occurs once for each C–H bond in the precursor. The cell potential is maintained near 5–6 V. The anode is nickel–plated. Simons discovered the process in the 1930s at Pennsylvania State College (U.S.), under the sponsorship of the 3M Corporation. The results were not published until after WWII because the work was classified due to its relevance to the manufacture of uranium hexafluoride. In 1949 Simons and his co-workers published a long paper in the Journal of the Electrochemical Society. The Simons process is used for the production of perfluorinated amines, ethers, carboxylic acids, and sulfonic acids. For carboxylic and sulfonic acids, the products are the corresponding acyl and sulfonyl fluorides. The method has been adapted to laboratory–scale preparations. Two noteworthy considerations are (i) the hazards associated with hydrogen fluoride (the solvent and fluorine source) and (ii) the requirement for anhydrous conditions.

Phillips Petroleum Process

This method is similar to the Simons Process but is typically applied to the preparation from volatile hydrocarbons and chlorohydrocarbons. In this process, electro-fluorination is conducted at porous graphite anodes in molten potassium fluoride in hydrogen fluoride. The species KHF_2 is relatively low melting, a good electrolyte, and an effective source of fluorine. The technology is sometimes called "CAVE" for Carbon Anode Vapour Phase Electrochemical Fluorination and was widely used at manufacturing sites of the 3M Corporation. The organic compound is fed through a porous anode leading to exchange of fluorine for hydrogen but not chlorine.

Other Methods

ECF has also been conducted in organic media, using for example organic salts of fluoride and acetonitrile as the solvent. A typical fluoride source is $(C_2H_5)_3N : 3HF$. In some cases, acetonitrile is omitted, and the solvent and electrolyte are the triethylamine–HF mixture. Representative products of this method are fluorobenzene (from benzene) and 1,2–difluoroalkanes (from alkenes).

Chapter 7

HALOCARBON

Halocarbon compounds are chemicals in which one or more carbon atoms are linked by covalent bonds with one or more halogen atoms (fluorine, chlorine, bromine or iodine–group 17) resulting in the formation of organofluorine compounds, organochlorine compounds, organobromine compounds, and organoiodine compounds. Chlorine halocarbons are the most common and are called organochlorides.

Many synthetic organic compounds such as plastic polymers, and a few natural ones, contain halogen atoms; they are known as *halogenated* compounds or *organohalogens*. Organochlorides are the most common industrially used organohalides, although the other organohalides are used commonly in organic synthesis. Except for extremely rare cases, organohalides are not produced biologically, but many pharmaceuticals are organohalides. Notably, many pharmaceuticals such as Prozac have trifluoromethyl groups.

CHEMICAL FAMILIES

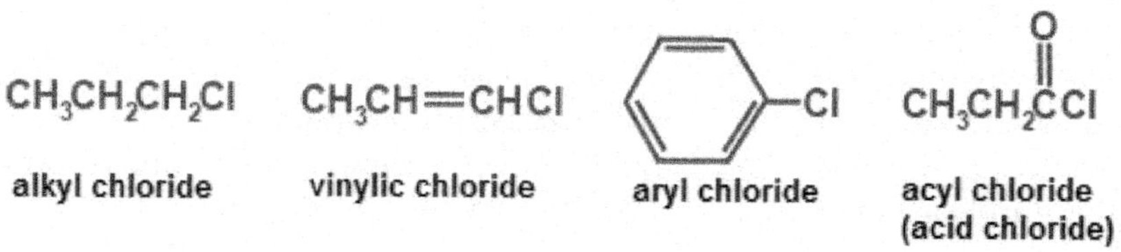

$CH_3CH_2CH_2Cl$	$CH_3CH{=}CHCl$	◯—Cl	$CH_3CH_2\overset{\text{O}}{\overset{\|}{C}}Cl$
alkyl chloride	vinylic chloride	aryl chloride	acyl chloride (acid chloride)

Fig. : Examples of organohalogens–chlorides.

Halocarbons are typically classified in the same ways as the similarly structured organic compounds that have hydrogen atoms occupying the molecular sites of the halogen atoms in halocarbons.

HALOALKANE

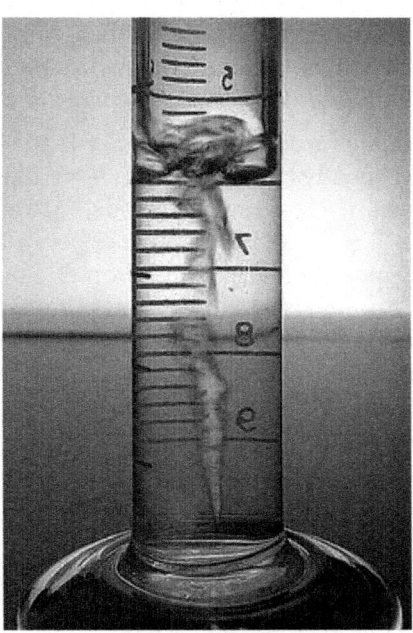

Fig. : Tetrafluoroethane (a haloalkane) is a colourless liquid that boils well below room temperature and can be extracted from common canned air canisters by simply inverting them during use.

The **haloalkanes** (also known as **halogenoalkanes** or **alkyl halides**) are a group of chemical compounds derived from alkanes containing one or more halogens. They are a subset of the general class of halocarbons, although the distinction is not often made. Haloalkanes are widely used commercially and, consequently, are known under many chemical and commercial names. They are used as flame retardants, fire extinguishants, refrigerants, propellants, solvents, and pharmaceuticals. Subsequent to the widespread use in commerce, many halo-carbons have also been shown to be serious pollutants and toxins. For example, the chloro-fluorocarbons have been shown to lead to ozone depletion. Methyl bromide is a controversial fumigant. Only haloalkanes which contain chlorine, bromine, and iodine are a threat to the ozone layer, but fluorinated volatile haloalkanes in theory may have activity as greenhouse gases. Methyl iodide, a naturally oc-curring substance, however, does not have ozone–depleting properties and the United States Environmental Protection Agency has designated the compound a non–ozone layer depleter. For more information. Haloalkane or alkyl halides are the compounds which have the general formua "RX" where R is an alkyl or substituted alkyl group and X is a halogen (F,Cl,Br,I).

Haloalkanes have been known for centuries. Chloroethane was produced synthetically in the 15th century. The systematic synthesis of such compounds developed in the 19th century in step with the development of organic chemistry and the understanding of the structure of alkanes. Methods were developed for the selective formation of C–halogen bonds. Especially versatile methods included the addition of halogens to alkenes, hydrohalogenation of alkenes, and the conversion of alcohols to alkyl halides. These methods are so reliable and so easily implemented that haloalkanes became cheaply available for use in industrial chemistry because the halide could be further replaced by other functional groups.

While most haloalkanes are human–produced, non–artificial–source haloalkanes do occur on Earth, mostly through enzyme–mediated synthesis by bacteria, fungi, and especially sea macroalgae (seaweeds). More than 1600 halogenated organics have been identified, with bromoalkanes being the most common haloalkanes. Brominated organics in biology range from biologically produced methyl bromide to non–alkane aromatics and unsaturates (indoles, terpenes, acetogenins, and phenols). Halogenated alkanes in land plants are more rare, but do occur, as for example the fluoroacetate produced as a toxin by at least 40 species of known plants. Specific dehalogenase enzymes in bacteria which remove halogens from haloalkanes, are also known.

Classes of Haloalkanes

From the structural perspective, haloalkanes can be classified according to the connectivity of the carbon atom to which the halogen is attached. In primary (1°) haloalkanes, the carbon that carries the halogen atom is only attached to one other alkyl group. An example is chloroethane (CH_3CH_2Cl). In secondary (2°) haloalkanes, the carbon that carries the halogen atom has two C–C bonds. In tertiary (3°) haloalkanes, the carbon that carries the halogen atom has three C–C bonds.

Haloalkanes can also be classified according to the type of halogen. Haloalkanes containing carbon bonded to fluorine, chlorine, bromine, and iodine results in organofluorine, organochlorine, organobromine and organoiodine compounds, respectively. Compounds containing more than one kind of halogen are also possible. Several classes of widely used haloalkanes are classified in this way chloro-fluorocarbons (CFCs), hydro-chloro-fluorocarbons (HCFCs) and hydro-fluorocarbons (HFCs). These abbreviations are particularly common in discussions of the environmental impact of haloalkanes.

Properties

Haloalkanes generally resemble the parent alkanes in being colourless, relatively odourless, and hydrophobic. Their boiling points are higher than the corresponding alkanes and scale with the atomic weight and number of halides. This is due to the increased strength of the inter-molecular forces – from London dispersion to dipole–dipole interaction because of the increased polarity. Thus carbon tetraiodide (CI_4) is a solid whereas carbon tetrafluoride (CF_4) is a gas. As they contain fewer C–H bonds, halocarbons are less flammable than alkanes, and some are used in fire extinguishers. Haloalkanes are better solvents than the corresponding alkanes because of their increased polarity. Haloalkanes containing halogens other than fluorine are more reactive than the parent alkanes – it is this reactivity that is the basis of most controversies. Many are alkylating agents, with primary haloalkanes and those containing heavier halogens being the most active (fluoroalkanes do not act as alkylating agents under normal conditions). The ozone–depleting abilities of the CFCs arises from the photolability of the C–Cl bond.

Occurrence

Haloalkanes are of wide interest because they are widespread and have diverse beneficial and detrimental impacts. The oceans are estimated to release 1–2 million tons of bromomethane annually.

A large number of pharmaceuticals contain halogens, especially fluorine. An estimated one fifth of pharmaceuticals contain fluorine, including several of the most widely used drugs. Examples include 5–fluorouracil, fluoxetine (Prozac), paroxetine (Paxil), ciprofloxacin (Cipro), mefloquine, and fluconazole. The beneficial effects arise because the C–F bond is relatively unreactive. Fluorine–substituted ethers are volatile anesthetics, including the commercial products methoxyflurane, enflurane, isoflurane, sevoflurane and desflurane. Fluorocarbon anesthetics reduce the hazard of flammability with diethyl ether and cyclopropane. Perfluorinated alkanes are used as blood substitutes.

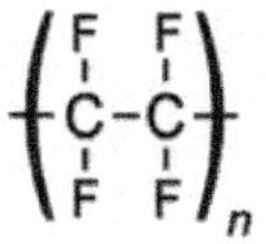

$$\left(\begin{array}{cc} F & F \\ | & | \\ C - C \\ | & | \\ F & F \end{array}\right)_n$$

Fig. : Teflon structure.

Chlorinated or fluorinated alkenes undergo polymerization. Important halogenated polymers include polyvinyl chloride (PVC), and polytetrafluoroethene (PTFE, or Teflon). The production of these materials releases substantial amounts of wastes.

Nomenclature

IUPAC

The formal naming of haloalkanes should follow IUPAC nomenclature, which put the halogen as a prefix to the alkane. For example, ethane with bromine becomes bromoethane, methane with four chlorine groups becomes tetrachloromethane. However, many of these compounds have already an established trivial name, which is endorsed by the IUPAC nomenclature, for example chloroform (trichloromethane) and methylene chloride (dichloromethane). For unambiguity, this article follows the systematic naming scheme throughout.

Production

Haloalkanes can be produced from virtually all organic precursors. From the perspective of industry, the most important ones are alkanes and alkenes.

From Alkanes

Alkanes react with halogens by free radical halogenation. In this reaction a hydrogen atom is removed from the alkane, then replaced by a halogen atom by reaction with a diatomic halogen molecule. The reactive intermediate in this reaction is a free radical and the reaction is called a *radical chain reaction*.

Free radical halogenation typically produces a mixture of compounds mono- or multi-halogenated at various positions. It is possible to predict the results of a halogenation reaction based on bond dissociation energies and the relative stabilities of the radical intermediates. Another factor to consider is the probability of reaction at each carbon atom, from a statistical point of view.

Due to the different dipole moments of the product mixture, it may be possible to separate them by distillation.

From Alkenes and Alkynes

In hydrohalogenation, an alkene reacts with a dry hydrogen halide (HX) like hydrogen chloride (HCl) or hydrogen bromide (HBr) to form a mono–haloalkane. The double bond of the alkene is replaced by two new bonds, one with the halogen and one with the hydrogen atom of the hydrohalic acid. Markovnikov's rule states that in this reaction, the halogen is more likely to become attached to the more substituted carbon. This is an electrophilic addition reaction. Water must be absent otherwise there will be a side product of a halohydrin. The reaction is necessarily to be carried out in a dry inert solvent such as CCl_4 or directly in the gaseous phase. The reaction of alkynes are similar, with the product being a geminal dihalide; once again, Markovnikov's rule is followed.

Alkenes also react with halogens (X_2) to form haloalkanes with two neighbouring halogen atoms in a halogen addition reaction. Alkynes react similarly, forming the tetrahalo compounds. This is sometimes known as "decolourizing" the halogen, since the reagent X_2 is coloured and the product is usually colourless and odourless.

From Alcohols

Tertiary alkanol reacts with hydrochloric acid directly to produce tertiary chloroalkane, but if primary or secondary alkanol is used, an activator such as zinc chloride is needed. This reaction is exploited in the Lucas test.

The most popular conversion is effected by reacting the alcohol with thionyl chloride ($SOCl_2$) in the "Darzen's Process," which is one of the most convenient laboratory methods because the byproducts are gaseous. Both phosphorus pentachloride (PCl_5) and phosphorus trichloride (PCl_3) also convert the hydroxyl group to the chloride.

Alcohols may likewise be converted to bromoalkanes using hydrobromic acid or phosphorus tribromide (PBr_3). A catalytic amount of PBr_3 may be used for the transformation using phosphorus and bromine; PBr_3 is formed *in situ*.

Iodoalkanes may similarly be prepared using red phosphorus and iodine (equivalent to phosphorus triiodide). The Appel reaction is also useful for preparing alkyl halides. The reagent is tetrahalomethane and triphenylphosphine; the co-products are haloform and triphenylphosphine oxide.

From Carboxylic Acids

Two methods for the synthesis of haloalkanes from carboxylic acids are the Hunsdiecker reaction and the Kochi reaction.

Bio-synthesis

Many chloro and bromolkanes are formed naturally. The principal pathways involve the enzymes chloroperoxidase and bromoperoxidase.

Reactions

Haloalkanes are reactive towards nucleophiles. They are polar molecules : the carbon to which the halogen is attached is slightly electropositive where the halogen is slightly electro-negative. This results in an electron deficient (electrophilic) carbon which, inevitably, attracts nucleophiles.

Substitution

Substitution reactions involve the replacement of the halogen with another molecule — thus leaving saturated hydrocarbons, as well as the halogenated product. Haloalkanes behave as the R^+ synthon, and readily react with nucleophiles.

Hydrolysis, a reaction in which water breaks a bond, is a good example of the nucleophilic nature of haloalkanes. The polar bond attracts a hydroxide ion, OH^-($NaOH_{(aq)}$ being a common source of this ion). This OH^- is a nucleophile with a clearly negative charge, as it has excess electrons it donates them to the carbon, which results in a covalent bond between the two. Thus C–X is broken by heterolytic fission resulting in a halide ion, X^-. As can be seen, the OH is now attached to the alkyl group, creating an alcohol. (Hydrolysis of bromoethane, for example, yields ethanol). Reaction with ammonia give primary amines.

Chloro–and bromoalkanes are readily substituted by iodide in the Finkelstein reaction. The iodoalkanes produced easily undergo further reaction. Sodium iodide is used thus as a catalyst.

Haloalkanes react with ionic nucleophiles (*e.g.* cyanide, thiocyanate, azide); the halogen is replaced by the respective group. This is of great synthetic utility : chloroalkanes are often inexpensively available. For example, after undergoing substitution reactions, cyanoalkanes may be hydrolyzed to carboxylic acids, or reduced to primary amines using lithium aluminium hydride. Azoalkanes may be reduced to primary amines by the Staudinger reduction or lithium aluminium hydride. Amines may also be prepared from alkyl halides in amine alkylation, the Gabriel synthesis and Delepine reaction, by undergoing nucleophilic substitution with potassium phthalimide or hexamine respectively, followed by hydrolysis.

In the presence of a base, haloalkanes alkylate alcohols, amines, and thiols to obtain ethers, N-substituted amines, and thioethers respectively. They are substituted by Grignard reagents to give magnesium salts and an extended alkyl compound.

Mechanism

Where the rate–determining step of a nucleophilic substitution reaction is unimolecular, it is known as an S_N1 reaction. In this case, the slowest (thus rate–determining step) is the heterolysis of a carbon–halogen bond to give a carbocation and the halide anion. The nucleophile (electron donor) attacks the carbocation to give the product.

S_N1 reactions are associated with the racemization of the compound, as the trigonal planar carbocation may be attacked from either face. They are favoured mechanism for tertiary haloalkanes, due to the stabilization of the positive charge on the carbocation by three electron–donating alkyl groups. They are also preferred where the substituents are sterically bulky, hindering the S_N2 mechanism.

Elimination

Rather than creating a molecule with the halogen substituted with something else, one can completely eliminate both the halogen and a nearby hydrogen, thus forming an alkene by dehydrohalogenation. For example, with bromoethane and sodium hydroxide (NaOH) in ethanol, the hydroxide ion HO⁻ abstracts a hydrogen atom. Bromide ion is then lost, resulting in ethylene, H_2O and NaBr. Thus, haloalkanes can be converted to alkenes. Similarly, dihaloalkanes can be converted to alkynes.

In related reactions, 1,2–dibromocompounds are debrominated by zinc dust to give alkenes and geminal dihalides can react with strong bases to give carbenes.

Other

Haloalkanes undergo free–radical reactions with elemental magnesium to give alkylmagnesium compounds : Grignard reagents. Haloalkanes also react with lithium metal to give organolithium compounds. Both Grignard reagents and organolithium compounds behave as the R⁻synthon. Alkali metals such as sodium and lithium are able to cause haloalkanes to couple in the Wurtz reaction, giving symmetrical alkanes. Haloalkanes, especially iodoalkanes, also undergo oxidative addition reactions to give organometallic compounds.

Applications

Haloalkanes are widely used as synthon equivalents to alkyl cation (R+) in organic synthesis. They can also participate in a wide variety of other organic reactions.

Short chain haloalkanes such as dichloromethane, trichloromethane (chloroform) and tetrachloromethane are commonly used as hydrophobic solvents in chemistry. They were formerly very common in industry; however, their use has been greatly curtailed due to their toxicity and harmful environmental effects.

Chloro-fluorocarbons were used almost universally as refrigerants and propellants due to their relatively low toxicity and high heat of vapourization. Starting in the 1980s, as their contribution to ozone depletion became known, their use was increasingly restricted, and they have now largely been replaced by HFCs.

VINYL HALIDE

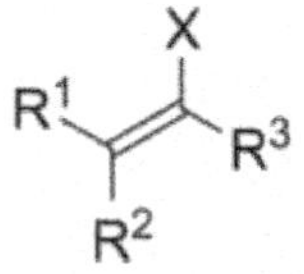

Fig. : General structure of a vinyl halide, where X is a halogen and R is a radical group.

In organic chemistry, a **vinyl halide** is any alkene with at least one halide substituent bonded directly on one of the alkene carbons. Vinyl chloride is one such substance.

Vinyl halides are very useful synthetic intermediates due to the vast number of reactions that make use of them. These include conversion to vinyl Grignard reagents, elimination to give the corresponding alkyne, and most importantly their use in cross–coupling reactions (*e.g.* Suzuki–Miyaura coupling, Stille coupling, Heck coupling, etc.).

As a result, there is a large number of reactions to form vinyl halides, which includes the reaction of vinyl organometallic species with halogens, and the Takai and Wittig olefination reactions.

Besides, some vinyl halides are useful for synthesizing polymers and co-polymers. The unsubstituted vinyl halides ($R_1 = R_2 = R_3 = H$) may polymerize spontaneously under certain conditions.

ARYL HALIDE

In organic chemistry, an **aryl halide** (also known as **haloarene** or **halogenoarene**) is an aromatic compound in which one or more hydrogen atoms directly bonded to an aromatic ring are replaced by a halide. The haloarene are distinguished from haloalkanes because they exhibit many differences in methods of preparation and properties. The most important members are the aryl chlorides, but the class of compounds is so broad that many derivatives enjoy niche applications.

Preparation

The two main preparatory routes to aryl halides are direct halogenation and via diazonium salts.

Direct Halogenation

In the Friedel–Crafts halogenation, a Lewis acid serve as catalysts. Many metal chlorides are used, examples include iron (III) chloride or aluminium chloride. The most important aryl halide, chlorobenzene is produced by this route. Monochlorination of benzene is always accompanied by formation of the dichlorobenzene derivatives.

Arenes with electron donating groups react with halogens even in the absence of Lewis acids. For example, phenols and anilines react quickly with chlorine and bromine water to give multiply halogenated products. The decolouration of bromine water by electron–rich arenes is used in the bromine test.

Direct halogenation of arenes are possible in the presence of light or at high temperature. For alkylbenzene derivatives, the alkyl positions tend to be halogenated first in the free radical halogenation. To halogenate the ring, Lewis acids are required, and light should be excluded to avoid the competing reaction.

Fig. : Reaction between benzene and halogen to form an halogenobenzene.

Sandmeyer, Schiemann and Gatterman Reactions

The second main route is the Sandmeyer reaction. Anilines (aryl amines) are converted to their diazonium salts using nitrous acid. For example, copper (I) chloride converts diazonium salts to the aryl chloride. Nitrogen gas is the leaving group, which makes this reaction very favourable. The similar Schiemann reaction uses the tetra-fluoroborate anion as the fluoride donor. Gatterman Reaction can also be used to convert Diazonium salt to chlorobenzene or bromobenzene by using copper powder instead of copper chloride or copper bromide. But this must be done in the presence of HCl and HBr respectively.

Halogenation in Nature

Aryl halides occur widely in nature, most commonly produced by marine organisms that utilize the chloride and bromide in ocean waters. Chlorinated and brominated aromatic compounds are also numerous, *e.g.* derivatives of tyrosine, tryptophan, and various pyrrole derivatives. Some of these naturally occurring aryl halides exhibit useful medicinal properties.

Structural Trends

The C–X distances for aryl halides follow the expected trend. These distances for fluorobenzene, chlorobenzene, bromobenzene, and methyl 4–iodobenzoate are 135.6(4), 173.90(23), 189.8(1), and 209.9 pm, respectively.

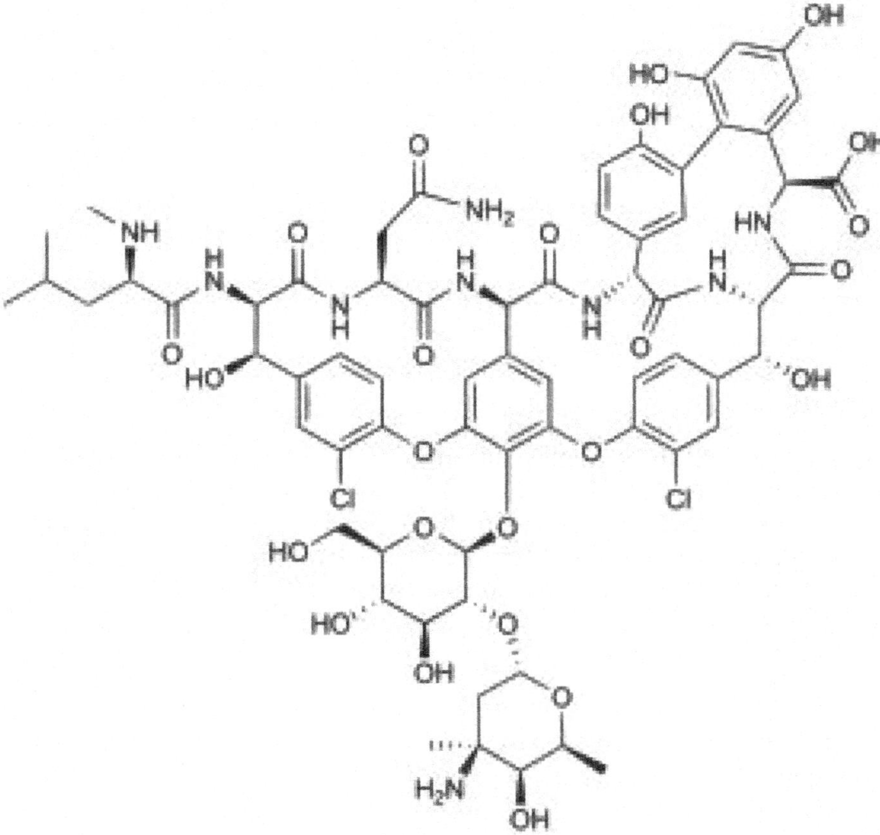

Fig. : Vancomycin, an important antibiotic, is an aryl chloride isolated from soil fungi.

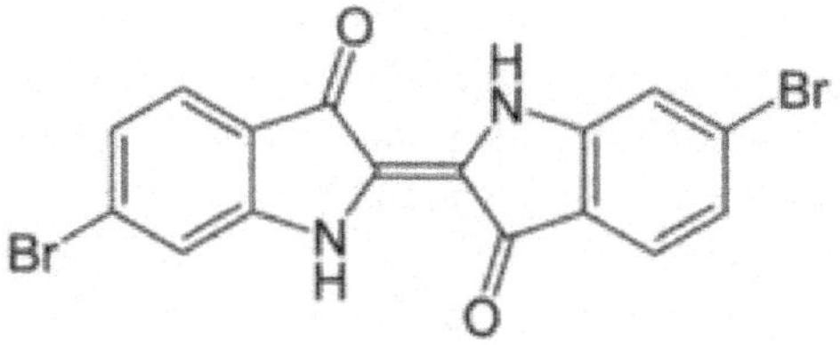

Fig. : The chemical structure of 6,6'-dibromoindigo, the main component of Tyrian Purple.

Reactions

Aryl halides do not participate in conventional SN_2 nucleophilic aromatic substitution reactions. Instead the halides are displaced by strong nucleophiles via reactions involving radical anions. Alternatively aryl halides, especially the bromides and iodides, undergo oxidative addition, and thus are subject to Buchwald–Hartwig amination–type reactions.

Aryl halides react with metals to give more reactive derivatives that behave as sources of aryl anions. Magnesium aryl halides are Grignard reagents, which are useful in organic synthesis of other aryl compounds.

Chlorobenzene was once the precursor to phenol, which is now made by oxidation of cumene. At high temperatures, aryl groups react with ammonia to give anilines.

Bio-degradation

Rhodococcus phenolicus is a bacterium species able to degrade dichlorobenzene as sole carbon sources.

Applications

The aryl halides produced on the largest scale are chlorobenzene and the isomers of dichlorobenzene. One major but discontinued application was the use of chlorobenzene as a solvent for dispersing the herbicide Lasso. Overall, production of aryl chlorides as well as related naphthyl derivatives has been declining since the 1980s, in part due to environmental concerns. Triphenylphosphine is produced from chlorobenzene :

$$3\ C_6H_5Cl + PCl_3 + 6\ Na \rightarrow P(C_6H_5)_3 + 6\ NaCl$$

Aryl bromides are widely used as fire-retardants. The most prominent member is tetra-bromobisphenol-A, which is prepared by direct bromination of the diphenol.

HISTORY AND CONTEXT

A few halocarbons are produced in massive amounts by microorganisms. For example, several million tons of methyl bromide are estimated to be produced by marine organisms annually. Most of the halocarbons encountered in everyday life — solvents, medicines, plastics are man-made. The synthesis of halocarbons in the early 1800s but accelerated when their useful properties as solvents and anesthetics were discovered. The development of plastics and synthetic elastomers led to greatly expanded scale of production. A substantial percentage of drugs are halocarbons.

NATURAL HALOCARBONS

A large amount of the naturally occurring halocarbons are created by wood fire, dioxine for example, or volcanic activities. A second large source are marine algae which produce several chlorinated methane and ethane derivates. There are several thousand more complex halocarbons known, produced mainly by marine species. Although chlorine compounds are the majority of the discovered compounds, bromides, iodides and fluorides have also been found. The tyrian purple, which is a dibromoindigo, is representative of the bromides, while the thyroxine secreted from the thyroid gland, is an iodide, and the highly toxic fluoroacetate is one of the rare organo-fluorides. These three representatives, thyroxine from humans, tyrian purple from snails and fluoroacetate from plants, also show that unrelated species use halocarbons for several purposes.Uses

The first halocarbon commercially used was Tyrian purple a natural organo-bromide of the *Murex brandaris* marine snail.

Common uses for halocarbons have been as solvents, pesticides, refrigerants, fire–resistant oils, ingredients of elastomers, adhesives and sealants, electrically insulating coatings, plasticizers, and plastics. Many halocarbons have specialized uses in industry. One halocarbon, sucralose, is a sweetener.

Before they became strictly regulated, the general public often encountered haloalkanes as paint and cleaning solvents such as trichloroethane (1,1,1–trichloroethane) and carbon tetrachloride (tetrachloromethane), pesticides like 1,2–dibromoethane (EDB, ethylene dibromide), and refrigerants like Freon–22 (duPont trademark for chlorodifluoromethane). Some haloalkanes are still widely used for industrial cleaning, such as methylene chloride (dichloromethane), and as refrigerants, such as R–134a (1,1,1,2–tetrafluoroethane).

Haloalkenes have also been used as solvents, including perchloroethylene (Perc, tetrachloroethene), widespread in dry cleaning, and trichloroethylene (TCE, 1,1,2–trichloroethene). Other haloalkenes have been chemical building blocks of plastics such as polyvinyl chloride ("vinyl" or PVC, polymerized chloroethene) and Teflon (duPont trademark for polymerized tetrafluoroethene, PTFE).

Haloaromatics include the former Aroclors (Monsanto Company trademark for polychlorinated biphenyls, PCBs), once widely used in power transformers and capacitors and in building caulk, the former Halowaxes (Union Carbide trademark for polychlorinated naphthalenes, PCNs), once used for electrical insulation, and the chlorobenzenes and their derivatives, used for disinfectants, pesticides such as dichloro–diphenyl–trichloroethane (DDT, 1,1,1–trichloro–2,2–bis (p–chlorophenyl) ethane), herbicides such as 2,4–D (2,4–dichlorophenoxyacetic acid), askarel dielectrics (mixed with PCBs, no longer used in most countries), and chemical feedstocks.

A few halocarbons, including acid halides like acetyl chloride, are highly reactive; these are rarely found outside chemical processing. The widespread uses of halocarbons were often driven by observations that most of them were more stable than other substances. They may be less affected by acids or alkalis; they may not burn as readily; they may not be attacked by bacteria or molds; or they may not be affected as much by sun exposure.

HAZARDS

The stability of halocarbons tended to encourage beliefs that they were mostly harmless, although in the mid–1920s physicians reported workers in polychlorinated naphthalene manufacturing suffering from chloracne (Teleky 1927), and by the late 1930s it was known that workers exposed to PCNs could die from liver disease (Flinn & Jarvik 1936) and that DDT would kill mosquitos and other insects (Müller 1948). By the 1950s, there had been several reports and investigations of workplace hazards. In 1956, for example, after testing hydraulic oils containing PCBs, the U.S. Navy found that skin contact caused fatal liver disease in animals and rejected them as "too toxic for use in a submarine" (Owens v. Monsanto 2001).

In 1962 a book by U.S. biologist Rachel Carson (Carson 1962) started a storm of concerns about environmental pollution, first focused on DDT and other pesticides, some of them also halocarbons. These concerns were amplified when in 1966 Swedish chemist Soren Jensen reported widespread residues of PCBs among Arctic and sub–Arctic fish and birds (Jensen 1966). In 1974, U.S. chemists Mario Molina and Sherwood Rowland predicted that common halocarbon refrigerants, the chloro-fluorocarbons (CFCs), would accumulate in the upper atmosphere and destroy protective ozone (Molina & Rowland 1974). Within a few years, ozone depletion was being observed above Antarctica, leading to bans on production and use of chloro-fluorocarbons in many countries. In 2007, the Inter-governmental Panel on Climate Change (IPCC) said halocarbons were a direct cause of global warming.

Since the 1970s there have been long-standing, unresolved controversies over potential health hazards of trichloroethylene (TCE) and other halocarbon solvents that had been widely used for industrial cleaning (Anderson v. Grace 1986) (Scott & Cogliano 2000) (U.S. National Academies of Science 2004) (United States 2004). More recently perfluorooctanoic acid (PFOA), a precursor in the most common manufacturing process for Teflon and also used to make coatings for fabrics and food packaging, has become a health and environmental concern (United States 2006), suggesting that halocarbons thought to be among the most inert may also present hazards.

Halocarbons, including those that might not be hazards in themselves, can present waste disposal issues. Because they do not readily degrade in natural environments, halocarbons tend to accumulate. Incineration and accidental fires can create corrosive by-products like hydrochloric acid and hydrofluoric acid and poisons like halogenated dioxins and furans. Species of Desulfitobacterium are being investigated for their potential in bio-remediating halogenic organic compounds.

Chapter 8

ORGANOIODINE COMPOUND

Organoiodine compounds are organic compounds that contain one or more carbon–iodine bonds. They occur widely in organic chemistry, but are relatively rare in nature. The thyroxine hormones are organoiodine compounds that are required for health and the reason for government-mandated iodization of salt.

STRUCTURE, BONDING, GENERAL PROPERTIES

Almost all organoiodine compounds feature iodide connected to one carbon center. These are usually classified as derivatives of I^-. Some organoiodine compounds feature iodine in higher oxidation states.

The C–I bond is the weakest of the carbon–halogen bonds. These bond strengths correlate with the electro-negativity of the halogen, decreasing in the order F > Cl > Br > I. This periodic order also follows the atomic radius of halogens and the length of the carbon-halogen bond. For example, in the molecules represented by CH_3X, where X is a halide, the carbon-X bonds have strengths, or bond dissociation energies, of 115, 83.7, 72.1, and 57.6 kcal/mol for X = fluoride, chloride, bromide, and iodide, respectively. Of the halides, iodide usually is the best leaving group. Because of the weakness of the C-I bond, samples of organoiodine compounds are often yellow due to an impurity of I_2.

A noteworthy aspect of organoiodine compounds is their high density, which arises from the high atomic weight of iodine. For example, one millilitre of methylene iodide weighs 3.325 g.

INDUSTRIAL APPLICATIONS

Few organoiodine compounds are important industrially, at least in terms of large scale production. Iodide containing intermediates are common in organic synthesis, because of the easy formation and cleavage of the C–I bond. Industrially significant organoiodine compounds, often used as disinfectants or pesticides, are iodoform (CHI_3), methylene iodide (CH_2I_2), and methyl iodide (CH_3I). Although methyl iodide is not an industrially important product, it is an important intermediate, being a transiently generated intermediate in the industrial production of acetic acid and acetic anhydride. The potential for methyl iodide to replace the ubiquitous dependence on methyl bromide as a soil fumigant has been considered, however limited information is available on environmental behaviour of the former. Ioxynil (3,5-diiodo-4-hydroxybenzonitrile), which inhibits photosynthesis at photo-system II, is among the very few organoiodine herbicides. A member of the hydroxybenzonitrile herbicide class, ioxynil is an iodinated analog of the brominated herbicide, bromoxynil (3,5-dibromo-4-hydroxybenzonitrile).

Iodinated and brominated organic compounds are of concern as environmental contaminants owing to very limited information available on environment fate behaviour. However, recent reports have shown promise in biological detoxification of these classes of contaminants. For example, Iodotyrosine deiodinase is a mammalian enzyme with the unusual function of aerobic reductive dehalogenation of iodine-or bromine-substituted organic substrates. Bromoxynil and ioxynil herbicides have been shown to undergo a variety of environmental transformations, including reductive dehalogenation by anaerobic bacteria.

Polyiodoorganic compounds are sometimes employed as X-ray contrast agents, in fluoroscopy, a type of medical imaging. This application exploits the X-ray absorbing ability of the heavy iodine nucleus. A variety of agents are available commercially, many are derivatives of 1,3,5-triiodobenzene and contain about 50% by weight iodine. For most applications, the agent must be highly soluble in water and, of course, non-toxic and readily excreted. A representative reagent is Ioversol, which has water-solubilizing diol substituents. Typical applications include urography and angiography.

Fig. : Structure of Ioversol, an organoiodine compound used as an X-ray contrast agent.

BIOLOGICAL ROLE

In terms of human health, the most important organoiodine compounds are the two thyroid hormones thyroxine ("T_4") and triiodothyronine ("T_3"). Marine natural products are rich sources of organoiodine compounds, like the recently discovered plakohypaphorines from the sponge *Plakortis simplex*.

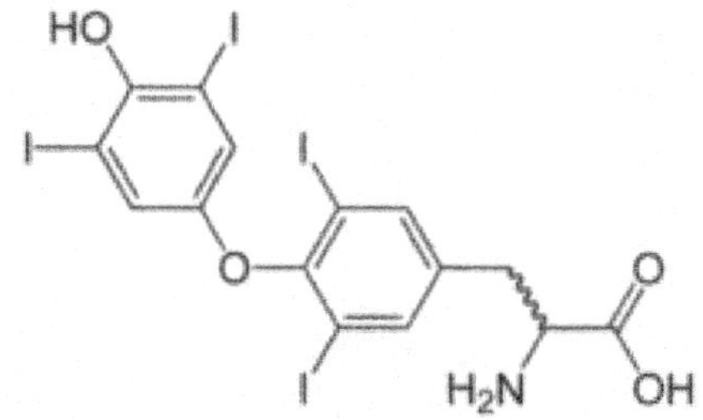

Fig. : Thyroxine (T$_4$).

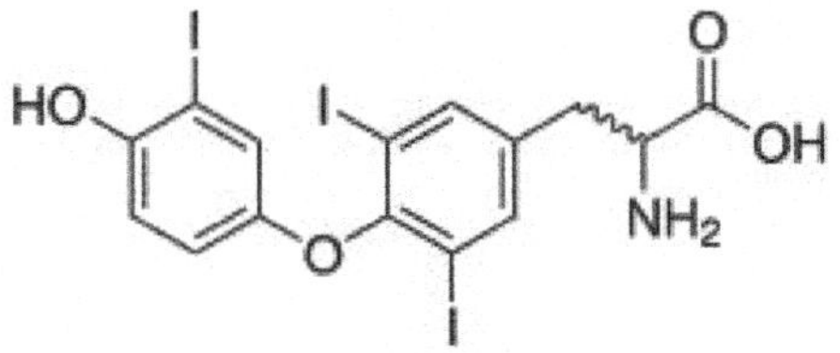

Fig. : Triiodothyronine (T$_3$).

The sum of iodomethane produced by the marine environment, microbial activitiy in rice paddies, and the burning of biological material is estimated to be 214 kilotonnes per year. The volatile iodomethane is broken up by oxidation reactions in the atmosphere and a global iodine cycle is established.

METHODS FOR PREPARATION OF THE C–I BOND

Organoiodine compounds are prepared by numerous routes, depending on the degree and regiochemistry of iodination sought and the nature of the precursors. The direct iodination with I$_2$ is employed with unsaturated substrates :

$$RHC=CH_2 + I_2 \rightarrow RHIC\text{-}CIH_2$$

The iodide anion is a good nucleophile and will displace chloride, tosylate, bromide and other leaving groups, as in the Finkelstein reaction. Aromatic iodides may be prepared via a diazonium salt in the Sandmeyer reaction.

Because of its low ionization energy, iodine is readily converted to reagents that deliver the equivalent of "I$^+$". A representative electrophilic iodination reagent is iodine monochloride.

ORGANO-FLUORINE COMPOUND

A : fluoromethane

B : isoflurane

C : a CFC

D : an HFC

E : triflic acid

F : Teflon

G : PFOS

H : fluorouracil

I : Prozac.

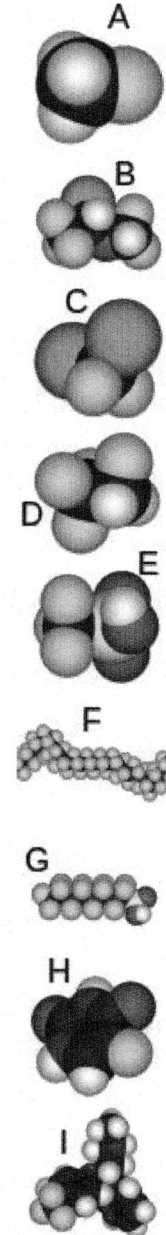

Fig. : Some important organo-fluorine compounds.

Organo-fluorine compounds are organic chemical compounds that contain carbon and fluorine bonded in the polarized and remarkably strong carbon–fluorine bond. Organo-fluorine compounds are diverse, they can be fluoro-carbons, perfluorinated, or biologically synthesized mono-fluorinated compounds, among other possibilities. These compounds have a wide range of function and can serve as refrigerants, pharmaceuticals, agrichemicals, surfactants, ozone depletors, poisons, or pollutants.

Bond

Fig. : C–F partial charges.

The carbon–fluorine bond is referred to as the strongest in organic chemistry because of stability added by its partial ionic character; it forms the strongest single bond to carbon. The ionic character is a result of the electro-negativity of fluorine. It induces partial charges on the carbon and fluorine atoms, leading to electrostatic attraction, making the bond short and strong.

Compounds

Organo-fluorine compounds that have the carbon–fluorine bond are diverse in their types. They can be fluoro-carbons, fluoro-carbon derivatives, fluorinated pharmaceuticals and agrichemicals, or mono-fluorinated biologically synthesized compounds, among others. Fluoro-carbons are compounds that contain only carbon and fluorine, while other molecules that contain many carbon–fluorine bonds are commonly referred to as fluoro-carbons. Pharmaceuticals and agrichemicals commonly contain only one fluorine or a trifluoromethyl group. However, some are more highly fluorinated, such as hexaflumuron, which has six fluorines, in large part to a tetrafluoroethoxy functional group. All known biologically synthesized organo-fluorines contain only one carbon–fluorine bond.

Fluoro-carbons

Fluoro-carbons are molecules that *only* contain carbon and fluorine. They can be gases, liquids, waxes, or solids, depending upon their molecular weight. The simplest fluoro-carbon is the gas tetrafluoromethane (CF_4). Liquids include perfluorooctane and perfluorodecalin. The fluoropolymer polytetrafluoroethylene (PTFE/Teflon) is a solid. While fluoro-carbons with single bonds are stable, unsaturated fluoro-carbons are more reactive, especially those with triple bonds.

Perfluorinated Compounds

Perfluorinated compounds are fluoro-carbon derivatives, as they are closely structurally related to fluoro-carbons. However, they also possess new atoms such as nitrogen, iodine, or ionic groups, such as perfluorinated carboxylic acids.

Alkyl Fluorides

Alkyl mono-fluorides can be obtained from alcohols and Olah reagent or another fluorinating agents.

Properties

Because of the varying degree of fluorination of organo-fluorine compounds, their properties are nearly impossible to compare as a group. "Every time you see a biologically active molecule that has fluorine in it, it could be in there for a different reason," says University of Florida chemistry professor William R. Dolbier Jr. By contrast, fluoro-carbon based compounds are fluorinated for specific chemical, physical, and sometimes biological reasons, as they have properties that are very distinct from hydrocarbons. Organo-fluorines with only one carbon–fluorine bond can simply behave as a hydrocarbon. Therefore, the one salient point for all organo-fluorine compounds is that the carbon–fluorine bond can dramatically alter biological properties. While both uracil and 5-fluorouracil are colourless, high-melting crystalline solids, the organo-fluorine is a potent anti-cancer drug. Similarly, fluoroacetate is a potent natural poison while dilute acetate in water is vinegar.

Biological Role

Biologically synthesized organo-fluorines have been found in micro-organisms and plants, but not animals. The most common example is fluoroacetate, which occurs as a plant defence against herbivores in at least 40 plants in Australia, Brazil and Africa. Other biologically synthesized organo-fluorines include ω-fluoro fatty acids, fluoroacetone, and 2-fluorocitrate which are all believed to be biosynthesized in bio-chemical pathways from the intermediate fluoroacetaldehyde. Adenosyl-fluoride synthase is an enzyme capable of biologically synthesizing the carbon–fluorine bond. Man made carbon–fluorine bonds are commonly found in pharmaceuticals and agrichemicals because it adds stability to the carbon framework; also, the relatively small size of fluorine is convenient as fluorine acts as an approximate bioisostere of the hydroxyl group. Introducing the carbon–fluorine bond to organic compounds is the major challenge for medicinal chemists using organo-fluorine chemistry, as the carbon–fluorine bond increases the probability of having a successful drug by about a factor of ten. An estimated 20% of pharmaceuticals, and 30–40% of agrichemicals are organo-fluorines, including several of the top drugs. Examples include 5-fluorouracil, fluoxetine (Prozac), paroxetine (Paxil), ciprofloxacin (Cipro), mefloquine, and fluconazole.

Environmental and Health Issues

Abiotic processes can also result in organo-fluorines considered as "problem molecules." Fluoro-carbon based CFCs and tetra-fluoro-methane have been reported in igneous and metamorphic rock. However, environmental and health issues still face many organo-fluorines. Because of the strength of the carbon–fluorine bond, many synthetic fluoro-carbons and fluorcarbon-based compounds are persistent in the environment. Others, such as CFCs, participate in ozone depletion. Fluoro-alkanes, commonly referred to as perfluoro-carbons, are potent greenhouse gases. The fluorosurfactants PFOS and PFOA, and other related chemicals, are persistent global contaminants. PFOS is a persistent organic pollutant and may be harming the health of wildlife; the potential health effects of PFOA to humans are under investigation by the C8 Science Panel.

ORGANOCHLORIDE

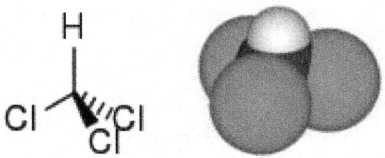

Fig. : Two representations of the organochloride chloroform.

An organochloride, organochlorine, chlorocarbon, chlorinated hydrocarbon, chloroalkane, or chlorinated solvent is an organic compound containing at least one covalently bonded atom of chlorine. Their wide structural variety and divergent chemical properties lead to a broad range of names and applications. Many derivatives are controversial because of the effects of these compounds on the environment and on human and animal health.

Physical Properties

Chloride substituents modify the physical properties of organic compounds in several ways. They are typically denser than water due to the presence of high atomic weight of chlorine. Chloride substituents induce stronger inter-molecular interactions than hydrogen substituents. The effect is illustrated by trends in boiling points : methane (−161.6 °C), methyl chloride (−24.2 °C), dichloromethane (40 °C), chloroform (61.2 °C), and carbon tetrachloride (76.72 °C). The increased intermolecular interactions is attributed to the effects of both van der Waals and polarity.

Natural Occurrence

Although rare compared to non-halogenated organic compounds, many organochlorine compounds have been isolated from natural sources ranging from bacteria to humans. Chlorinated organic compounds are found in nearly every class of biomolecules including alkaloids, terpenes, amino acids, flavonoids, steroids, and fatty acids. Organochlorides, including dioxins, are produced in the high temperature environment of forest fires, and dioxins have been found in the preserved ashes of lightning-ignited fires that predate synthetic dioxins. In addition, a variety of simple chlorinated hydrocarbons including dichloromethane, chloroform, and carbon tetrachloride have been isolated from marine algae. A majority of the chloromethane in the environment is produced naturally by biological decomposition, forest fires, and volcanoes. The natural organochloride epibatidine, an alkaloid isolated from tree frogs, has potent analgesic effects and has stimulated research into new pain medication.

Preparation

From Chlorine

Alkanes and arylalkanes may be chlorinated under free radical conditions, with UV light. However, the extent of chlorination is difficult to control. Aryl chlorides may be prepared by the Friedel-Crafts halogenation, using chlorine and a Lewis acid catalyst.

The haloform reaction, using chlorine and sodium hydroxide, is also able to generate alkyl halides from methyl ketones, and related compounds. Chloroform was formerly produced thus.

Chlorine adds to the multiple bonds on alkenes and alkynes as well, giving di-or tetra-chloro compounds.

Reaction with Hydrogen Chloride

Alkenes react with hydrogen chloride (HCl) to give alkyl chlorides. For example, the industrial production of chloroethane proceeds by the reaction of ethylene with HCl :

$$H_2C=CH_2 + HCl \rightarrow CH_3CH_2Cl$$

Secondary and tertiary alcohols react with the Lucas reagent (zinc chloride in concentrated hydrochloric acid) to give the corresponding alkyl halide; this reaction a method for classifying alcohols :

$$R-OH + HCl \xrightarrow[\Delta]{ZnCl_2} R-Cl + H_2O$$

Other Chlorinating Agents

In the laboratory, alkyl chlorides are most easily prepared by reacting alcohols with thionyl chloride ($SOCl_2$), phosphorus trichloride (PCl_3), or phosphorus pentachloride (PCl_5) :

$$ROH + SOCl_2 \rightarrow RCl + SO_2 + HCl$$

$$3\ ROH + PCl_3 \rightarrow 3\ RCl + H_3PO_3$$

$$ROH + PCl_5 \rightarrow RCl + POCl_3 + HCl$$

In the laboratory, thionyl chloride is especially convenient, because the by-products are gaseous.

Alternatively, the Appel reaction :

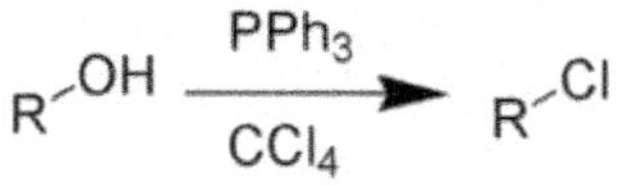

$$R{-}OH \xrightarrow[\text{CCl}_4]{\text{PPh}_3} R{-}Cl$$

Reactions

Alkyl chlorides are versatile building blocks in organic chemistry. While alkyl bromides and iodides are more reactive, alkyl chlorides tend to be less expensive and more readily available. Alkyl chlorides readily undergo attack by nucleophiles.

Heating alkyl halides with sodium hydroxide or water gives alcohols. Reaction with alkoxides or aroxides give ethers in the Williamson ether synthesis; reaction with thiols give thioethers. Alkyl chlorides readily react with amines to give substituted amines. Alkyl chlorides are substituted by softer halides such as the iodide in the Finkelstein reaction. Reaction with other pseudohalides such as azide, cyanide, and thiocyanate are possible as well. In the presence of a strong base, alkyl chlorides undergo dehydrohalogenation to give alkenes or alkynes.

Alkyl chlorides react with magnesium to give Grignard reagents, transforming an electrophilic compound into a nucleophilic compound. The Wurtz reaction reductively couples two alkyl halides to couple with sodium.

Applications

Vinyl Chloride

The largest application of organochlorine chemistry is the production of vinyl chloride. The annual production in 1985 was around 13 billion kilograms, almost all of which was converted into polyvinylchloride (PVC).

Chloromethanes

Most low molecular weight chlorinated hydrocarbons such as chloroform, dichloromethane, dichloroethene, and trichloroethane are useful solvents. These solvents tend to be relatively non-polar; they are therefore immiscible with water and effective in cleaning applications such as degreasing and dry cleaning. Several billion kilograms of chlorinated methanes are produced annually, mainly by chlorination of methane :

$$CH_4 + x\ Cl_2 \rightarrow CH_{4-x}Cl_x + x\ HCl$$

The most important is dichloromethane, which is mainly used as a solvent. Chloromethane is a precursor to chlorosilanes and silicones. Historically significant, but smaller in scale is chloroform, mainly a precursor to chlorodifluoromethane ($CHClF_2$) and tetrafluoroethene which is used in the manufacture of Teflon.

Pesticides

Many pesticides contain chlorine. Notable examples include DDT, dicofol, heptachlor, endosulfan, chlordane, aldrin, dieldrin, endrin, mirex, kepone and pentachlorophenol. These can be either hydrophilic or hydrophobic depending on their molecular structure. Many of these agents have been banned in various countries, *e.g.* mirex, aldrin.

Insulators

Polychlorinated biphenyls (PCBs) were once commonly used electrical insulators and heat transfer agents. Their use has generally been phased out due to health concerns. PCBs were replaced by polybrominated diphenyl ethers (PBDEs), which bring similar toxicity and bioaccumulation concerns.

Toxicity

Some types of organochlorides have significant toxicity to plants or animals, including humans. Dioxins, produced when organic matter is burned in the presence of chlorine, and some insecticides, such as DDT, are persistent organic pollutants which pose dangers when they are released into the environment. For example, DDT, which was widely used to control insects in the mid 20th century, also accumulates in food chains, and causes reproductive problems (*i.e.*, eggshell thinning) in certain bird species.

When chlorinated solvents, such as carbon tetrachloride, are not disposed of properly, they accumulate in groundwater. Some highly reactive organochlorides such as phosgene have even been used as chemical warfare agents.

However, the presence of chlorine in an organic compound does not ensure toxicity. Some organochlorides are considered safe enough for consumption in foods and medicines. For example, peas and broad beans contain the natural chlorinated plant hormone 4-chloroindole-3-acetic acid (4-Cl-IAA); and the sweetener sucralose (Splenda) is widely used in diet products. As of 2004, there were at least 165 organochlorides approved worldwide for use as pharmaceutical drugs, including the natural antibiotic vancomycin, the anti-histamine loratadine (Claritin), the anti-depressant sertraline (Zoloft), the anti-epileptic lamotrigine (Lamictal), and the inhalation anesthetic isoflurane.

Rachel Carson brought the issue of DDT pesticide toxicity to public awareness with her 1962 book *Silent Spring*. While many countries have phased out the use of some types of organochlorides such as the US ban on DDT, persistent DDT, PCBs, and other organochloride residues continue to be found in humans and mammals across the planet many years after production and use have been limited. In Arctic areas, particularly high levels are found in marine mammals. These chemicals concentrate in mammals, and are even found in human breast milk. Males typically have far higher levels, as females reduce their concentration by transfer to their offspring through breast feeding.

ORGANOBROMINE COMPOUND

Organobromine compounds are organic compounds that contain carbon bonded to bromine. The most pervasive is the naturally produced bromomethane. One prominent application is the use of polybrominated diphenyl ethers as fire-retardants. A variety of minor organobromine compounds are found in nature, but none are bio-synthesized or required by mammals. Organobromine compounds have fallen under increased scrutiny for their environmental impact.

General Properties

Most organobromine compounds, like most organohalide compounds, are relatively non-polar. Bromine is more electro-negative than carbon (2.8 *vs* 2.5). Consequently, the carbon in a carbon–bromine bond is electrophilic, *i.e.* alkyl bromides are alkylating agents.

Carbon–halogen bond strengths, or bond dissociation energies are of 115, 83.7, 72.1, and 57.6 kcal/mol for bonded to fluorine, chlorine, bromine, or iodine, respectively.

The reactivity of organobromine compounds resembles but is intermediate between the reactivity of organochlorine and organoiodine compounds. For many applications, organobromides represent a compromise of reactivity and cost. The principal reactions for organobromides include dehydrobromination, Grignard reactions, reductive coupling, and nucleophilic substitution.

Synthetic Methods

From Bromine

Alkenes reliably add bromine without catalysis to give the vicinal dibromides :

$$RCH{=}CH_2 + Br_2 \rightarrow RCHBrCH_2Br$$

Aromatic compounds undergo bromination simultaneously with evolution of hydrogen bromide. Catalysts such as AlBr3 or FeBr3 are needed for the reaction to happen on aromatic rings. Chlorine-based catalysts (FeCl3, AlCl3) could be used, but yield would drop slightly as dihalogens(BrCl) could form. The reaction details following the usual patterns of electrophilic aromatic substitution :

$$RC_6H_5 + Br_2 \rightarrow RC_6H_4Br + HBr$$

A prominent application of this reaction is the production of tetrabromobisphenol-A from bisphenol-A.

Free-radical substitution with bromine is commonly used to prepare organobromine compounds. Carbonyl-containing, benzylic, allylic substrates are especially prone to this reactions. For example, the commercially significant bromoacetic acid is generated directly from acetic acid and bromine in the presence of phosphorus tribromide catalyst :

$$CH_3CO_2H + Br_2 \rightarrow BrCH_2CO_2H + HBr$$

Bromine also converts fluoroform to bromotrifluoromethane.

From Hydrogen Bromide

Hydrogen bromide adds across double bonds to give alkyl bromides, following the Markovnikov rule :

$$RCH=CH_2 + HBr \rightarrow RCHBrCH_3$$

Under free radical conditions, the direction of the addition can be reversed. Free-radical addition is used commercially for the synthesis of 1-bromoalkanes, precursors to tertiary amines and quaternary ammonium salts. 2-Phenethyl bromide ($C_6H_5CH_2CH_2Br$) is produced via this route from styrene.

Hydrogen bromide can also be used to convert alcohols to alkyl bromides. This reaction, that must be done under low temperature conditions, is employed in the industrial synthesis of allyl bromide :

$$HOCH_2CH=CH_2 + HBr \rightarrow BrCH_2CH=CH_2 + H_2O$$

Methyl bromide, another fumigant, is generated from methanol and hydrogen bromide.

From Bromide Salts

Bromide ions, as provided by salts like sodium bromide, function as a nucleophiles in the formation of organobromine compounds by displacement.

Industrially Significant Organobromine Compounds

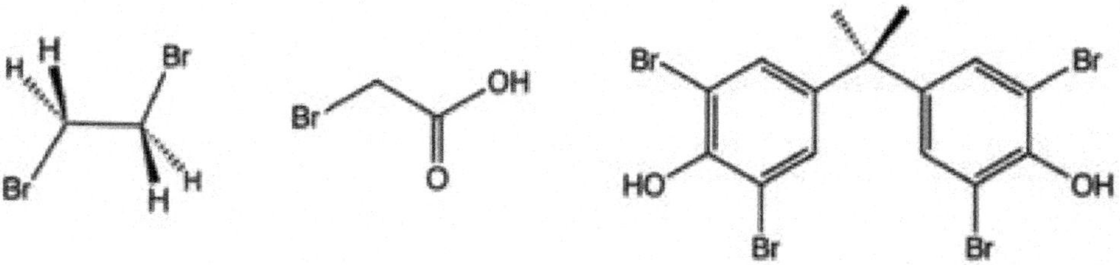

Fig. : Structure of three industrially significant organobromine compounds. From left : ethylene bromide, bromoacetic acid, and tetrabromobisphenol-A.

Fire-retardants

Organobromine compounds are widely used as fire-retardants. It and tetrabromophthalic anhydride are precursors to polymers wherein the backbone features covalent carbon-bromine bonds. Other fire retardants, such as hexabromocyclododecane and the bromodiphenyl ethers, are additives and are not chemically attached to the material they protect. The use of organobromine fire-retardants is growing but is also controversial because they are persistent pollutants.

Fumigants and Biocides

Ethylene bromide, obtained by addition of bromine to ethylene, was once of commercial significance as a component of leaded gasoline. It was also a popular fumigant in agriculture, displacing 1,2-dibromo-3-chloropropane ("DBCB"). Both applications are declining owing to environmental and health considerations. Methyl bromide is also an effective fumigant, but its production and use are controlled by the Montreal Protocol. Growing in use are organobromine biocides used in water treatment. Representative agents include bromoform and dibromodimethylhydantoin ("DBDMH"). Some herbicides, such as bromoxynil, contain also bromine moieties. Like other halogenated pesticides, bromoxynil is subject to reductive dehalogenation under anaerobic conditions, and can be debrominated by organisms originally isolated for their ability to reductively dechlorinate phenolic compounds.

Dyes

Many dyes contain carbon-bromine bonds. The naturally occurring Tyrian purple (6,6'-dibromoindigo) was a valued dye before the development of the synthetic dye industry in the late 19th century. Several brominated anthroquinone derivatives are used commercially. Bromothymol blue is a popular indicator in analytical chemistry.

Pharmaceuticals

Commercially available organobromine pharmaceuticals include the vasodilator nicergoline, the sedative brotizolam, the anticancer agent pipobroman, and the antiseptic merbromin. Otherwise, organobromine compounds are rarely pharmaceutically useful, in contrast to the situation for organo-fluorine compounds. Several drugs are produced as the bromide (or equivalents, hydrobromide) salts, but in such cases bromide serves as an innocuous counterion of no biological significance.

Organobromine Compounds in Nature

Organobromine compounds are the most common organohalides in nature. Even though the concentration of bromide is only 0.3% of that for chloride in sea-water, organobromine compounds are more prevalent in marine organisms than organochlorine derivatives. Their abundance reflects the easy oxidation of bromide to the equivalent of Br^+, a potent electrophile. The enzyme bromoperoxidase catalyzes this reaction. The oceans are estimated to release 1–2 million tons of bromoform and 56,000 tons of bromomethane annually. Red algae, such as the edible Asparagopsis taxiformis, eaten in Hawaii as "limu kohu", concentrate organobromine and organoiodine compounds in "vesicle cells"; 95% of the essential volatile oil of Asparagopsis, prepared by drying the seaweed in a vacuum and condensing using dry ice, is organohalogen compounds, of which bromoform comprises 80% by weight. Bromoform, produced by several algae, is a known toxin, though the small amounts present in edible algae do not appear to pose human harm. Some of these organobromine compounds are employed in a form of interspecies "chemical warfare." 5-Bromouracil and 3-Bromo-tyrosine have been identified in human white blood cells as products of myeloperoxidase-induced halogenation on invading pathogens.

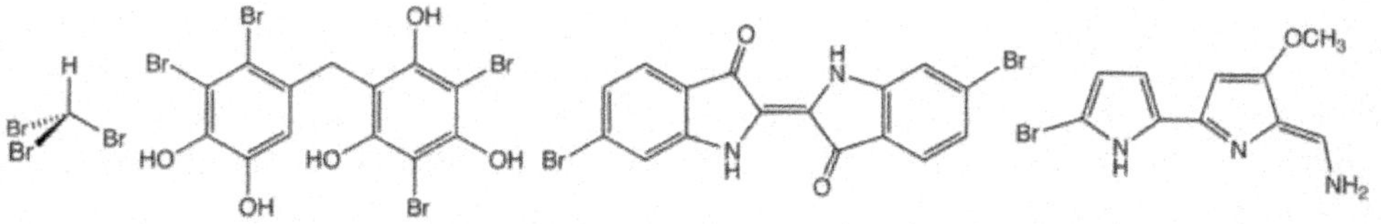

Fig. : Structure of naturally-occurring organobromine compounds. From left : bromoform, a brominated bisphenol, dibromoindigo (Tyrian purple), and the antifeedant tambjamine B.

In addition to conventional brominated natural products, a variety of organo-bromine compounds result from the biodegradation of fire-retardants. Metabolites include methoxylated and hydroxylated aryl bromides as well as brominated dioxin derivatives. Such compounds are considered persistent organic pollutants and have been found in mammals.

Safety

Alkyl bromine compounds are often alkylating agents and the brominated aromatic derivatives are implicated as hormone disruptors. Of the commonly produced compounds, ethylene dibromide is of greatest concern as it is both highly toxic and highly carcinogenic.

PERIODINANE

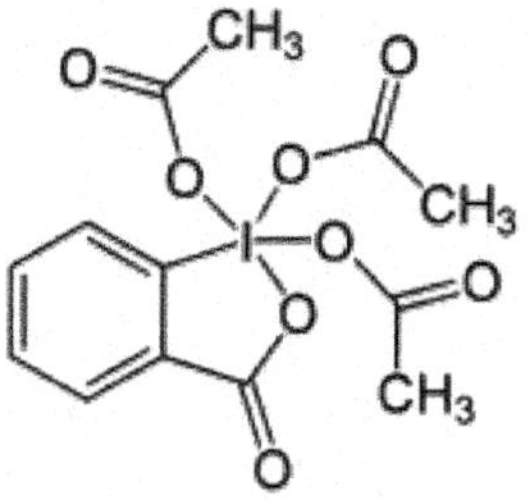

Fig. : The Dess-Martin periodinane.

Periodinanes are chemical compounds containing **hypervalent iodine**. These iodine compounds are hypervalent because the iodine atom in it contains more than the 8 electrons in the valence shell required for the octet rule. When iodine is complexed with a monodentate electro-negative ligand such as chlorine, iodine compounds occur with a +3 oxidation number as iodine (III) or λ^3-**iodanes** or as a +5 oxidation number as iodine(V) or λ^5-**iodanes**. Iodine itself contains 7 valence electrons and in a λ^3-iodane three more are donated by the ligands making it a **decet** structure. λ^5-iodanes are **dodecet** molecules. In an ordinary iodine compound such as iodobenzene the number of valence electrons is eight as expected. In order to get from iodine to a hypervalent iodine compound it gets oxidized with removal of first 3 electrons and then 5 electrons. The ligands in turn contribute electrons pairs and form co-ordinate covalent bonds by adding a total of 6 or 10 electrons back to iodine. In the **L-I-N notation** L stands for the number of electrons donated by ligands and N the number of ligands.

Periodinane Compounds

The concept of hypervalent iodine was developed by J.J. Musher in 1969. In order to accommodate the excess of electrons in hypervalent compounds the 3-center-4-electron bond was introduced in analogy with the 3-center-2-electron bond observed in electron deficient compounds. One such bond exists in iodine (III) compounds and two such bonds reside in iodine (V) compounds.

The first hypervalent iodine compound, (dichloroiodo) benzene ($C_6H_5Cl_2I$) was prepared in 1886 by the German chemist Conrad Willgerodt by passing chlorine gas through iodobenzene in a cooled solution of chloroform.

$$C_6H_5I + Cl_2 \rightarrow C_6H_5ICl_2$$

λ^3-iodanes such as diarylchloroiodanes have a pseudotrigonal bipyramidal geometry displaying apicophilicity with a phenyl group and a chlorine group at the apical positions and other phenyl group with two lone pair electrons in the equatorial positions. The λ^5-iodanes such as the Dess-Martin periodinane have square pyramidal geometries with 4 heteroatoms in basal positions and one apical phenyl group.

Classical organic procedures exist for the preparation of **iodosobenzene diacetate** from peracetic acid and acetic acid.

$$C_6H_5I + CH_3COOOH \rightarrow C_6H_5I(OOCCH_3)_2$$

Phenyliodine bis(trifluoroacetate), PIFA, or (bis (trifluoroacetoxy) iodo) benzene is a related compound based on trifluoroacetic acid.

Phenyliodine diacetate, PIDA, is an organic reagent used as an oxidizing agent. This is suitable for cleaving glycols and alpha-hydroxy ketones in ring opening and Mannich reaction.

The acetate can be hydrolysed with water to **iodoxybenzene** or **iodylbenzene** $C_6H_5O_2I$

This compound was first prepared by Willgerodt by disproportionation of iodosylbenzene under steam distillation to iodylbenzene and iodobenzene :

$$2\ PhIO \rightarrow PhIO_2 + PhI$$

is a known oxidizing agent. Iodosobenzene diacetate can also be hydrolyzed to **iodosylbenzene** with sodium hydroxide which is actually a polymer with the molecular formula $(C_6H_5OI)_n$. Iodosylbenzene is used in organic oxidations. Dess-Martin periodinane (1983) is another powerful oxidant and an improvement of the IBX acid already in existence in 1983. The IBX acid is prepared from 2-iodobenzoic acid and potassium bromate and sulfuric acid and is insoluble in most solvents whereas the Dess-Martin reagent prepared from reaction of the IBX acid with acetic anhydride is very soluble. The oxidation mechanism ordinarily consists of a ligand exchange reaction followed by a reductive elimination.

Diaryliodonium Salts

Diaryliodonium salts are compounds of the type $[Ar-I^+-Ar]X^-$. The term salt is misleading and IUPAC prefers diaryl-λ^3-iodane. The first such compound was synthesised in 1894 (**Meyer and Hartmann reaction**). With halogen counterions diaryliodonium salts show poor solubility in many organic solvents but solubility is improved with triflate and tetrafluoroborate counterions.

Diaryliodonium salts can be prepared in a number of ways. In one method an aryl iodide is first oxidized to an aryliodine (III) compound (ArIO, $ArIO_2$) with as a second step a ligand exchange with an arene (AES), an arylstannane or an arylsilane. In another method diaryliodonium salts a prepared from preformed hypervalent iodine compounds such as iodic acid, iodosyl sulfate or iodosyl triflate.

diaryliodonium salts react with nucleophiles at iodine replacing one ligand and then form the substituted arene ArNu and iodobenzen ArI by reductive elimination or by substitution by ligand. diaryliodonium salts also react with metals M through ArMX intermediates in cross-coupling reactions.

Periodinane Uses

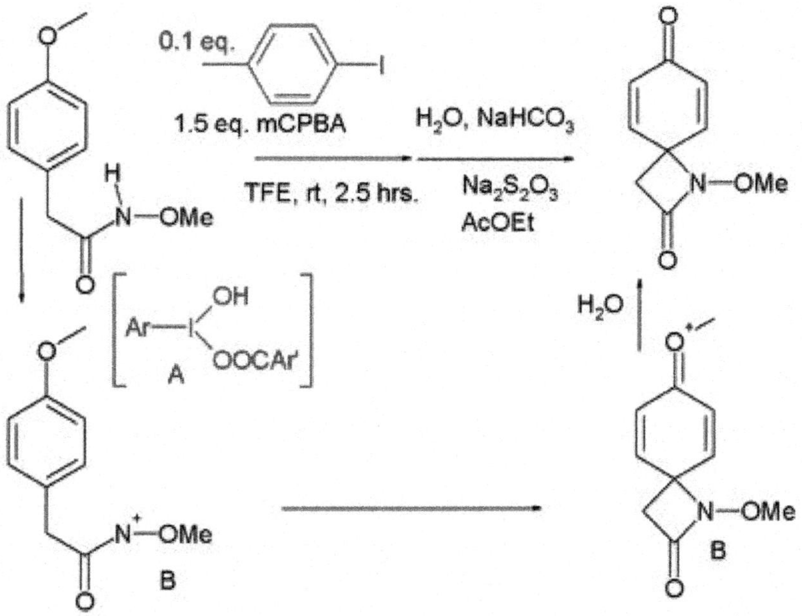

The predominant use of hypervalent iodine compounds is that of oxidizing reagent replacing many toxic reagents based on heavy metals. Thus, a hypervalent iodine (III) reagent was used, as oxidant, together with ammonium acetate, as the nitrogen source, to provide 2-Furonitrile, a pharmaceutical intermediate and potential artificial sweetener, in aqueous acetonitrile at 80 °C in 90% yield.

Current research focuses on their use in carbon-carbon and carbon-heteroatom bond forming reactions. In one study such reaction, an intra-molecular C-N coupling of an alkoxyhydroxylamine to its anisole group is accomplished with a catalytic amount of aryliodide in trifluoroethanol :

In this reaction the periodinane (depicted as intermediate A) is formed by oxidation of the aryliodide with the sacrificial catalyst mCPBA which in turn converts the hydroxylamine group to a nitrenium ion **B**. This ion is the electrophile in ipso addition to the aromatic ring forming a lactam with an enone group.

Chapter 9

REDUCTION OF NITRO COMPOUNDS

The chemical reactions described as **reduction of nitro compounds** can be facilitated by many different reagents and reaction conditions. Historically, the nitro-group was one of the first functional groups to be reduced, due to the ease of nitro-group reduction.

Nitro-groups behave differently whether a neighbouring hydrogen is present or not. Thus, reduction conditions can be initially classified by starting materials : aliphatic nitro compounds or aromatic nitro compounds. Secondary classifications are based upon reaction products.

ALIPHATIC NITRO COMPOUNDS

Reduction to Hydrocarbons

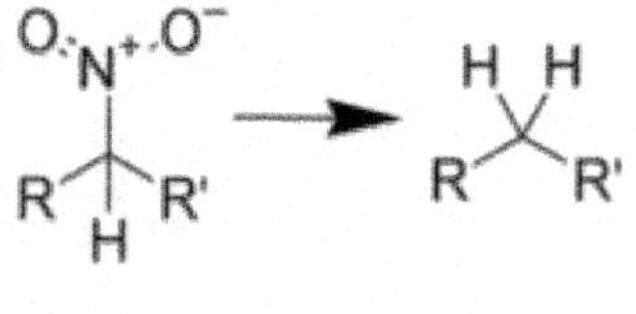

Hydrodenitration (replacement of a nitro-group with hydrogen) is difficult to achieve, but can be completed by catalytic hydrogenation over platinum on silica gel at high temperatures.

Reduction to Amines

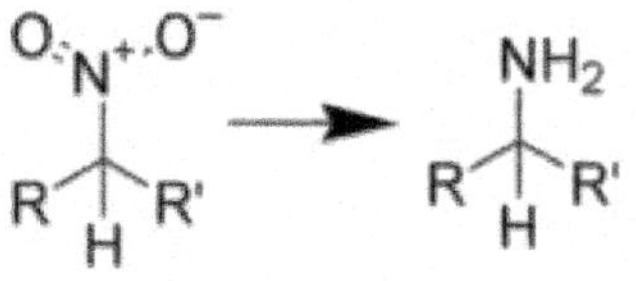

Aliphatic nitro compounds can be reduced to aliphatic amines using several different reagents :

- Catalytic hydrogenation using platinum (IV) oxide (PtO_2) or Raney nickel
- Iron metal in refluxing acetic acid
- Samarium diiodide

    α,β-Unsaturated nitro compounds can be reduced to saturated amines using :

- Catalytic hydrogenation over palladium-on-carbon
- Iron metal
- Lithium aluminium hydride (Note : Hydroxylamine and oxime impurities are typically found.)

Reduction to Hydroxylamines

Aliphatic nitro compounds can be reduced to aliphatic hydroxylamines using diborane.

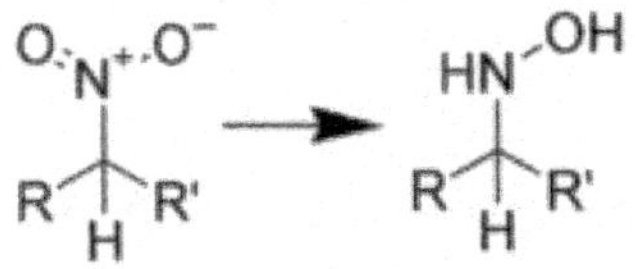

The reaction can also be carried out with zinc dust and ammonium chloride :

$$R\text{-}NO_2 + 4\,NH_4Cl + 2\,Zn \rightarrow R\text{-}NH\text{-}OH + 2\,ZnCl_2 + 4\,NH_3 + H_2O$$

Reduction to Oximes

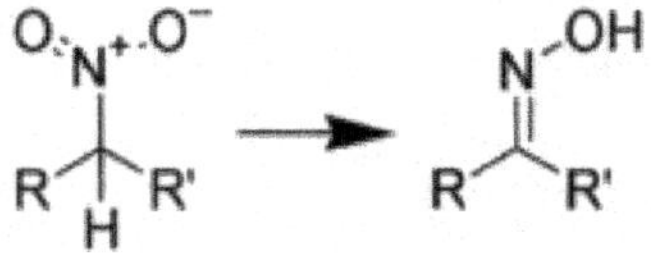

Nitro compounds are typically reduced to oximes using metal salts, such as stannous chloride or chromium (II) chloride. Additionally, catalytic hydrogenation using a controlled amount of hydrogen can generate oximes.

AROMATIC NITRO COMPOUNDS

The reduction of aryl nitro compounds can be finely tuned to obtain a different products typically in high yields.

Reduction to Anilines

Many methods for the production of anilines from aryl nitro compounds exist, such as :

- Catalytic hydrogenation using palladium-on-carbon, platinum (IV) oxide, or Raney nickel
- Iron in acidic media (Note : Iron is particularly well suited for this reduction as the reaction conditions are typically gentle and also because iron has a high functional group tolerance.)
- Sodium sulfide (or hydrogen sulfide and base)
- Tin (II) chloride
- Titanium (III) chloride
- Zinc
- Samarium

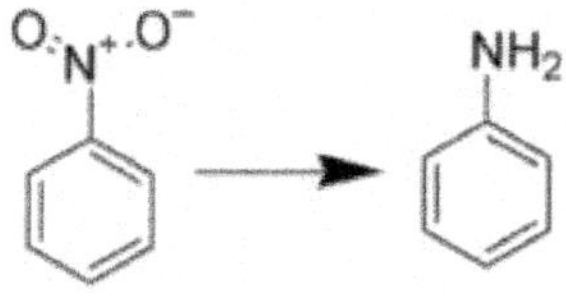

It is also possible to form a 3-nitroaniline by reduction of a 1,3-dinitrobenzene using sodium sulfide.

Metal hydrides are typically not used to reduce aryl nitro compounds to anilines because they tend to produce azo compounds.

Reduction to Hydroxylamines

Several methods for the production of aryl hydroxylamines from aryl nitro compounds exist :

- Raney nickel and hydrazine at 0-10°C
- Electrolytic reduction
- Zinc metal in aqueous ammonium chloride.

Reduction to Hydrazino Compounds

Treatment of nitroarenes with excess zinc metal results in the formation of N,N'-diarylhydrazine.

Reduction to Azo Compounds

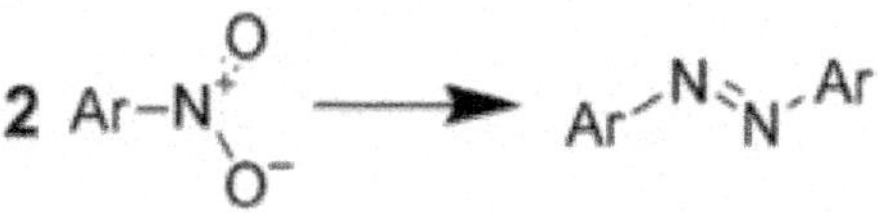

Treatment of aromatic nitro compounds with metal hydrides gives good yields of azo compounds. For example, one could use :

- Lithium aluminium hydride

- Zinc metal with sodium hydroxide. (Excess zinc will reduce the azo group to a hydrazino compound.)

NITRILE REDUCTION

In **nitrile reduction** a nitrile is reduced to either an amine or an aldehyde with a suitable chemical reagent.

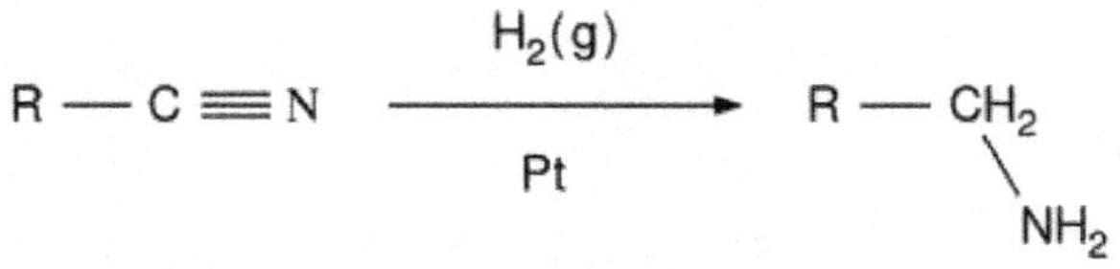

$$R - C \equiv N \quad \xrightarrow[\text{Pt}]{\text{H}_2(\text{g})} \quad R - \underset{\underset{NH_2}{\displaystyle\diagdown}}{CH_2}$$

Reagents for the conversion to amines are lithium aluminium hydride, Raney nickel/hydrogen/or diborane This organic reaction is one of several nitrogen-hydrogen bond forming reactions.

Nitriles can also be reduced to aldehydes. One method is called the Stephen aldehyde synthesis (Tin (II) chloride, hydrochloric acid and hydrolysis of the iminum salt). Aldehydes also form by reduction with hydrogen with *in-situ* hydrolysis of the imine. Reagents are Raney nickel, lithium aluminium hydride and sodium borohydride.

Nitriles can also be reduced electro-chemically.

Chapter 10

CARBONYL

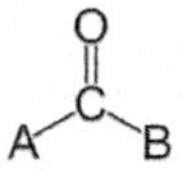

Fig. : Carbonyl group.

In organic chemistry, a **carbonyl group** is a functional group composed of a carbon atom double-bonded to an oxygen atom : C=O. It is common to several classes of organic compounds, as part of many larger functional groups.

The term carbonyl can also refer to carbon monoxide as a ligand in an inorganic or organometallic complex (a metal carbonyl, *e.g.* nickel carbonyl).

The remainder of this article concerns itself with the organic chemistry definition of carbonyl, where carbon and oxygen share a double bond.

CARBONYL COMPOUNDS

A carbonyl group characterizes the following types of compounds :

Compound	Aldehyde	Ketone	Carboxylic acid	Ester	Amide
Structure	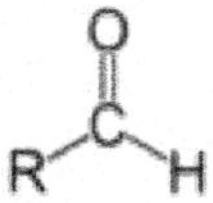	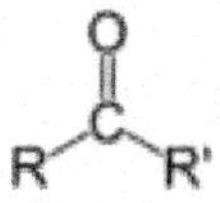	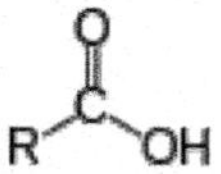	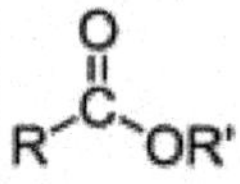	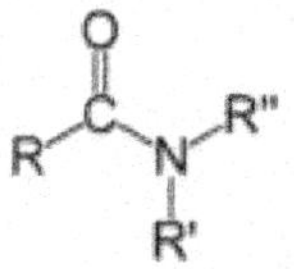
General formula	RCHO	RCOR′	RCOOH	RCOOR′	RCONR′R″

Compound	Enone	Acyl halide	Acid anhydride	Imide
Structure	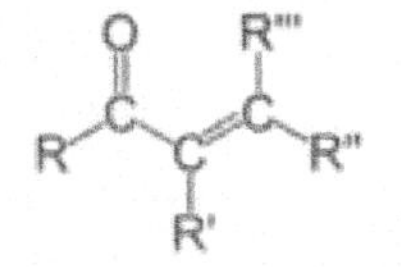	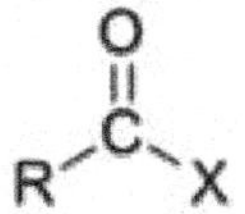	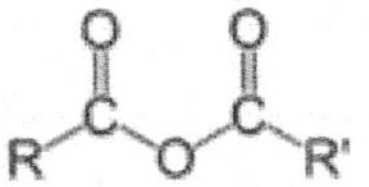	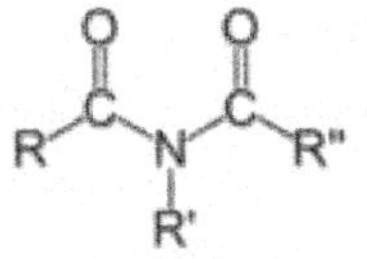
General formula	RC(O)C(R')CR''R'''	RCOX	(RCO)$_2$O	RC(O)N(R')C(O)R'''

Note that the most specific labels are usually employed. For example, R(CO)O(CO)R' structures are known as acid anhydride rather than the more generic ester, even though the ester motif is present.

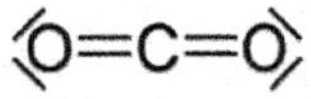

Fig. : Carbon dioxide.

Other organic carbonyls are urea and the carbamates, the derivatives of acyl chlorides chloroformates and phosgene, carbonate esters, thioesters, lactones, lactams, hydroxamates, and isocyanates. Examples of inorganic carbonyl compounds are carbon dioxide and carbonyl sulfide.

A special group of carbonyl compounds are **1,3-dicarbonyl compounds** that have acidic protons in the central methylene unit. Examples are Meldrum's acid, diethyl malonate and acetylacetone.

REACTIVITY

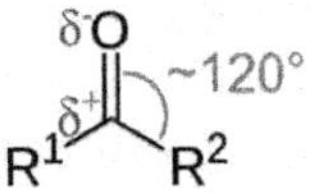

Fig. : Carbonyl group.

Oxygen is more electro-negative than carbon, and thus draws electron density away from carbon to increase the bond's polarity. Therefore, the carbonyl carbon becomes electrophilic, and thus more reactive with nucleophiles. Also, the electro-negative oxygen can react with an electrophile; for example a proton in an acidic solution or other Lewis acid forming an oxocarbenium ion.

The alpha hydrogens of a carbonyl compound are *much* more acidic (roughly 10^{30} times more acidic) than typical sp^3 C-H bonds, such as those in methane. For example, the pK$_a$ values of acetaldehyde and acetone are 16.7 and 19 respectively, while the pK$_a$ value of methane is extrapolated to be approximately 50. This is because a carbonyl is in tautomeric resonance with an enol. The deprotonation of the enol with a strong base produces an enolate, which is a powerful nucleophile and can alkylate electrophiles such as other carbonyls.

Amides are the most stable of the carbonyl couplings due to their high resonance stabilization between the nitrogen-carbon and carbon-oxygen bonds.

Carbonyl groups can be reduced by reaction with hydride reagents such as $NaBH_4$ and $LiAlH_4$, or catalytically by hydrogen and a catalyst such as copper chromite, Raney nickel, rhenium, ruthenium or even rhodium. Ketones give secondary alcohols; aldehydes, esters and carboxylic acids give primary alcohols.

Carbonyls would be alkylated by nucleophilic attack by organometallic reagents such as organolithium reagents and Grignard reagents. Carbonyls also may be alkylated by enolates as in aldol reactions. Carbonyls are also the prototypical groups with vinylogous reactivity, *e.g.* the Michael reaction where an unsaturated carbon in conjugation with the carbonyl is alkylated instead of the carbonyl itself.

In case of multiple carbonyl types in one molecule, one can expect the most electrophilic one to react first. Acyl chlorides and carboxylic anhydrides react fastest, followed by aldehydes and ketones. Esters react much slower and amides barely react due to resonance of the amide nitrogen towards the carbonyl group. This effect is mainly due to the decreasing electrophilic character of the carbonyl carbon atom. This reactivity difference permits to induce chemoselectivity. An instructive example is found in the last part of the total synthesis of monensin by Kishi in 1979 :

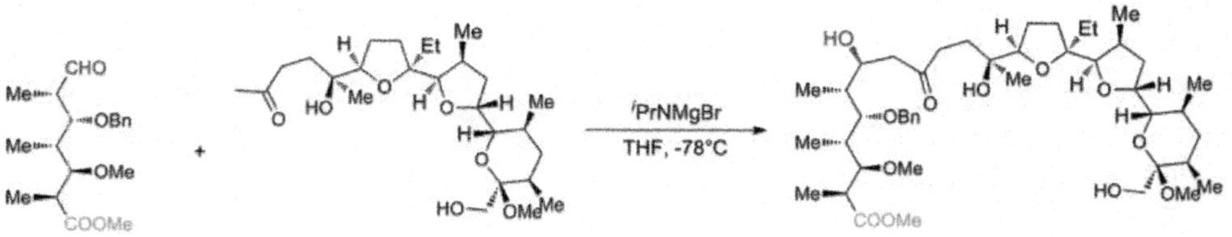

The left wing of the molecule possesses two potential electrophilic sites : an aldehyde (indicated in blue) and an ester (indicated in green). Only the aldehyde, which is more electrophilic, will react with the enolate of the methyl ketone in the other part of the molecule. The methyl ester remains untouched. Of course, other effects can play a role in this selectivity process : electronic effects, steric effects, kinetic versus thermodynamic effect,...

Other important reactions include :

*   Carbonyl Alpha-Substitution Reactions
*   Wittig Reaction a phosphonium ylid is used to create an alkene
*   Wolff-Kishner reduction into a hydrazone and further into a saturated alkane
*   Clemmensen reduction into a saturated alkane
*   Mozingo reduction into a saturated alkane
*   Conversion into thioacetals
*   Hydration to hemiacetals and hemiketals, and then to acetals and ketals
*   Reaction with ammonia and primary amines to form imines
*   Reaction with hydroxylamines to form oximes
*   Reaction with cyanide anion to form cyanohydrins
*   Oxidation with oxaziridines to acyloins

- Reaction with Tebbe's reagent and phosphonium ylides to alkenes.
- Perkin reaction, an aldol reaction variant
- Aldol condensation, a reaction between an enolate and a carbonyl
- Cannizzaro reaction, a disproportionation of aldehydes into alcohols and acids
- Tishchenko reaction, another disproportionation of aldehydes that gives a dimeric ester
- Nucleophilic abstraction is used to produce carbon dioxide.

A,B-UNSATURATED CARBONYL COMPOUNDS

Fig. : Acrolein, an α,β-unsaturated carbonyl compound.

α,β-Unsaturated carbonyl compounds are an important class of carbonyl compounds with the general structure $-(O=C)-C^{\alpha}=C^{\beta}-$. In these compounds the carbonyl group is conjugated with an alkene (hence the adjective *unsaturated*), from which they derive special properties. Unlike the case for simple carbonyls, α,β-unsaturated carbonyl compounds are often attacked by nucleophiles at the β carbon. This pattern of reactivity is called vinylogous. Examples of unsaturated carbonyls are acrolein (propenal), mesityl oxide, acrylic acid, and maleic acid. Unsaturated carbonyls can be prepared in the laboratory in an aldol reaction and in the Perkin reaction.

The carbonyl group draws electrons away from the alkene, and the alkene group is, therefore, deactivated towards an electrophile, such as bromine or hydrochloric acid. As a general rule with asymmetric electrophiles, hydrogen attaches itself at the α-position in an electrophilic addition. On the other hand, these compounds are activated towards nucleophiles in nucleophilic conjugate addition.

Since α,β-unsaturated compounds are electrophiles, many α,β-unsaturated carbonyl compounds are toxic, mutagenic and carcinogenic. DNA can attack the β carbon and thus be alkylated. However, the endogenous scavenger compound glutathione naturally protects from toxic electrophiles in the body.

SPECTROSCOPY

- Infrared spectroscopy : the C=O double bond absorbs infrared light at wavenumbers between approximately 1600–1900 cm^{-1}. The exact location of the absorption is well understood with respect to the geometry of the molecule. This absorption is known as the "carbonyl stretch" when displayed on an infrared absorption spectrum.

- Nuclear magnetic resonance : the C=O double-bond exhibits different resonances depending on surrounding atoms, generally a downfield shift. The ^{13}C NMR of a carbonyl carbon is in the range of 160-220 ppm.

THE CARBONYL GROUP

A carbonyl group is a chemically organic functional group composed of a carbon atom double-bonded to an oxygen atom--> [C=O] The simplest carbonyl groups are aldehydes and ketones usually attached to another carbon compound. These structures can be found in many aromatic compounds contributing to smell and taste.

Introduction

Before going into anything in depth be sure to understand that the **C=O** entity itself is known as the "Carbonyl group" while the members of this group are called "carbonyl compounds"--> **X-C=O**. The carbon and oxygen are usually sp^2 hybridized and planar.

Carbonyl Group Double Bonds

The double bonds in alkenes and double bonds in carbonyl groups are VERY different in terms of reactivity. The C=C is *less* reactive due to C=O electro-negativity attributed to the oxygen and its two lone pairs of electrons. One pair of the oxygen lone pairs are located in 2s while the other pair are in 2p orbital where its axis is directed perpendicular to the direction of the pi orbitals. The Carbonyl groups properties are directly tied to its electronic structure as well as geometric positioning. For example, the electro-negativity of oxygen also polarizes the pi bond allowing the single bonded substituent connected to become electron withdrawing.

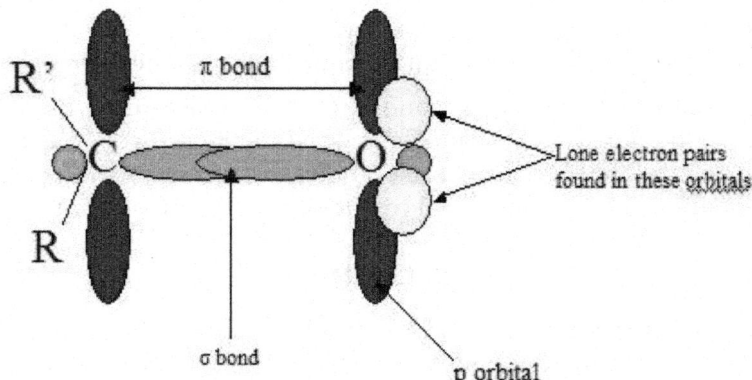

Note : Both the pi bonds are in phase (top and botom blue ovals).

The double bond lengths of a carbonyl group is about 1.2 angstroms and the strength is about 176-179 kcal/mol). It is possible to correlate the length of a carbonyl bond with its polarity; the longer the bond meaing the lower the polarity. For example, the bond length in C=O is larger in acetaldehyde than in formaldehyde (this of course takes into account the inductive effect of CH_3 in the compound).

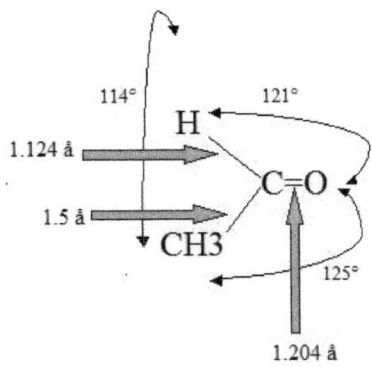

Polarization

As discussed before, we understand that oxygen has two lone pairs of electrons hanging around. These electrons make the oxygen more electro-negative than carbon. The carbon is then partially postive (electrophillic) and the oxygen partially negative (nucleophillic). The polarizability is denoted by a lowercase delta and a positive or negative superscript depending. For example, carbon would have d+ and oxygen delta^(-). The polarization of carbonyl groups also effects the boiling point of aldehydes and ketones to be higher than those of hydrocarbons in the same amount. The larger the carbonyl compound the less soluble it is in water. If the compound exceeds six carbons it then becomes insoluble.

For more information about carbonyl solubility, look in the "outside links" section

Amides are the most stable of the carbonyl couplings due to the high-resonance stabilization between nitrogen-carbon and carbon-oxygen.

Some Carbonyl Compounds

Compound	Aldehyde	Ketone	Formaldehyde	Carboxylic Acid	Ester	Amide	Enone	Acyl Halide	Acid Anhydride
Structure	O‖C R H	O‖C R R'	O‖C H H	O‖C R OH	O‖C R OR'	O‖C R N R" R'	O‖C R C R" R' R'''	O‖C R X	O O‖ ‖C C R O R'
General Formula	RCHO	RCOR'	CH2O	RCOOH	RCOOR'	RCONR'R"	RC(O)C(R')CR"R'''	RCOX	(RCO)2O

Nucleophile Addition to a Carbonyl Group

C=O is prone to additions and nucleophillic attack because or carbon's positive charge and oxygen's negative charge. The resonance of the carbon partial positive charge allows the negative charge on the nucleophile to attack the Carbonyl group and become a part of the structure and a positive charge (usually a proton hydrogen) attacks the oxygen. Just a reminder, the nucleophile is a good acid therefore "likes protons" so it will attack the side with a positive charge.

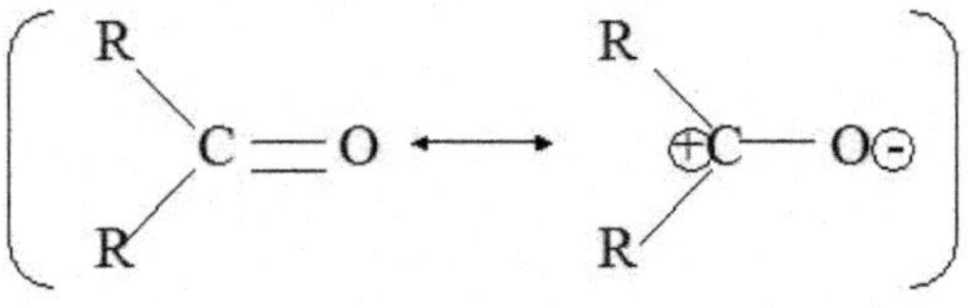

Remember : due to the electro-negative nature of oxygen the carbon is partially positive and oxygen is partially negative

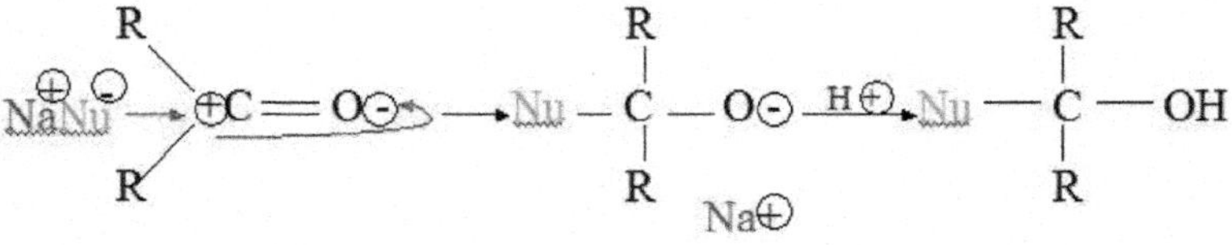

| 1 | 2 | 3 |

1. The Nucleophile (Nu) attacks the positively charged carbon and pushes one of the double bond electrons onto oxygen to give it a negative charge.
2. The Nucleophile is now a part of the carbonl structure with a negatively cahrged oxygen and a Na⁺ "floating" around.
3. The negatively charged oxygen attacks the proton (H⁺) to give the resulting product above.

CARBONYL OXIDATION WITH HYPERVALENT IODINE REAGENTS

Carbonyl oxidation with hypervalent iodine reagents involves the functionalization of the α position of carbonyl compounds through the intermediacy of a hypervalent iodine (III) enolate species. This electrophilic intermediate may be attacked by a variety of nucleophiles or undergo re-arrangement or elimination.

Introduction

Hypervalent iodine (III) compounds are attractive oxidizing agents because of their stability and selectivity. In the presence of enolizable carbonyl compounds, they are able to accomplish oxidative functionalization of the α position. A key iodine (III) enolate intermediate forms, which then undergoes either nucleophilic substitution (α-functionalization), elimination (dehydrogenation), or re-arrangement. Common hypervalent iodine reagents used to effect these transformations include iodosylbenzene (PhIO), iodobenzene diacetate (PhI(OAc)$_2$), Koser's reagent (PhI(OTs)OH), and (dichloroiodo)benzene (Cl$_2$IPh).

(1)

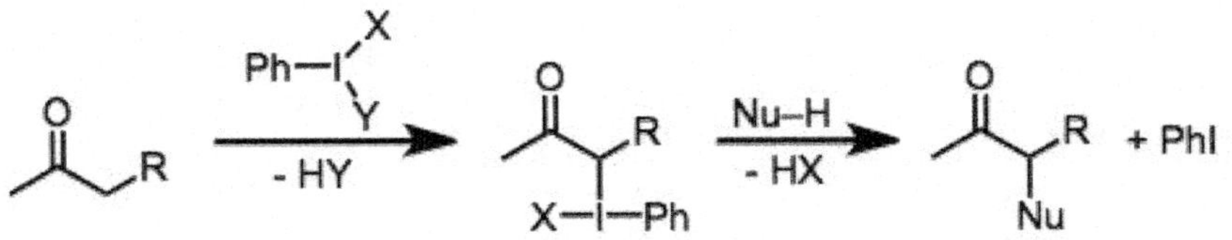

Mechanism and Stereochemistry

Prevailing Mechanism

The mechanism of carbonyl oxidation by iodine (III) reagents varies as a function of substrate structure and reaction conditions, but some generalizations are possible. Under basic conditions, the active iodinating species are iodine (III) compounds in which any relatively acidic ligands on iodine (such as acetate) have been replaced by alkoxide. In all cases, the α carbon forms a bond to iodine. Reduction of iodine (III) to iodine (I) then occurs via attack of a nucleophile on the now electrophilic α carbon. Under basic conditions, nucleophilic attack at the carbonyl carbon is faster than attack at the α carbon. Iodine displacement is actually accomplished intra-molecularly by the carbonyl oxygen, which becomes the α-hydroxyl oxygen in the product.

(2)

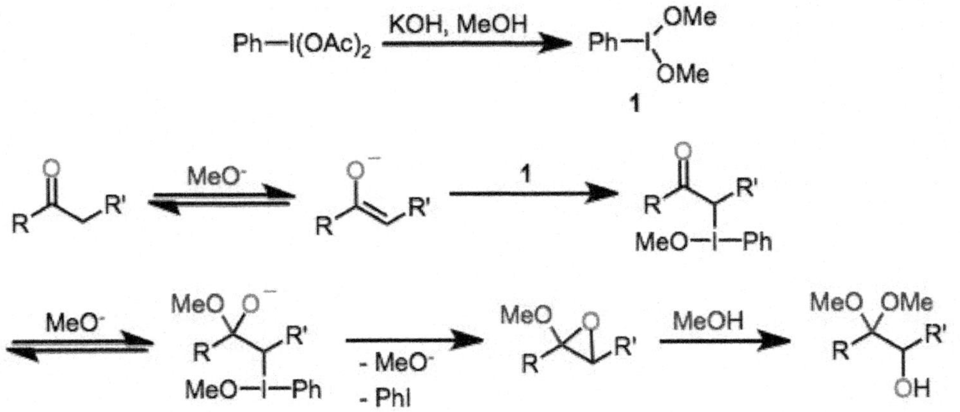

Re-arrangements of the iodine (III) enolate species have been observed. Under acidic conditions, oxidations of aryl enol ethers lead to α-aryl esters via 1,2-aryl migration. Ring-contractive Favorskii re-arrangements may take place under basic conditions.

(3)

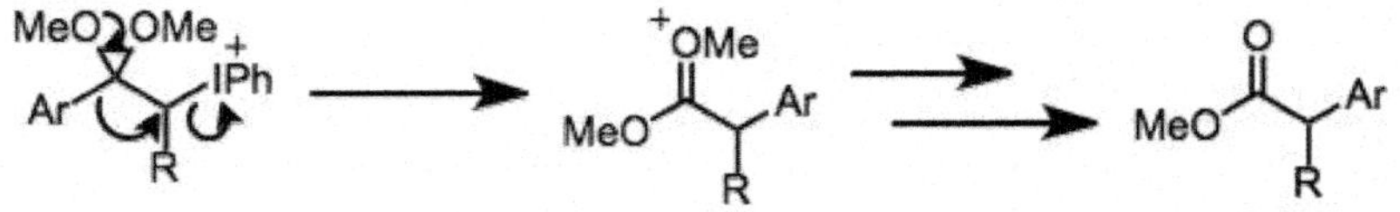

Stereo-chemistry

Using a chromium carbonyl complex, it was shown that displacement of iodine likely occurs with inversion of configuration. Iodine approaches on the side opposite the chromium tricarbonyl unit due to steric hindrance. Invertive displacement leads to a *syn* relationship between chromium and the α hydroxyl group.

(4)

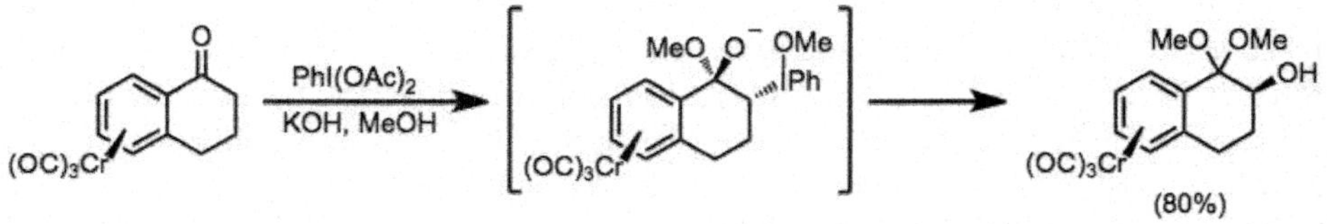

Studies on the oxidation of unsaturated carbonyl compounds also provide stereo-chemical insight. Only the isomer with a *syn* relationship between the α-hydroxy and β-methoxy groups was observed. After nucleophilic attack by methoxide, iodine approaches the face opposite the methoxide. Invertive displacement by hydroxide then leads to the *syn* isomer.

(5)

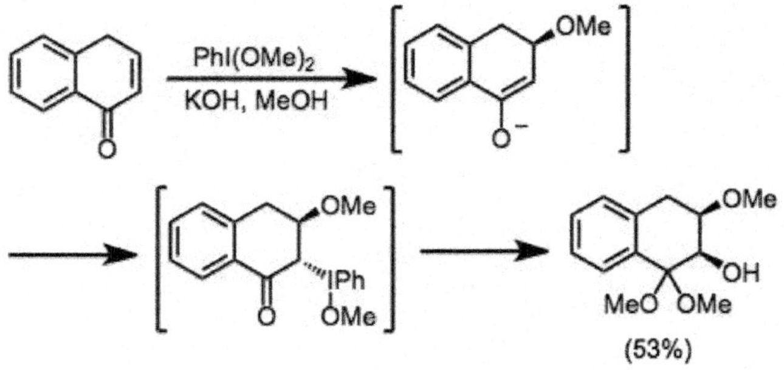

Scope and Limitations

Under protic conditions, ketones undergo α-hydroxylation and dimethyl acetal formation. Both iodosylbenzene and iodobenzene diacetate (IBD) can effect this transformation. This method can be used to synthesize α-hydroxy ketones after acidic hydrolysis of the ketal functionality.

(6)

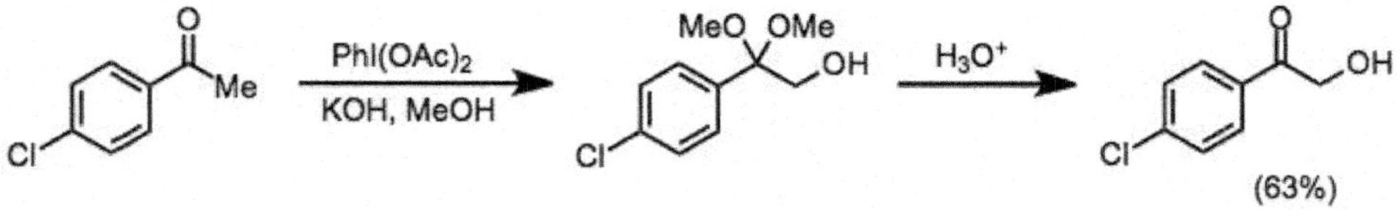

(63%)

In the presence of diaryliodonium salts, enolates undergo α-arylation. Bulky diaryliodoniums react more slowly, and enolate homocoupling begins to compete as the aromatic ring is substituted.

(7)

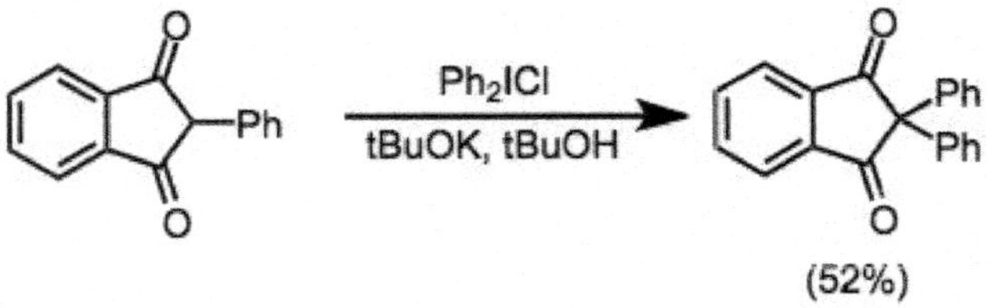

(52%)

α-Oxytosylation facilitates the elaboration of carbonyl compounds into a variety of α-functionalized products. The α-tosyloxycarbonyl compounds that result are more stable than α-halocarbonyl compounds and are not lachrymators.

(8)

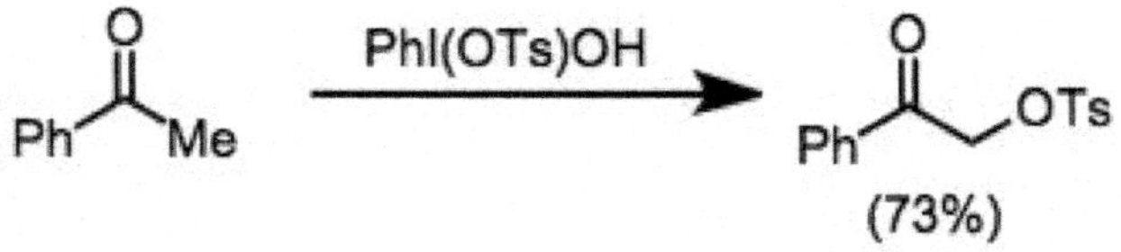

(73%)

Silyl enol ethers undergo many of the same reactions as carbonyl compounds in the presence of iodine (III) reagents. α-Alkoxylation is possible in the presence of an external alcohol nucleophile, although yields are somewhat variable.

(9)

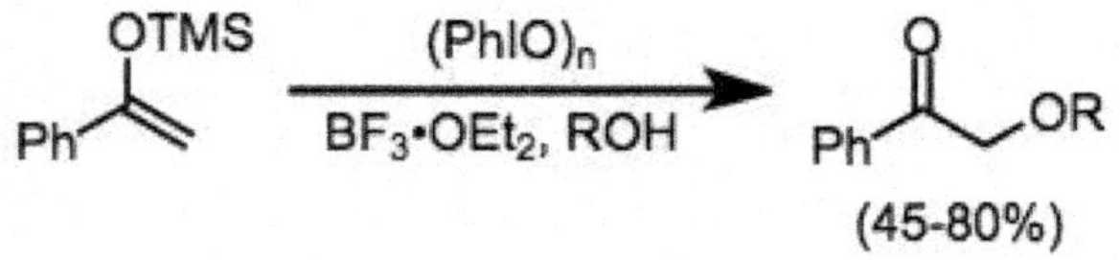

(45-80%)

When no external or internal nucleophile is present, oxidative homocoupling occurs, yielding 1,4-dicarbonyl compounds.

(10)

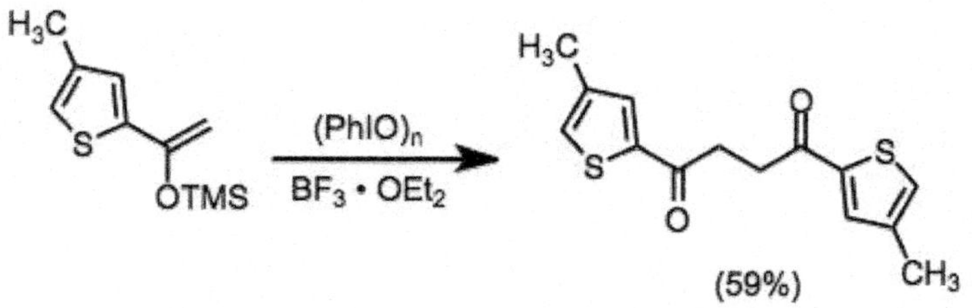

(59%)

Intra-molecularly tethered nucleophiles may displace iodobenzene to afford lactones or other heterocycles. If acidic hydrogens are present in the cyclic product, over-oxidation can occur under the reaction conditions.

(11)

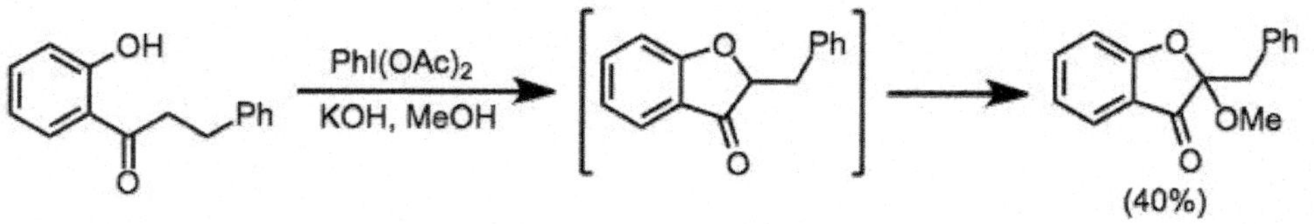

(40%)

In some cases, re-arrangements complicate hypervalent iodine oxidations of carbonyl compounds. Aryl migration may occur under acidic conditions, yielding α-aryl esters from enol ethers. Favorskii re-arrangements have also been observed, and these have been particularly useful for steroid synthesis.

(12)

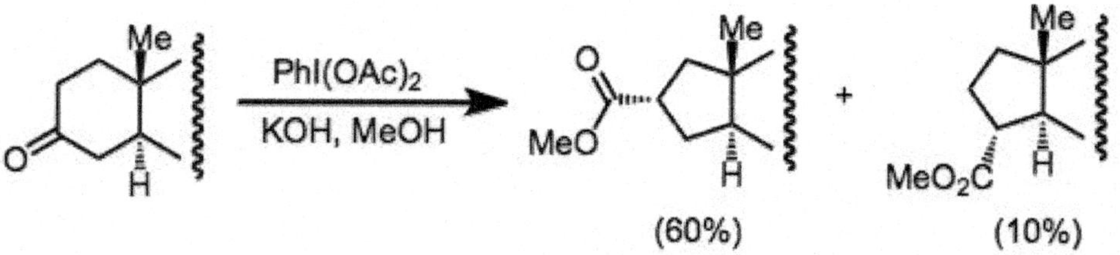

(60%)   (10%)

Synthetic Applications

Oxidative functionalization of silyl enol ethers in low concentration (to avoid homocoupling) without an external nucleophile leads to dehydrogenation. This can be a useful way to generate α, β-unsaturated carbonyl compounds in the absence of functional handles. For instance, dehydrogenation is employed in steroid synthesis to form unsaturated ketones.

(13)

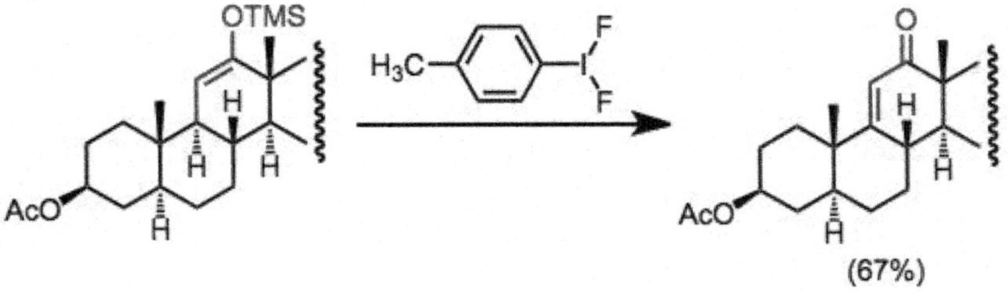

(67%)

Comparison with Other Methods

Few compounds that oxidize carbonyl compounds rival the safety, selectivity, and versatility of hypervalent iodine reagents. Other methods for the α-hydroxylation of carbonyl compounds may employ toxic organometallic compounds (such as lead tetra-acetate or osmium tetroxide). One alternative to hypervalent iodine oxidation that does not employ heavy metals is the attack of a metal enolate on dioxygen, followed by reduction of the resulting peroxide (equation (14)). The most popular method for the α-hydroxylation of carbonyl compounds is the Rubottom oxidation, which employs silyl enol ethers as substrates and peracids as oxidants.

(14)

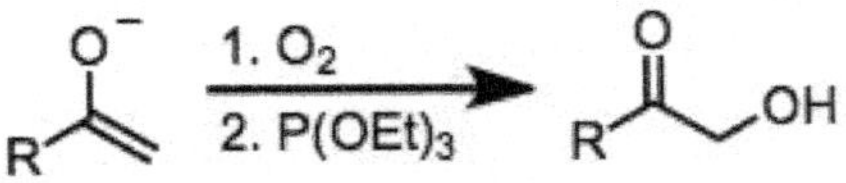

Oxidative re-arrangements are generally easier to accomplish using hypervalent iodine reagents than other oxidizing agents. The Willgerodt-Kindler reaction of alkyl aryl ketones, for instance, requires forcing conditions and often gives low yields of amide products.

(15)

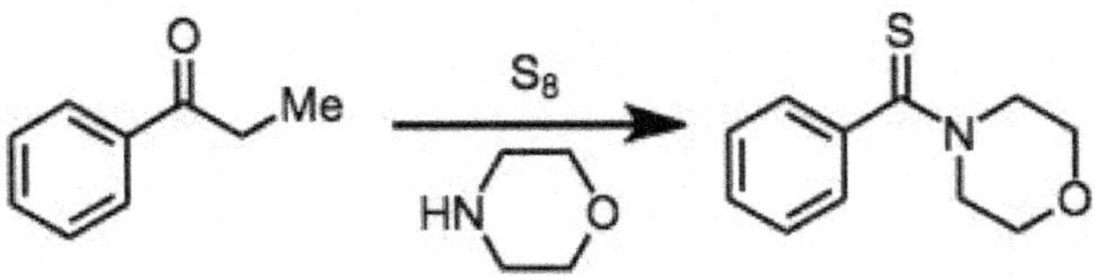

Experimental Conditions and Procedure

Example Procedure

(16)

Hydroxy (mesyloxy) iodobenzene (3.16 g, 10 mmol) was added to a solution of the methyl trimethylsilyl phenylketene acetal derived from methyl phenylacetate

(3.33 g, 15 mmol) in dry dichloromethane (50 mL). The mixture was stirred at room temperature for 2 hours and then washed with aqueous sodium bicarbonate solution (3 × 50 mL). The organic phase was dried (MgSO$_4$) and concentrated in vacuo to yield the crude mesyloxyester which was purified by column chromatography on silica gel (hexane-dichloromethane, 1 : 1) to give 1.58 g (65%) of the title compound, mp 91–92°; IR (KBr) 1760 cm^{-1} (CO); ^{1}H NMR (CDCl$_3$) : δ 3.10 (s, 3H), 3.80 (s, 3H), 6.00 (s, H), 7.40–7.80 (m, 5H); ^{13}C NMR (CDCl$_3$) : δ 168.2 (s), 132.2 (s), 130.0 (s), 129.0 (s), 127.7 (s), 78.9 (s), 53.0 (s), 39.45 (s); MS, m/z 185 (53), 165 (15), 145 (15), 107 (100), 90 (12), 79 (65), 51 (17).

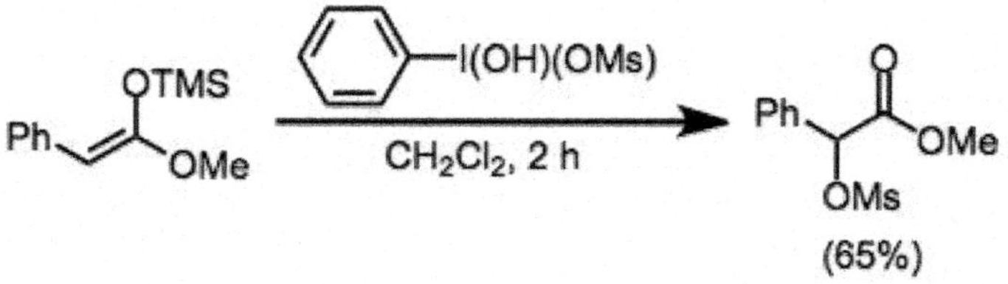

ALDEHYDE

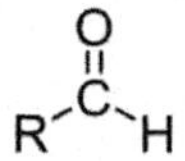

Fig. : An aldehyde.

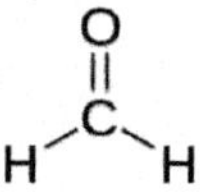

Fig. : Formaldehyde, the simplest aldehyde.

An **aldehyde**/'ældɪhaɪd/is an organic compound containing a formyl group. This functional group, with the structure R-CHO, consists of a carbonyl center (a carbon double bonded to oxygen) bonded to hydrogen and an R group, which is any generic alkyl or side chain. The group without R is called the **aldehyde group** or **formyl group**. Aldehydes differ from ketones in that the carbonyl is placed at the end of a carbon skeleton rather than between two carbon atoms. Aldehydes are common in organic chemistry. Many fragrances are aldehydes.

Structure and Bonding

Aldehydes feature an sp^2-hybridized, planar carbon center that is connected by a double bond to oxygen and a single bond to hydrogen. The C-H bond is not acidic. Because of resonance stabilization of the conjugate base, an α-hydrogen in an aldehyde (not shown in the picture above) is far more acidic, with a pKa near 17, than a C-H bond in a typical alkane (pKa about 50). This acidification is attributed to (i) the electron-withdrawing quality of the formyl center and (ii) the fact that the conjugate base, an enolate anion, delocalizes its negative charge. Related to (i), the aldehyde group is somewhat polar.

Aldehydes (except those without an alpha carbon, or without protons on the alpha carbon, such as formaldehyde and benzaldehyde) can exist in either the keto or the enol tautomer. Keto-enol tautomerism is catalyzed by either acid or base. Usually the enol is the minority tautomer, but it is more reactive.

Nomenclature

IUPAC Names for Aldehydes

The common names for aldehydes do not strictly follow official guidelines, such as those recommended by IUPAC but these rules are useful. IUPAC prescribes the following nomenclature for aldehydes :

1. Acyclic aliphatic aldehydes are named as derivatives of the longest carbon chain containing the aldehyde group. Thus, HCHO is named as a derivative of methane, and $CH_3CH_2CH_2CHO$ is named as a derivative of butane. The name is formed by changing the suffix-e of the parent alkane to-al, so that HCHO is named *methanal*, and $CH_3CH_2CH_2CHO$ is named *butanal*.

2. In other cases, such as when a-CHO group is attached to a ring, the suffix-*carbaldehyde* may be used. Thus, $C_6H_{11}CHO$ is known as *cyclohexanecarbaldehyde*. If the presence of another functional group demands the use of a suffix, the aldehyde group is named with the prefix *formyl-*. This prefix is preferred to *methanoyl-*.

3. If the compound is a natural product or a carboxylic acid, the prefix *oxo*-may be used to indicate which carbon atom is part of the aldehyde group; for example, $CHOCH_2COOH$ is named *3-oxopropanoic acid*.

4. If replacing the aldehyde group with a carboxyl group (-COOH) would yield a carboxylic acid with a trivial name, the aldehyde may be named by replacing the suffix-*ic acid* or-*oic acid* in this trivial name by-*aldehyde*.

Etymology

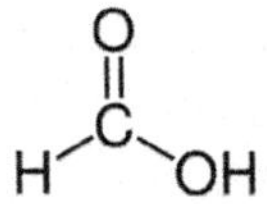

$$\underset{H}{\overset{\displaystyle \overset{\textstyle O}{\|}}{}}\underset{}{\overset{C}{}}\ OH$$

Fig. : Formic acid.

The word *aldehyde* was coined by Justus von Liebig as a contraction of the Latin *alcohol dehydrogenatus* (dehydrogenated alcohol). In the past, aldehydes were sometimes named after the corresponding alcohols, for example, *vinous aldehyde* for acetaldehyde. (*Vinous* is from Latin *vinum* = wine (the traditional source of ethanol), cognate with *vinyl*.)

The term *formyl group* is derived from the Latin and/or Italian word *formica* = ant. This word can be recognized in the simplest aldehyde, formaldehyde (methanal), and in the simplest carboxylic acid, formic acid (methanoic acid, an acid).

Physical Properties and Characterization

Aldehydes have properties that are diverse and that depend on the remainder of the molecule. Smaller aldehydes are more soluble in water, formaldehyde and acetaldehyde completely so. The volatile aldehydes have pungent odours. Aldehydes degrade in air *via* the process of autoxidation.

The two aldehydes of greatest importance in industry, formaldehyde and acetaldehyde, have complicated behaviour because of their tendency to oligomerize or polymerize. They also tend to hydrate, forming the geminal diol. The oligomers/polymers and the hydrates exist in equilibrium with the parent aldehyde.

Aldehydes are readily identified by spectroscopic methods. Using IR spectroscopy, they display a strong v_{CO} band near 1700 cm^{-1}. In their ^{1}H NMR spectra, the formyl hydrogen center absorbs near δ9, which is a distinctive part of the spectrum. This signal shows the characteristic coupling to any protons on the alpha carbon.

Applications and Occurrence

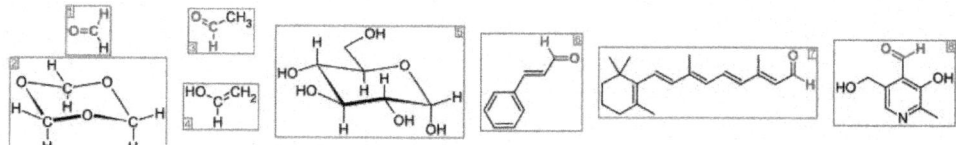

Important aldehydes and related compounds. The **aldehyde group** (or **formyl group**) is coloured red. From the left : (1) formaldehyde and (2) its trimer 1,3,5-trioxane, (3) acetaldehyde and (4) its enol vinyl alcohol, (5) glucose (pyranose form as α-D-glucopyranose), (6) the flavorant cinnamaldehyde, (7) the visual pigment retinal, and (8) the vitamin pyridoxal.

Naturally Occurring Aldehydes

Traces of many aldehydes are found in essential oils and often contribute to their favourable odors, *e.g.* cinnamaldehyde, cilantro, and vanillin. Possibly because of the high reactivity of the formyl group, aldehydes are not common in several of the natural building blocks : amino acids, nucleic acids, lipids. Most sugars, however, are derivatives of aldehydes. These "aldoses" exist as hemiacetals, a sort of masked form of the parent aldehyde. For example, in aqueous solution only a tiny fraction of glucose exists as the aldehyde.

Synthesis

There are several methods for preparing aldehydes, but the dominant technology is hydroformylation. Illustrative is the generation of butyraldehyde by hydroformylation of propene :

$$H_2 + CO + CH_3CH=CH_2 \rightarrow CH_3CH_2CH_2CHO$$

Oxidative Routes

Aldehydes are commonly generated by alcohol oxidation. In industry, formaldehyde is produced on a large scale by oxidation of methanol. Oxygen is the reagent of choice, being "green" and cheap. In the laboratory, more specialized oxidizing agents are used, but chromium (VI) reagents are popular. Oxidation can be achieved by heating the alcohol with an acidified solution of potassium dichromate. In this case, excess dichromate will further oxidize the aldehyde to a carboxylic acid, so either the aldehyde is distilled out as it forms (if volatile) or milder reagents such as PCC are used.

$$[O] + CH_3(CH_2)_9OH \rightarrow CH_3(CH_2)_8CHO + H_2O$$

Oxidation of primary alcohols to form aldehydes and can be achieved under milder, chromium-free conditions by employing methods or reagents such as IBX acid, Dess-Martin periodinane, Swern oxidation, TEMPO, or the Oppenauer oxidation.

Another oxidation route significant in industry is the Wacker process, whereby ethylene is oxidized to acetaldehyde in the presence of copper and palladium catalysts (acetaldehyde is also produced on a large scale by the hydration of acetylene).

Specialty Methods

Reaction name	Substrate	Comment
Ozonolysis	alkene	ozonolysis of non-fully-substituted alkenes yield aldehydes upon reductive work-up.
Organic reduction	ester	Reduction of an ester with diisobutylaluminium hydride (DIBAL-H) or sodium aluminium hydride
Rosenmund reaction	acid chloride	or using lithium tri-t-butoxyaluminium hydride (LiAlH(O^{t}Bu)$_3$).
Wittig reaction	ketone	reagent methoxymethylenetriphenylphosphine in a modified Wittig reaction.
Formylation reactions	nucleophilic arenes	various reactions for example the Vilsmeier-Haack reaction
Nef reaction	Nitro compound	
Zincke reaction	pyridines	Zincke aldehydes form in a variation
Stephen aldehyde synthesis	nitriles	reagents tin (II) chloride and hydrochloric acid.
Meyers synthesis	oxazine	oxazine hydrolysis
McFadyen-Stevens reaction	hydrazide	is a base-catalyzed thermal decomposition of acylsulfonylhydrazides.

Common Reactions

Aldehydes are highly reactive and participate in many reactions." From the industrial perspective, important reactions are condensations, *e.g.* to prepare plasticizers and polyols, and reduction to produce alcohols, especially "oxo-alcohols." From the biological perspective, the key reactions involve addition of nucleophiles to the formyl carbon in the formation of imines (oxidative deamination) and hemiacetals (structures of aldose sugars).

Reduction

The formyl group can be readily reduced to a primary alcohol ($-CH_2OH$). Typically this conversion is accomplished by catalytic hydrogenation either directly or by transfer hydrogenation. Stoichiometric reductions are also popular, as can be effected with sodium borohydride.

Oxidation

The formyl group readily oxidizes to the corresponding carboxylic acid (-COOH). The preferred oxidant in industry is oxygen or air. In the laboratory, popular oxidizing agents include potassium permanganate, nitric acid, chromium(VI) oxide, and chromic acid. The combination of manganese dioxide, cyanide, acetic acid and methanol will convert the aldehyde to a methyl ester.

Another oxidation reaction is the basis of the *silver mirror test*. In this test, an aldehyde is treated with Tollens' reagent, which is prepared by adding a drop of sodium hydroxide solution into silver nitrate solution to give a precipitate of silver (I) oxide, and then adding just enough dilute ammonia solution to redissolve the precipitate in aqueous ammonia to produce $[Ag(NH_3)_2]^+$ complex. This reagent will convert aldehydes to carboxylic acids without attacking carbon-carbon double-bonds. The name *silver mirror test* arises because this reaction will produce a precipitate of silver whose presence can be used to test for the presence of an aldehyde.

A further oxidation reaction involves Fehling's reagent as a test. The Cu^{2+} complex ions are reduced to a red brick coloured Cu_2O precipitate.

If the aldehyde cannot form an enolate (*e.g.*, benzaldehyde), addition of strong base induces the Cannizzaro reaction. This reaction results in disproportionation, producing a mixture of alcohol and carboxylic acid.

Nucleophilic Addition Reactions

Nucleophiles add readily to the carbonyl group. In the product, the carbonyl carbon becomes sp^3 hybridized, being bonded to the nucleophile, and the oxygen center becomes protonated :

$$RCHO + Nu^- \rightarrow RCH(Nu)O^-$$
$$RCH(Nu)O^- + H^+ \rightarrow RCH(Nu)OH$$

In many cases, a water molecule is removed after the addition takes place; in this case, the reaction is classed as an addition-elimination or addition-condensation reaction. There are many variations of nucleophilic addition reactions.

Oxygen Nucleophiles

In the acetalisation reaction, under acidic or basic conditions, an alcohol adds to the carbonyl group and a proton is transferred to form a hemiacetal. Under acidic conditions, the hemiacetal and the alcohol can further react to form an acetal and water. Simple hemiacetals are usually unstable, although cyclic ones such as glucose can be stable. Acetals are stable, but revert to the aldehyde in the presence of acid. Aldehydes can react with water to form hydrates, R-C(H)(OH)(OH). These diols are stable when strong electron withdrawing groups are present, as in chloral hydrate. The mechanism of formation is identical to hemiacetal formation.

Nitrogen Nucleophiles

In alkylimino-de-oxo-bisubstitution, a primary or secondary amine adds to the carbonyl group and a proton is transferred from the nitrogen to the oxygen atom to create a carbinolamine. In the case of a primary amine, a water molecule can be eliminated from the carbinolamine to yield an imine. This reaction is catalyzed by acid. Hydroxylamine (NH_2OH) can also add to the carbonyl group. After the elimination of water, this will result in an oxime. An ammonia derivative of the form H_2NNR_2 such as hydrazine (H_2NNH_2) or 2,4-dinitrophenylhydrazine can also be the nucleophile and after the elimination of water, resulting in the formation of a hydrazone, which are usually orange crystalline solids. This reaction forms the basis of a test for aldehydes and ketones.

Carbon Nucleophiles

The cyano group in HCN can add to the carbonyl group to form cyanohydrins, R-C(H)(OH)(CN). In this reaction the CN− ion is the nucleophile that attacks the partially positive carbon atom of the carbonyl group. The mechanism involves a pair of electrons from the carbonyl group double bond transferring to the oxygen atom, leaving it single bonded to carbon and giving the oxygen atom a negative charge. This intermediate ion rapidly reacts with H+, such as from the HCN molecule, to form the alcohol group of the cyanohydrin.

In the Grignard reaction, a Grignard reagent adds to the group, eventually yielding an alcohol with a substituted group from the Grignard reagent. Related reactions are the Barbier reaction and the Nozaki-Hiyama-Kishi reaction. In organostannane addition tin replaces magnesium.

In the aldol reaction, the metal enolates of ketones, esters, amides, and carboxylic acids will add to aldehydes to form β-hydroxycarbonyl compounds (aldols). Acid or base-catalyzed dehydration will then lead to α,β-unsaturated carbonyl compounds. The combination of these two steps is known as the aldol condensation. The Prins reaction occurs when a nucleophilic alkene or alkyne reacts with an aldehyde as electrophile. The product of the Prins reaction varies with reaction conditions and substrates employed.

Bisulphite Reaction

Aldehydes characteristically form "addition compounds" with sodium bisulphite :

$$RCHO + HSO_3^- \rightarrow RCH(OH)(SO_3)^-$$

This reaction is used as a test for aldehydes.

More Complex Reactions

Reaction name	Product	Comment
Wolff-Kishner reduction	alkane	If an aldehyde is converted to a simple hydrazone $(RCH=NHNH_2)$ and this is heated with a base such as KOH, the terminal carbon is fully reduced to a methyl group. The Wolff-Kishner reaction may be performed as a one-pot reaction, giving the overall conversion $RCH=O \rightarrow RCH_3$.
Pinacol coupling reaction	diol	with reducing agents such as magnesium
Wittig reaction	alkene	reagent an ylide
Takai reaction	alkene	diorganochromium reagent
Corey-Fuchs reactions	alkyne	phosphine-dibromomethylene reagent
Ohira–Bestmann reaction	alkyne	reagent dimethyl (diazomethyl)phosphonate
Johnson-Corey-Chaykovsky reaction	epoxide	reagent a sulfonium ylide
Oxo Diels Alder reaction	pyran	Aldehydes can, typically in the presence of suitable catalysts, serve as partners in cycloaddition reactions. The aldehyde serves as the dienophile component, giving a pyran or related compound.
Hydroacylation	ketone	In hydroacylation an aldehyde is added over an unsaturated bond to form a ketone.
decarbonylation	alkane	catalysed by transition metals.

Dialdehydes

A **dialdehyde** is an organic chemical compound with two aldehyde groups. The nomenclature of dialdehydes have the ending-dial or sometimes-dialdehyde. Short aliphatic dialdehydes are sometimes named after the diacid from which they can de derived. An example is butanedial, which is also called succinaldehyde (from succinic acid).

KETONE

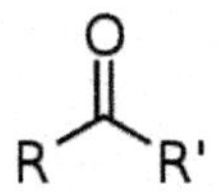

Fig. : Ketone group.

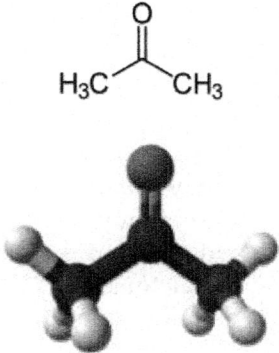

Fig. : Acetone.

In chemistry, a **ketone (alkanone)**/'ki:toʊn/is an organic compound with the structure RC(=O)R', where R and R' can be a variety of carbon-containing substituents. Ketones feature a carbonyl group (C=O) bonded to two other carbon atoms. Many ketones are known and many are of great importance in industry and in biology. Examples include many sugars (ketoses) and the industrial solvent acetone.

Nomenclature and Etymology

The word *ketone* derives its name from *Aketon*, an old German word for acetone.

According to the rules of IUPAC nomenclature, ketones are named by changing the suffix-*ane* of the parent alkane to-*anone*. For the most important ketones, however, traditional non-systematic names are still generally used, for example acetone and benzophenone. These non-systematic names are considered retained IUPAC names, although some introductory chemistry textbooks use names such as 2-propanone or propan-2-one instead of acetone, the simplest ketone (C H_3-CO-CH_3). The position of the carbonyl group is usually denoted by a number.

Although used infrequently, *oxo* is the IUPAC nomenclature for a ketone functional group. Other prefixes, however, are also used. For some common chemicals (mainly in bio-chemistry), *keto* or *oxo* refer to the ketone functional group. The term *oxo* is used widely through chemistry. For example, it also refers to an oxygen atom bonded to a transition metal (a metal oxo).

Structure and Properties

Representative ketones, from the left : acetone, a common solvent; oxaloacetate, an intermediate in the metabolism of sugars; acetylacetone in its (mono) enol form (the enol highlighted in blue); cyclohexanone, precursor to nylon; muscone, an animal scent; and tetracycline, an antibiotic.

The ketone carbon is often described as "sp² hybridized," a description that includes both their electronic and molecular structure. Ketones are trigonal planar around the ketonic carbon, with C-C-O and C-C-C bond angles of approximately 120°. Ketones differ from aldehydes in that the carbonyl group (CO) is bonded to two carbons within a carbon skeleton. In aldehydes, the carbonyl is bonded to one carbon and one hydrogen and are located at the ends of carbon chains. Ketones are also distinct from other carbonyl-containing functional groups, such as carboxylic acids, esters and amides.

The carbonyl group is polar as a consequence of the fact that the electronegativity of the oxygen is greater than that for carbon. Thus, ketones are nucleophilic at oxygen and electrophilic at carbon. Because the carbonyl group interacts with water by hydrogen bonding, ketones are typically more soluble in water than the related methylene compounds. Ketones are hydrogen-bond acceptors. Ketones are not usually hydrogen-bond donors and cannot hydrogen-bond to themselves. Because of their inability to serve both as hydrogen-bond donors and acceptors, ketones tend not to "self-associate" and are more volatile than alcohols and carboxylic acids of comparable molecular weights. These factors relate to the pervasiveness of ketones in perfumery and as solvents.

Classes of Ketones

Ketones are classified on the basis of their substituents. One broad classification sub-divides ketones into symmetrical and asymmetrical derivatives, depending on the equivalency of the two organic substituents attached to the carbonyl center. Acetone and benzophenone ($C_6H_5C(O)C_6H_5$) are symmetrical ketones. Acetophenone ($C_6H_5C(O)CH_3$) is an asymmetrical ketone. In the area of stereo-chemistry, asymmetrical ketones are known for being prochiral.

Diketones

Many kinds of diketones are known, some with unusual properties. The simplest is diacetyl ($CH_3C(O)C(O)CH_3$), once used as butter-flavoring in popcorn. Acetylacetone (pentane-2,4-dione) is virtually a misnomer (inappropriate name) because this species exists mainly as the monoenol $CH_3C(O)CH=C(OH)CH_3$. Its enolate is a common ligand in coordination chemistry.

Unsaturated Ketones

Ketones containing alkene and alkyne units are often called unsaturated ketones. The most widely used member of this class of compounds is methyl vinyl ketone, $CH_3C(O)CH=CH_2$, which is useful in the Robinson annulation reaction. Lest there be confusion, a ketone itself is a site of unsaturation; that is, it can be hydrogenated.

Cyclic Ketones

Many ketones are cyclic. The simplest class have the formula $(CH_2)_nCO$, where n varies from 3 for cyclopropanone to the teens. Larger derivatives exist. Cyclohexanone, a symmetrical cyclic ketone, is an important intermediate in the production of nylon. Isophorone, derived from acetone, is an unsaturated, asymmetrical ketone that is the precursor to other polymers. Muscone, 3-methylpentadecanone, is an animal pheromone. Another cyclic ketone is cyclobutanone, having the formula C_4H_6O.

Keto-enol Tautomerization

Fig. : Keto-enol tautomerism. **1** is the keto form; **2** is the enol.

Ketones that have at least one alpha-hydrogen, undergo keto-enol tautomerization; the tautomer is an enol. Tautomerization is catalyzed by both acids and bases. Usually, the keto form is more stable than the enol. This equilibrium allows ketones to be prepared via the hydration of alkynes.

Acidity of Ketones

Ketones are far more acidic ($pK_a \approx 20$) than a regular alkane ($pK_a \approx 50$). This difference reflects resonance stabilization of the enolate ion that is formed upon deprotonation. The relative acidity of the α-hydrogen is important in the enolization reactions of ketones and other carbonyl compounds. The acidity of the α-hydrogen also allows ketones and other carbonyl compounds to undergo nucleophilic reactions at that position, with either stoichiometric and catalytic base.

Characterization

Spectroscopy

Ketones and aldehydes absorb strongly in infra-red spectrum near 1700 cm^{-1}. The exact position of the peak depends on the substituents.

Whereas ^1H NMR spectroscopy is, in general, not useful for establishing the presence of a ketone, ^{13}C NMR spectra exhibit signals somewhat downfield of 200 ppm depending on structure. Such signals are typically weak due to the absence of nuclear Overhauser effects. Since aldehydes resonate at similar chemical shifts, multiple resonance experiments are employed to definitively distinguish aldehydes and ketones.

Qualitative Organic Tests

Ketones give positive results in Brady's test, the reaction with 2,4-dinitrophenyl-hydrazine to give the corresponding hydrazone. Ketones may be distinguished from aldehydes by giving a negative result with Tollens' reagent or with Fehling's solution. Methyl ketones give positive results for the iodoform test.

Synthesis

Many methods exist for the preparation of ketones in industrial scale and academic laboratories. Ketones are also produced in various ways by organisms.

In industry, the most important method probably involves oxidation of hydrocarbons, often with air. For example, a billion kilograms of cyclohexanone are produced annually by aerobic oxidation of cyclohexane. Acetone is prepared by air-oxidation of cumene.

For specialized or small scale organic synthetic applications, ketones are often prepared by oxidation of secondary alcohols :

$$R_2CH(OH) + O \rightarrow R_2C{=}O + H_2O$$

Typical strong oxidants (source of "O" in the above reaction) include potassium permanganate or a Cr (VI) compound. Milder conditions make use of the Dess–Martin periodinane or the Moffatt–Swern methods.

Many other methods have been developed including :

- By geminal halide hydrolysis.
- By hydration of alkynes. Such processes occur *via* enols and require the presence of an acid and $HgSO_4$. Subsequent enol–keto tautomerization gives a ketone. This reaction always produces a ketone, even with a terminal alkyne.
- From Weinreb Amides using stoichiometric organometallic reagents.
- Aromatic ketones can be prepared in the Friedel–Crafts acylation, the related Houben–Hoesch reaction and the Fries re-arrangement.
- Ozonolysis, and related dihydroxylation/oxidative sequences, cleave alkenes to give aldehydes and/or ketones, depending on alkene substitution pattern.
- In the Kornblum–DeLaMare re-arrangement ketones are prepared from peroxides and base.
- In the Ruzicka cyclization, cyclic ketones are prepared from dicarboxylic acids.
- In the Nef reaction, ketones form by hydrolysis of salts of secondary nitro compounds.
- In the Fukuyama coupling, ketones form from a thioester and an organozinc compound.
- By the reaction of an acid chloride with organocadmium compounds or organocopper compounds.
- The Dakin–West reaction provides an efficient method for preparation of certain methyl ketones from carboxylic acids.

- Ketones can also be prepared by the reaction of Grignard reagents with nitriles, followed by hydrolysis.
- By decarboxylation of carboxylic anhydride.
- Ketones can be prepared from haloketones in reductive dehalogenation of halo ketones.
- In ketonic decarboxylation symmetrical ketones are prepared from carboxylic acids.

Reactions

Ketones engage in many organic reactions. The most important reactions follow from the susceptibility of the carbonyl carbon toward nucleophilic addition and the tendency for the enolates to add to electrophiles. Nucleophilic additions include in approximate order of their generality :

- With water (hydration) gives geminal diols, which are usually not formed in appreciable (or observable) amounts.
- With an acetylide to give the α-hydroxyalkyne.
- With ammonia or a primary amine gives an imine.
- With secondary amine gives an enamine.
- With Grignard and organolithium reagents to give, after aqueous workup, a tertiary alcohol.
- With an alcohols or alkoxides to gives the hemiketal or its conjugate base. With a diol to the ketal. This reaction is employed to protect ketones.
- With sodium amide resulting in C–C bond cleavage with formation of the amide $RCONH_2$ and the alkane R'H, a reaction called the Haller–Bauer reaction.
- With strong oxidising agents to give carboxylic acids.

  Electrophilic addition, reaction with an electrophile gives a resonance stabilized cation :

- With phosphonium ylides in the Wittig reaction to give the alkenes.
- With thiols to give the thioacetal.
- With hydrazine or 1-disubstituted derivatives of hydrazine to give hydrazones.
- With a metal hydride gives a metal alkoxide salt, hydrolysis of which gives the alcohol, an example of ketone reduction.
- With halogens to form α-haloketone, a reaction that proceeds via an enol.
- With heavy water to give a α-deuterated ketone.
- Fragmentation in photochemical Norrish reaction.
- Reaction of 1,4-aminodiketones to oxazoles by dehydration in the Robinson–Gabriel synthesis.
- In the case of aryl–alkyl ketones, with sulfur and an amine give amides in the Willgerodt reaction.
- With hydroxylamine to produce oximes.

Bio-chemistry

Ketones are pervasive in nature. The formation of organic compounds in photosynthesis occurs *via* the ketone ribulose-1,5-bisphosphate. Many sugars are ketones, known collectively as ketoses. The best known ketose is fructose, which exists as a cyclic hemiketal, which masks the ketone functional group. Fatty acid synthesis proceeds *via* ketones. Acetoacetate is an intermediate in the Krebs cycle which releases energy from sugars and carbohydrates.

In medicine, acetone, acetoacetate, and beta-hydroxybutyrate are collectively called ketone bodies, generated from carbohydrates, fatty acids, and amino acids in most vertebrates, including humans. Ketone bodies are elevated in the blood (ketosis); after fasting, including a night of sleep; in both blood and urine in starvation; in hypoglycemia, due to causes other than hyperinsulinism; in various inborn errors of metabolism, and intentionally induced via a ketogenic diet, and in ketoacidosis (usually due to diabetes mellitus). Although ketoacidosis is characteristic of decompensated or untreated type 1 diabetes, ketosis or even ketoacidosis can occur in type 2 diabetes in some circumstances as well.

Applications

Ketones are produced on massive scales in industry as solvents, polymer precursors, and pharmaceuticals. In terms of scale, the most important ketones are acetone, methylethyl ketone, and cyclohexanone. They are also common in biochemistry, but less so than in organic chemistry in general. The combustion of hydrocarbons is an uncontrolled oxidation process that gives ketones as well as many other types of compounds.

Toxicity

Although it is difficult to generalize on the toxicity of such a broad class of compounds, simple ketones are, in general, not highly toxic. This characteristic is one reason for their popularity as solvents. Exceptions to this rule are the unsaturated ketones such as methyl vinyl ketone with LD_{50} of 7 mg/kg (oral).

Chapter 11

SELENIUM

Selenium is a chemical element with symbol Se and atomic number 34. It is a non-metal with properties that are intermediate between those of its periodic table column-adjacent chalcogen elements sulfur and tellurium. It rarely occurs in its elemental state in nature, or as pure ore compounds. Selenium (Greek σελήνη *selene* meaning "Moon") was discovered in 1817 by Jöns Jacob Berzelius, who noted the similarity of the new element to the previously known tellurium (named for the Earth).

Selenium is found impurely in metal sulfide ores, where it partially replaces the sulfur. Commercially, selenium is produced as a by-product in the refining of these ores, most often during copper production. Minerals that are pure selenide or selenate compounds are known, but are rare. The chief commercial uses for selenium today are in glass-making and in pigments. Selenium is a semi-conductor and is used in photo cells. Uses in electronics, once important, have been mostly supplanted by silicon semi-conductor devices. Selenium continues to be used in a few types of DC power surge protectors, baby formula, and one type of fluorescent quantum dot.

Selenium salts are toxic in large amounts, but trace amounts are necessary for cellular function in many organisms, including all animals. Selenium is a component of the antioxidant enzymes glutathione peroxidase and thioredoxin reductase (which indirectly reduce certain oxidized molecules in animals and some plants). It is also found in three deiodinase enzymes, which convert one thyroid hormone to another. Selenium requirements in plants differ by species, with some plants requiring relatively large amounts, and others apparently requiring none.

CHARACTERISTICS

Physical Properties

Selenium exists in several allotropes that interconvert upon heating and cooling carried out at different temperatures and rates. As prepared in chemical reactions, selenium is usually amorphous, brick-red powder. When rapidly melted, it forms the black, vitreous form, which is usually sold industrially as beads. The structure of black selenium is irregular and complex and consists of polymeric rings with up to 1000 atoms per ring. Black Se is a brittle, lustrous solid that is slightly soluble in CS_2. Upon heating, it softens at 50°C and converts to gray selenium at 180°C; the transformation temperature is reduced by presence of halogens and amines.

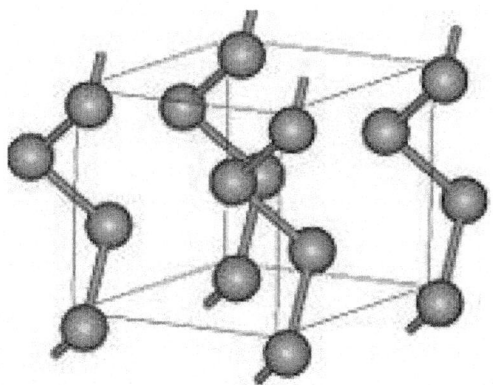

Fig. : Structure of hexagonal (gray) selenium.

The red-coloured α, β and γ forms are produced from solutions of black selenium by varying evaporation rates of the solvent (usually CS_2). They all have relatively low, monoclinic crystal symmetries and contain nearly identical puckered Se_8 rings arranged in different fashions, as in sulfur. The packing is most dense in the α form. In the Se_8 rings, the Se-Se distance is 233.5 pm and Se-Se-Se angle is 105.7 degrees. Other selenium allotropes may contain Se_6 or Se_7 rings.

The most stable and dense form of selenium has a gray colour and hexagonal crystal lattice consisting of helical polymeric chains, wherein the Se-Se distance is 237.3 pm and Se-Se-Se angle is 130.1 degrees. The minimum distance between chains is 343.6 pm. Gray Se is formed by mild heating of other allotropes, by slow cooling of molten Se, or by condensing Se vapors just below the melting point. Whereas other Se forms are insulators, gray Se is a semi-conductor showing appreciable photo-conductivity. Contrary to other allotropes, it is insoluble in CS_2. It resists oxidation by air and is not attacked by non-oxidizing acids. With strong reducing agents, it forms polyselenides. Selenium does not exhibit the unusual changes in viscosity that sulfur undergoes when gradually heated.

Isotopes

Selenium has six naturally occurring isotopes, five of which are stable : [74]Se, [76]Se, [77]Se, [78]Se, and [80]Se. The last three also occur as fission products, along with [79]Se, which has a half-life of 327,000 years. The final naturally occurring isotope, [82]Se, has a very long half-life (~10^{20} yr, decaying via double beta decay to [82]Kr), which, for practical purposes, can be considered to be stable. Twenty-three other unstable isotopes have been characterized.

CHEMICAL COMPOUNDS

Selenium compounds commonly exist in the oxidation states −2, +2, +4, and +6.

Chalcogen Compounds

Selenium forms two oxides : selenium dioxide (SeO_2) and selenium trioxide (SeO_3). Selenium dioxide is formed by the reaction of elemental selenium with oxygen :

$$Se_8 + 8\ O_2 \rightarrow 8\ SeO_2$$

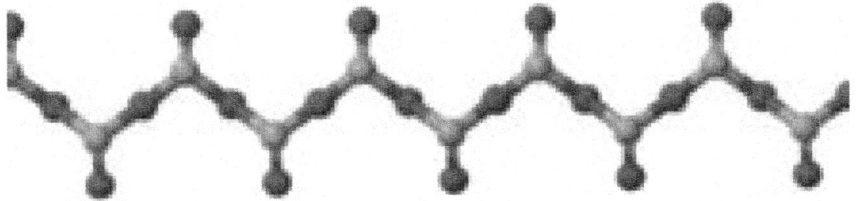

Fig. : Structure of the polymer SeO_2. The (pyramidal) Se atoms are yellow.

It is a polymeric solid that forms monomeric SeO_2 molecules in the gas phase. It dissolves in water to form selenous acid, H_2SeO_3. Selenous acid can also be made directly by oxidizing elemental selenium with nitric acid :

$$3\ Se + 4\ HNO_3 + H_2O \rightarrow 3\ H_2SeO_3 + 4\ NO$$

Unlike sulfur, which forms a stable trioxide, selenium trioxide is thermodynamically unstable and decomposes to the dioxide above 185°C :

$$2\ SeO_3 \rightarrow 2\ SeO_2 + O_2\ (\Delta H = -54\ kJ/mol)$$

Selenium trioxide is produced in the laboratory by the reaction of anhydrous potassium selenate (K_2SeO_4) and sulfur trioxide (SO_3).

Salts of selenous acid are called *selenites*. These include silver selenite (Ag_2SeO_3) and sodium selenite (Na_2SeO_3).

Hydrogen sulfide reacts with aqueous selenous acid to produce selenium disulfide :

$$H_2SeO_3 + 2\ H_2S \rightarrow SeS_2 + 3\ H_2O$$

Selenium disulfide consists of 8-membered rings of a nearly statistical distribution of sulfur and selenium atoms. It has an approximate composition of SeS_2, with individual rings varying in composition, such as Se_4S_4 and Se_2S_6. Selenium disulfide has been use in shampoo as an anti-dandruff agent, an inhibitor in polymer chemistry, a glass dye, and a reducing agent in fireworks.

Selenium trioxide may be synthesized by dehydrating selenic acid, H_2SeO_4, which is itself produced by the oxidation of selenium dioxide with hydrogen peroxide :

$$SeO_2 + H_2O_2 \rightarrow H_2SeO_4$$

Hot, concentrated selenic acid is capable of dissolving gold, forming gold (III) selenate.

Halogen Compounds

Iodides of selenium are not well known. The only stable chloride is selenium monochloride (Se_2Cl_2), which might be better known as selenium (I) chloride; the corresponding bromide is also known. These species are structurally analogous to the corresponding disulfur dichloride. Selenium dichloride is an important reagent in the preparation of selenium compounds (*e.g.* the preparation of Se_7). It is prepared by treating selenium with sulfuryl chloride (SO_2Cl_2). Selenium reacts with fluorine to form selenium hexafluoride :

$$Se_8 + 24\ F_2 \rightarrow 8\ SeF_6$$

In comparison with its sulfur counterpart (sulfur hexafluoride), selenium hexafluoride (SeF_6) is more reactive and is a toxic pulmonary irritant. Some of the selenium oxyhalides, such as selenium oxyfluoride ($SeOF_2$) and selenium oxychloride ($SeOCl_2$) have been used as specialty solvents.

Selenides

Analogous to the behaviour of other chalcogens, selenium forms a dihydride H_2Se. It is a strongly odiferous, toxic, and colourless gas. It is more acidic than H_2S. In solution it ionizes to HSe^-. The selenide dianion Se^{2-} forms a variety of compounds, including the minerals from which selenium is obtained commercially. Illustrative selenides include mercury selenide (HgSe), lead selenide (PbSe), zinc selenide (ZnSe), and copper indium gallium diselenide ($Cu(Ga,In)Se_2$). These materials are semi-conductors. With highly electro-positive metals, such as aluminium, these selenides are prone to hydrolysis :

$$Al_2Se_3 + 6\ H_2O \rightarrow Al_2O_3 + 6\ H_2Se$$

Alkali metal selenides react with selenium to form polyselenides, $Se2-x$, which exist as chains.

Other Compounds

Tetraselenium tetranitride, Se_4N_4, is an explosive orange compound analogous to tetra-sulfur tetra-nitride (S_4N_4). It can be synthesized by the reaction of selenium tetra-chloride ($SeCl_4$) with $[((CH3)3Si)_2N]_2Se$.

Selenium reacts with cyanides to yield selenocyanates :

$$8\ KCN + Se_8 \rightarrow 8\ KSeCN$$

Organoselenium Compounds

Selenium, especially in the II oxidation state, forms stable bonds to carbon, which are structurally analogous to the corresponding organo-sulfur compounds. Especially common are selenides (R_2Se, analogues of thioethers), diselenides (R_2Se_2, analogues of disulfides), and selenols (RSeH, analogues of thiols). Representatives of selenides, diselenides, and selenols include respectively selenomethionine, diphenyldiselenide, and benzeneselenol. The sulfoxide in sulfur chemistry is represented in selenium chemistry by the selenoxides (formula RSe(O)R), which are intermediates in organic synthesis, as illustrated by the selenoxide elimination reaction. Consistent with trends indicated by the double bond rule, selenoketones, R(C=Se)R, and selenaldehydes, R(C=Se)H, are rarely observed.

HISTORY

Selenium was discovered in 1817 by Jöns Jakob Berzelius and Johan Gottlieb Gahn. Both chemists owned a chemistry plant near Gripsholm, Sweden producing sulfuric acid by the lead chamber process. The pyrite from the Falun mine created a red precipitate in the lead chambers which was presumed to be an arsenic compound, and so the pyrite's use to make acid was discontinued. Berzelius and Gahn wanted to use the pyrite and they also observed that the red precipitate gave off a smell like horseradish when burned. This smell was not typical of arsenic, but a similar odour was known from tellurium compounds. Hence, Berzelius's first letter to Alexander Marcet stated that this was a tellurium compound. However, the lack of tellurium compounds in the Falun mine minerals eventually led Berzelius to reanalyze the red precipitate, and in 1818 he wrote a second letter to Marcet describing a newly found element similar to sulfur and tellurium. Because of its similarity to tellurium, named for the Earth, Berzelius named the new element after the Moon.

In 1873, Willoughby Smith found that the electrical resistance of grey selenium was dependent on the ambient light. This led to its use as a cell for sensing light. The first of commercial products using selenium were developed by Werner Siemens in the mid-1870s. The selenium cell was used in the photophone developed by Alexander Graham Bell in 1879. Selenium transmits an electric current proportional to the amount of light falling on its surface. This phenomenon was used in the design of light meters and similar devices. Selenium's semi-conductor properties found numerous other applications in electronics. The development of selenium rectifiers began during the early 1930s, and these replaced copper oxide rectifiers because of their superior efficiencies. These lasted in commercial applications until the 1970s, following which they were replaced with less expensive and even more efficient silicon rectifiers.

Selenium came to medical notice later because of its toxicity to human beings working in industries. Selenium was also recognized as an important veterinary toxin, which is seen in animals that have eaten high-selenium plants. In 1954, the first hints of specific biological functions of selenium were discovered in microorganisms. Its essentiality for mammalian life was discovered in 1957. In the 1970s, it was shown to be present in two independent sets of enzymes. This was followed by the discovery of selenocysteine in proteins. During the 1980s, it was shown that selenocysteine is encoded by the codon UGA. The recoding mechanism was worked out first in bacteria and then in mammals.

OCCURRENCE

Native (*i.e.*, elemental) selenium is a rare mineral, which does not usually form good crystals, but, when it does, they are steep rhombohedra or tiny acicular (hair-like) crystals. Isolation of selenium is often complicated by the presence of other compounds and elements.

Selenium occurs naturally in a number of inorganic forms, including selenide-, selenate-, and selenite-containing minerals, but these minerals are rare. The common mineral selenite is *not* a selenium mineral, and contains no selenite ion, but is rather a type of gypsum (calcium sulfate hydrate) named like selenium for the moon well before the discovery of selenium. Selenium is most commonly found quite impurely, replacing a small part of the sulfur in sulfide ores of many metals.

In living systems, selenium is found in the amino acids selenomethionine, selenocysteine, and methylselenocysteine. In these compounds, selenium plays a role analogous to that of sulfur. Another naturally occurring organoselenium compound is dimethyl selenide.

Certain solids are selenium-rich, and selenium can be bio-concentrated by certain plants. In soils, selenium most often occurs in soluble forms such as selenate (analogous to sulfate), which are leached into rivers very easily by runoff. Ocean water contains significant amounts of selenium.

Anthropogenic sources of selenium include coal burning and the mining and smelting of sulfide ores.

PRODUCTION

Selenium is most commonly produced from selenide in many sulfide ores, such as those of copper, nickel, or lead. Electrolytic metal refining is particularly conducive to producing selenium as a by-product, and it is obtained from the anode mud of copper refineries. Another source was the mud from the lead chambers of sulfuric acid plants but this method to produce sulfuric acid is no longer used. These muds can be processed by a number of means to obtain selenium. However, most elemental selenium comes as a by-product of refining copper or producing sulfuric acid. Since the invention of solvent extraction and electro-winning (SX/EW) for the production of copper this method takes an increasing share of the world wide copper production. This changes the availability of selenium because only a comparably small part of the selenium in the ore is leached together with the copper.

Industrial production of selenium usually involves the extraction of selenium dioxide from residues obtained during the purification of copper. Common production from the residue then begins by oxidation with sodium carbonate to produce selenium dioxide. The selenium dioxide is then mixed with water and the solution is acidified to form selenous acid (oxidation step). Selenous acid is bubbled with sulfur dioxide (reduction step) to give elemental selenium.

About 2,000 tonnes of selenium has been produced in 2011 worldwide, mostly in Germany (650 t), Japan (630 t), Belgium (200 t) and Russia (140 t), and the total reserves were estimated at 93,000 tonnes. These data however exclude two major producers, the United States and China. The price has been relatively stable during 2004-2010 at ~30 US dollars per pound (per 100-pound lot) but has increased to 65 $/lb in 2011. A previous sharp increase was observed in 2004 from 4-5 to 27 $/lb. The consumption in 2010 was divided as follows : metallurgy–30%, glass manufacturing–30%, agriculture–10%, chemicals and pigments–10%, electronics–10%. China is the dominant consumer of selenium at 1,500-2,000 tonnes/year.

APPLICATIONS

Manganese Electrolysis

During the electro winning of manganese an addition of selenium dioxide decreases the power necessary to operate the electrolysis cells. China is the largest consumer of selenium dioxide for this purpose. For every tonne of manganese an average of 2 kg selenium oxide is used.

Glass Production

The largest commercial use of Se, accounting for about 50% of consumption, is for the production of glass. Se compounds confer a red colour to glass. This colour cancels out the green or yellow tints that arise from iron impurities that are typical for most glass. For this purpose various selenite and selenate salts are added. For other applications, the red colour may be desirable, in which case mixtures of CdSe and CdS are added.

Alloys

Selenium is used with bismuth in brasses to replace more toxic lead. The regulation of lead in drinking water applications with the Safe Drinking Water Act of 1974 made a reduction of lead in brass necessary. The new brass is marketed under the name EnviroBrass. Like lead and sulfur, selenium improves the machinability of steel at concentrations of about 0.15%. The same improvement is also observed in copper alloys and therefore selenium is also used in machinable copper alloys.

Solar Cells

Copper indium gallium selenide is a material used in the production of solar cells.

Other Uses

Small amounts of organoselenium compounds are used to modify the vulcanization catalysts used in the production of rubber.

The demand for selenium by the electronics industry is declining, despite a number of continuing applications. Because of its photo-voltaic and photo-conductive properties, selenium is used in photocopying, photocells, light meters and solar cells. Its use as a photo-conductor in plain-paper copiers once was a leading application but in the 1980s, the photo-conductor application declined (although it was still a large end-use) as more and more copiers switched to the use of organic photo-conductors. It was once widely used in selenium rectifiers. These uses have mostly been replaced by silicon-based devices or are in the process of being replaced. The most notable exception is in power DC surge protection, where the superior energy capabilities of selenium suppressors make them more desirable than metal oxide varistors.

Zinc selenide was the first material for blue LEDs but gallium nitride is dominating the market now. Cadmium selenide has recently played an important part in the fabrication of quantum dots. Sheets of amorphous selenium convert x-ray images to patterns of charge in xeroradiography and in solid-state, flat-panel x-ray cameras.

Selenium is a catalyst in some chemical reactions but it is not widely used because of issues with toxicity. In X-ray crystallography, incorporation of one or more selenium atoms in place of sulfur helps with Multi-wavelength anomalous dispersion and Single wavelength anomalous dispersion phasing.

Selenium is used in the toning of photographic prints, and it is sold as a toner by numerous photographic manufacturers. Its use intensifies and extends the tonal range of black-and-white photographic images and improves the permanence of prints.

^{75}Se is used as a gamma source in industrial radiography.

BIOLOGICAL ROLE

NFPA 704

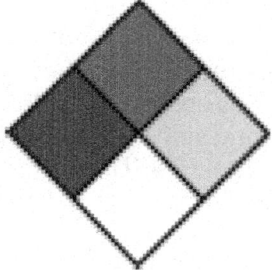

0 2 0
Fig. : Fire diamond for elemental selenium.

Although it is toxic in large doses, selenium is an essential micro-nutrient for animals. In plants, it occurs as a bystander mineral, sometimes in toxic proportions in forage (some plants may accumulate selenium as a defense against being eaten by animals, but other plants such as locoweed require selenium, and their growth indicates the presence of selenium in soil).

Selenium is a component of the unusual amino acids selenocysteine and selenomethionine. In humans, selenium is a trace element nutrient that functions as co-factor for reduction of antioxidant enzymes, such as glutathione peroxidases and certain forms of thioredoxin reductase found in animals and some plants (this enzyme occurs in all living organisms, but not all forms of it in plants require selenium).

The glutathione peroxidase family of enzymes (GSH-Px) catalyze certain reactions that remove reactive oxygen species such as hydrogen peroxide and organic hydroperoxides :

$$2 \, GSH + H_2O_2 \text{----} GSH\text{-}Px \rightarrow GSSG + 2 \, H_2O$$

Selenium also plays a role in the functioning of the thyroid gland and in every cell that uses thyroid hormone, by participating as a co-factor for the three of the four known types of thyroid hormone deiodinases, which activate and then deactivate various thyroid hormones and their metabolites : the iodothyronine deiodinases are the sub-family of deiodinase enzymes that use selenium as the otherwise rare amino acid selenocysteine. (Only the deiodinase iodotyrosine deiodinase, which works on the last break-down products of thyroid hormone, does not use selenium).

Selenium may inhibit Hashimoto's disease, in which the body's own thyroid cells are attacked as alien. A reduction of 21% on TPO anti-bodies was reported with the dietary intake of 0.2 mg of selenium.

Increased dietary selenium intakes reduce the effects of mercury toxicity and it is now recognized that the molecular mechanism of mercury toxicity involves irreversible inhibition of selenoenzymes that are required to prevent and reverse oxidative damage in brain and endocrine tissues.

Evolution in Biology

From about three billion years ago, prokaryotic selenoprotein families drive the evolution of selenocysteine, an amino acid. Selenium is incorporated into several prokaryotic selenoprotein families in bacteria, archaea and eukaryotes as selenocysteine, where selenoprotein peroxiredoxins protect bacterial and eukaryotic cells against oxidative damage. Selenoprotein families of GSH-Px and the deiodinases of eukaryotic cells seem to have a bacterial phylogenetic origin. The selenocysteine-containing form occurs in species as diverse as green algae, diatoms, sea urchin, fish and chicken. Selenium enzymes are involved in utilization of the small reducing molecules glutathione and thioredoxin. One family of selenium-containing molecules (the glutathione peroxidases) destroy peroxide and repair damaged peroxidized cell membranes, using glutathione. Another selenium-containing enzyme in some plants and in animals (thioredoxin reductase) generates reduced thioredoxin, a dithiol that serves as an electron source for peroxidases and also the important reducing enzyme ribonucleotide reductase that makes DNA precursors from RNA precursors.

Trace elements involved in GSH-Px and superoxide dismutase enzymes activities, *i.e.* selenium, vanadium, magnesium, copper, and zinc, may have been lacking in some terrestrial mineral-deficient areas. Marine organisms retained and sometimes expanded their seleno-proteomes, whereas the seleno-proteomes of some terrestrial organisms were reduced or completely lost. These findings suggest that, with the exception of vertebrates, aquatic life supports selenium utilization, whereas terrestrial habitats lead to reduced use of this trace element. Marine fishes and vertebrate thyroid glands have the highest concentration of selenium and iodine. From about 500 Mya, freshwater and terrestrial plants slowly optimized the production of "new" endogenous antioxidants such as ascorbic acid (Vitamin C), polyphenols (including flavonoids), tocopherols, etc. A few of these appeared more recently, in the last 50–200 million years, in fruits and flowers of angiosperm plants. In fact, the angiosperms (the dominant type of plant today) and most of their antioxidant pigments evolved during the late Jurassic period.

The deiodinase isoenzymes constitute another family of eukaryotic seleno-proteins with identified enzyme function. Deiodinases are able to extract electrons from iodides, and iodides from iodothyronines. They are, thus, involved in thyroid-hormone regulation, participating in the protection of thyrocytes from damage by H_2O_2 produced for thyroid-hormone bio-synthesis. About 200 Mya, new selenoproteins were developed as mammalian GSH-Px enzymes.

Nutritional Sources of Selenium

Dietary selenium comes from nuts, cereals, meat, mushrooms, fish, and eggs. Brazil nuts are the richest ordinary dietary source (though this is soil-dependent, since the Brazil nut does not require high levels of the element for its own needs). In descending order of concentration, high levels are also found in kidney, tuna, crab, and lobster.

The human body's content of selenium is believed to be in the 13–20 milligram range.

Indicator Plant Species

Certain species of plants are considered indicators of high selenium content of the soil, since they require high levels of selenium to thrive. The main selenium indicator plants are *Astragalus* species (including some locoweeds), prince's plume (*Stanleya* sp.), woody asters (*Xylorhiza* sp.), and false goldenweed (*Oonopsis* sp.)

Medical Use

The substance loosely called selenium sulfide (approximate formula SeS_2) is the active ingredient in some anti-dandruff shampoos. The selenium compound kills the scalp fungus *Malassezia*, which causes shedding of dry skin fragments. The ingredient is also used in body lotions to treat Tinea versicolour due to infection by a different species of *Malassezia* fungus.

Detection in Biological Fluids

Selenium may be measured in blood, plasma, serum or urine to monitor excessive environmental or occupational exposure, confirm a diagnosis of poisoning in hospitalized victims or to assist in a forensic investigation in a case of fatal overdosage. Some analytical techniques are capable of distinguishing organic from inorganic forms of the element. Both organic and inorganic forms of selenium are largely converted to monosaccharide conjugates (selenosugars) in the body prior to being eliminated in the urine. Cancer patients receiving daily oral doses of selenothionine may achieve very high plasma and urine selenium concentrations.

Toxicity

Although selenium is an essential trace element, it is toxic if taken in excess. Exceeding the Tolerable Upper Intake Level of 400 micrograms per day can lead to selenosis. This 400 microgram (μg) Tolerable Upper Intake Level is based primarily on a 1986 study of five Chinese patients who exhibited overt signs of selenosis and a follow up study on the same five people in 1992. The 1992 study actually found the maximum safe dietary Se intake to be approximately 800 micrograms per day (15 micrograms per kilogram body weight), but suggested 400 micrograms per day to not only avoid toxicity, but also to avoid creating an imbalance of nutrients in the diet and to account for data from other countries. In China, people who ingested corn grown in extremely selenium-rich stony coal (carbonaceous shale) have suffered from selenium toxicity. This coal was shown to have selenium content as high as 9.1%, the highest concentration in coal ever recorded in literature.

Symptoms of selenosis include a garlic odour on the breath, gastrointestinal disorders, hair loss, sloughing of nails, fatigue, irritability, and neurological damage. Extreme cases of selenosis can result in cirrhosis of the liver, pulmonary edema, and death. Elemental selenium and most metallic selenides have relatively low toxicities because of their low bio-availability. By contrast, selenates and selenites are very toxic, having an oxidant mode of action similar to that of arsenic trioxide. The chronic toxic dose of selenite for humans is about 2400 to 3000 micrograms of selenium per day for a long time. Hydrogen selenide is an extremely toxic, corrosive gas. Selenium also occurs in organic compounds, such as dimethyl selenide, selenomethionine, selenocysteine and methylselenocysteine, all of which have high bio-availability and are toxic in large doses.

On 19 April 2009, 21 polo ponies died shortly before a match in the United States Polo Open. Three days later, a pharmacy released a statement explaining that the horses had received an incorrect dose of one of the ingredients used in a vitamin/mineral supplement compound that had been incorrectly compounded by a compounding pharmacy. Analysis of blood levels of inorganic compounds in the supplement indicated the selenium concentrations were ten to fifteen times higher than normal in the horses' blood samples, and 15 to 20 times higher than normal in their liver samples. It was later confirmed that selenium was the ingredient in question.

Selenium poisoning of water systems may result whenever new agricultural runoff courses through normally dry, undeveloped lands. This process leaches natural soluble selenium compounds (such as selenates) into the water, which may then be concentrated in new "wetlands" as the water evaporates. High selenium levels produced in this fashion have been found to have caused certain congenital disorders in wetland birds.

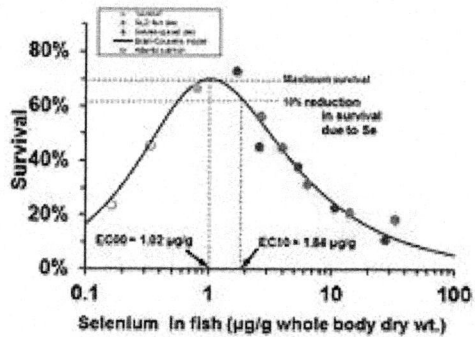

Relationship between survival of juvenile salmon and concentration of selenium in their tissues after 90 days (Chinook salmon) or 45 days (Atlantic salmon) exposure to dietary selenium. The 10% lethality level (LC10=1.84 µg/g) was derived by applying the biphasic model of Brain and Cousens to only the Chinook salmon data. The Chinook salmon data comprise two series of dietary treatments, combined here because the effects on survival are in distinguishable.

In fish and other wildlife, low levels of selenium cause deficiency while high levels cause toxicity. For example, in salmon, the optimal concentration of selenium in the fish tissue (whole body) is about 1 microgram selenium per gram of tissue (dry weight). At levels much below that concentration, young salmon die from selenium deficiency; much above that level they die from toxic excess.

DEFICIENCY

Selenium deficiency is rare in healthy, well-nourished individuals. It can occur in patients with severely compromised intestinal function, those undergoing total parenteral nutrition, and in those of advanced age (over 90). Also, people dependent on food grown from selenium-deficient soil are at risk. Although New Zealand has low levels of selenium in its soil, adverse health effects have not been detected.

Selenium deficiency as defined by low (<60% of normal) selenoenzyme activity levels in brain and endocrine tissues only occurs when a low selenium status is linked with an additional stress, such as high exposures to mercury or as a result of increased oxidant stress due to vitamin E deficiency.

There are interactions between selenium and other nutrients, such as iodine and vitamin E. The effect of selenium deficiency on health remains uncertain, particularly in relation to Kashin-Beck disease. Also, there are interactions between selenium and other minerals, such as zinc and copper. It seems that a high dose of Se supplements to pregnant animals might disturb the Zn : Cu ratio which, in turn, leads to Zn reduction. It can be concluded that the Zn status should be monitored when high doses of Se are supplemented to pregnant animals. Further studies need to be done with higher levels of Se supplement to confirm these interactions.

In some regions (*e.g.* various regions within North America) where low available selenium levels in soil lead to low concentrations in dry matter of plants, Se deficiency in some animal species may occur unless dietary (or injected) selenium supplementation is done. Ruminants are particularly susceptible. In general, absorption of dietary selenium is lower in ruminants than in non-ruminants, and is lower from forages than from grain. Ruminants grazing certain forages, *e.g.* some white clover varieties containing cyanogenic glycosides, may have higher selenium requirements, presumably because of cyanide from the aglycone released by glucosidase activity in the rumen and inactivation of glutathione peroxidases due to absorbed cyanide's effect on the glutathione moiety. Neonate ruminants at risk of WMD (white muscle disease) may be administered both selenium and vitamin E by injection; some of the WMD myopathies respond only to selenium, some only to vitamin E, and some to either.

Controversial Health Effects

A number of correlative epidemiological studies have implicated selenium deficiency (as measured by blood levels) in a number of serious or chronic diseases, such as cancer, diabetes, HIV/AIDS, and tuberculosis. In addition, selenium supplementation has been found to be a chemopreventive for some types of cancer in some types of rodents. However, in randomized, blinded, controlled prospective trials in humans, selenium supplementation has not succeeded in reducing the incidence of any disease, nor has a meta-analysis of such selenium supplementation studies detected a decrease in overall mortality.

Chapter 12

ELECTROLYTIC FLUORINATION OF HETEROCYCLIC COMPOUNDS

Partly fluorinated heterocyclic compounds are in focus of researchers due to their high biological activity. They are used as intermediate products when creating preparations for medicine and agriculture. Their synthesis is rather non-selective and limited in many cases. Continuous search for fluorinating reagents for introduction of the fluorine atoms and reagents for introduction of perfluoroalkyl groups into organic molecules is going on because of this. Unlike this electrochemical fluorination method is the ideal method of direct fluorination and can be carried out at one go when conditions are mild. However there are few examples of such fluorination, though high yield and selectivity make this method rather attractive.

Having a carbonyl group in α-position to nitrogen atom N-methylpyrazoles are subject to electrolytic fluorination in the Py/9HF system at 20°C with formation of fluoropyridizinone. Et_3N*3HF system can also be used as electrolyte.

In the case of 4-thiazolydinone **10** $Et_3N*3HF/MeCN$ system proved to be rather effective and convenient. The formation of mono-fluoroderivative **11** with high regioselectivity occurs when carrying out anodic fluorination. These are the first examples of effective anodic fluorination of sulfur-containing heterocyclic compounds. Subsequent oxidization using peracids affected sulfur atom and compound **12** was formed.

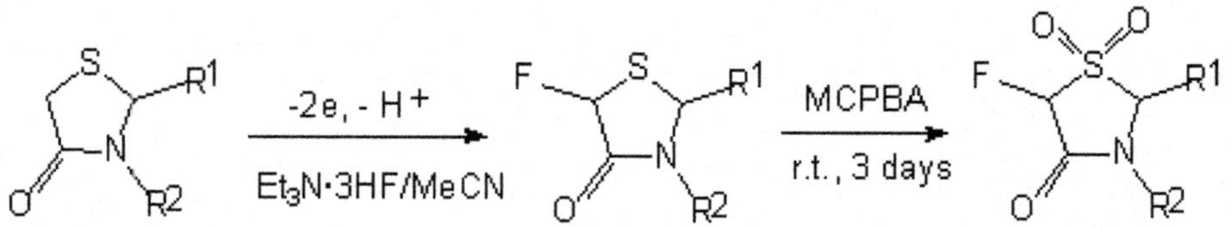

R^1	R^2	Yield 11,%	cis : trans	Yield 12,%	cis :trans
1-naphthyl	Me	79	29 : 71		
Ph	H	14	35 : 65		
Ph	Me			42	35 : 65
Ph	CMe₃			59	45 : 55
Ph	Ph	84	43 : 57		
PhCH₂	Ph			65	46 : 54

Other heterocyclic compounds react similarly.

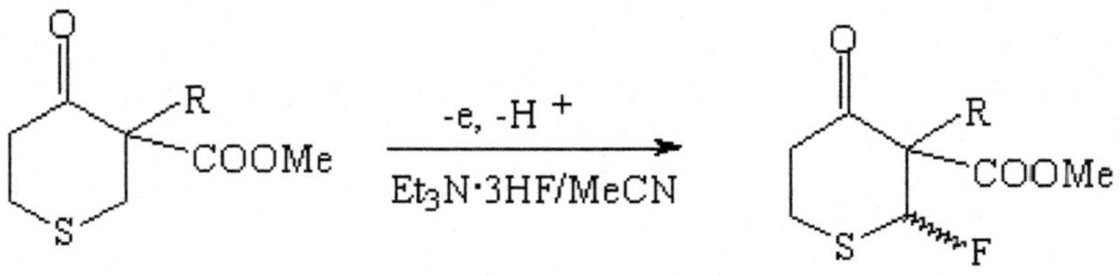

R	Yield,%	Stereoselectivity,%
Me	7	100
CH₂Ph	27	76

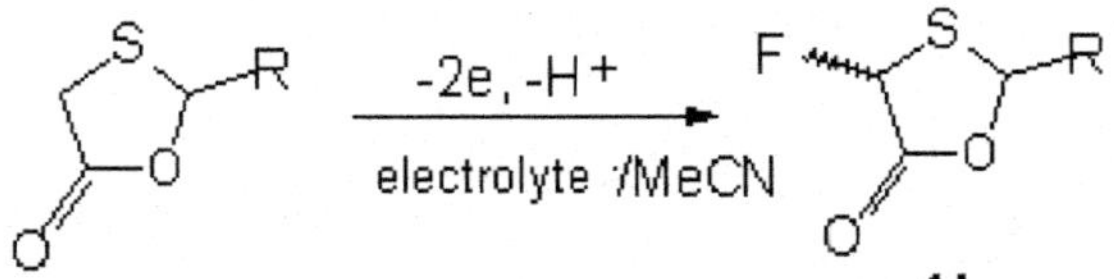

13   **14**

So, during anodic fluorination of 1,3-dithiolane-4-ones and 1,3-oxathiolan-5-ones **13**, which have the substitutes in position 2, appropriate fluoro-derivatives are formed.

In table there is information on anodic fluorination of 2-substituted 1, 3-oxothiolane-5-ones, during which mono-fluoroderivative 14 is formed. We'll notice, that fluoride-ion matters a lot in this process.

Table : Anodic mono-fluorination of 2-substituted 1,3-oxothiolane-5-ones.

R	E^{ox}_p/B vs SSSE*	Electrolyte	Potential of oxidation, V	F/mol	Yield of 14,%	cis/trans
Et	2.34	Et₃N*3HF Et₄NF*4HF	2,3 2,1	3,4 2,6	0 86	47/53
n-Pr	2,32	Et₃N*3HF Et₄NF*4HF	2,2-2,3 2,1	3,6 2,2	5,4 67	d 45/55
i-Pr	2,44	Et₃N*3HF Et₄NF*4HF	2,2 2,1	2,2 2,4	0 78	43/57

(Contd...)

(Contd...)

R	E^{ox}_p/B vs SSSE*	Electrolyte	Potential of oxidation, V	F/mol	Yield of 14,%	cis/trans
Ph	1,88	Et$_3$N*3HF	1,6	4,0	0	
		Et$_4$NF*4HF	2,0	2,3	70	45/55
4-CNC$_6$H$_4$	2,17	Et$_3$N*3HF	2,2	3,4	0	39/61
		Et$_4$NF*4HF	2,0	3,2	66	

* Pt-electrodes, 0,1 M NaClO$_4$/MeCN

Taking into attention the data listed in the table the system Et$_3$N*3HF is unusable for carrying out of the anodic fluorination processes for this heterocycle.

4-Arylthio-1,3-dioxalane-2-one in dimethoxyethane undergoes the fluorodesulfurization and fluorination with formation of 4-fluoro-1,3-dioxolane-2-one.

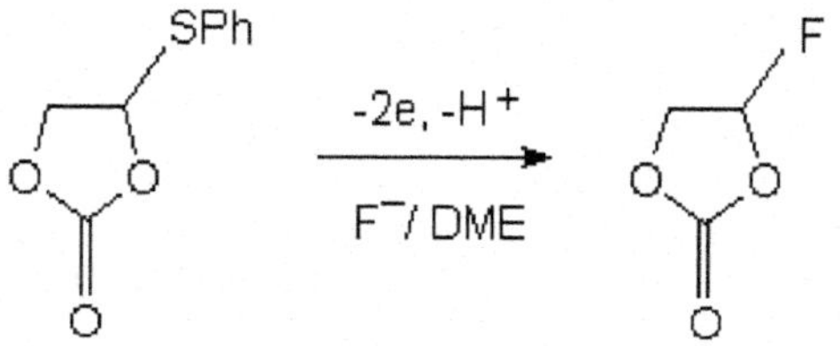

Electro-chemical oxidization of 2,2,-diphenyl-1,3-dithiane in Et$_3$N*4HF system results in formation of difluorodiphenylmethane.

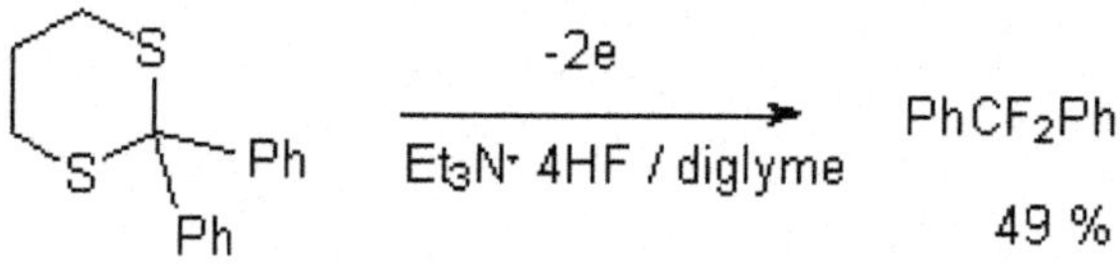

Mono-fluorinated β-lactams are of interest of fluoro-containing antibiotics synthesis as intermediate products and they also are of interest of creating amino-acids and sugars as block. Thus, regioselective anodic mono-fluorination of 3-phenylthio-2-azetidinones and 4-phenylthio-2-azetidinones results in formation of 3-and 4-substituted derivatives of 2-azetidinones. The product yield of mono-fluorination essentially depends on electrolyte used. It is interesting that in case Py/(HF)n system doesn't really work.

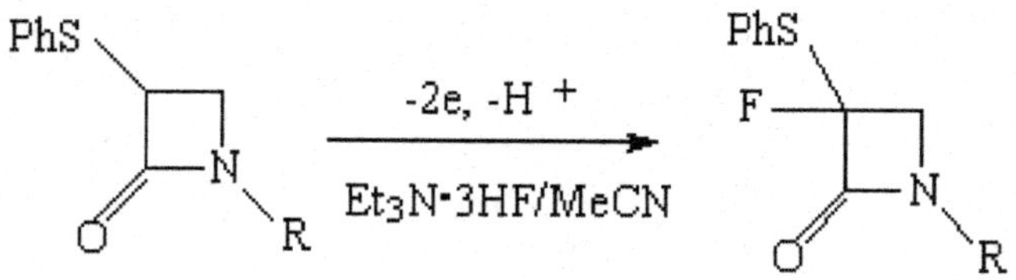

R	Electrolyte	B, vs. SCE	F/mol	Yield, %
Et			2,5	65
Pr^i	$Et_3N. 3HF$ $Et_4NF*4HF$ $Py(HF)n$ $Et_3N. 2HF$	1,8 1,5 1,9 1,7	2,3 2,0 2,2 13,2	77 64 traces 33
Bu^n		1,8	2,4	92
Bu^t		1,9	2,5	67
$c-C_6H_{11}$		1,8	2,5	84
$PhCH_2$		2,0	4,0	68

In general case electro-chemical fluorination of N-substituted lactams in electrolyte $Et_3N*3HF/MeCN$ goes selectively and affects the carbon atom, being in α-position to nitrogen atom of lactam. In this case the mono-fluoroderivative product was formed, moreover the yield is rather high.

If SR group is not in α-position to C=O heterocycle group, then during anodic fluorination its replacement for fluorine atom occurs.

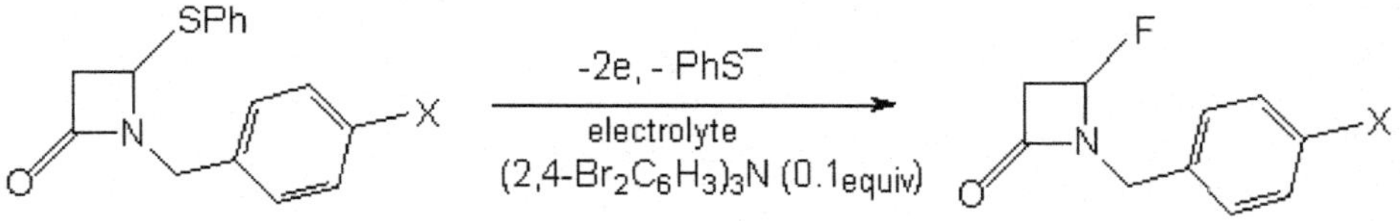

X	Electrolyte	Yield, %
H	$Et_3N. 3HF/MeCN$	83
Me	$Et_3N. 3HF/MeCN$	71
Br	$Et_3N. 3HF/MeCN$	quantitatively
H	$Et_4NF. 3HF/MeCN$	43

The presence of other substituents in heterocyclic ring doesn't change the reaction direction-fluorodesulfurization occurs.

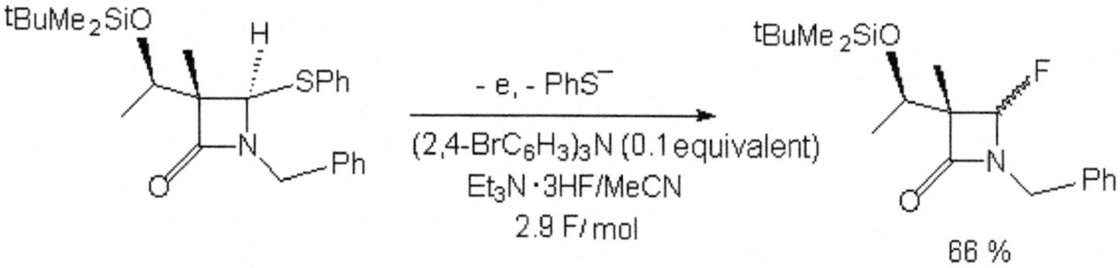

If in position 4 of azetidine-2-one there is $SiMe_3$ group, then the formation of 4-substituted fluoroderivative also takes place.

Heterocyclic compounds with carbonyl groups are subject to selective fluorination too. For example, the formation of mono-and difluoro-derivatives of

(4-methoxyphenyl) acetone, ethyl (4-methoxyphenyl) acetate and fluoro-derivatives of indan is observed when carrying out anodic fluorination of hydrocarbon substrate in electrolyte-complex Et_3N*3HF. Anodic fluorination of 1-naphthalene derivatives in Et_3N*3HF system, oxyindole and 3-oxy-1,2,3,4-tetrahydroisoquinoline in $Et_4NF.3HF$ system goes without problems when groups CN, COOEt or SPh are available.

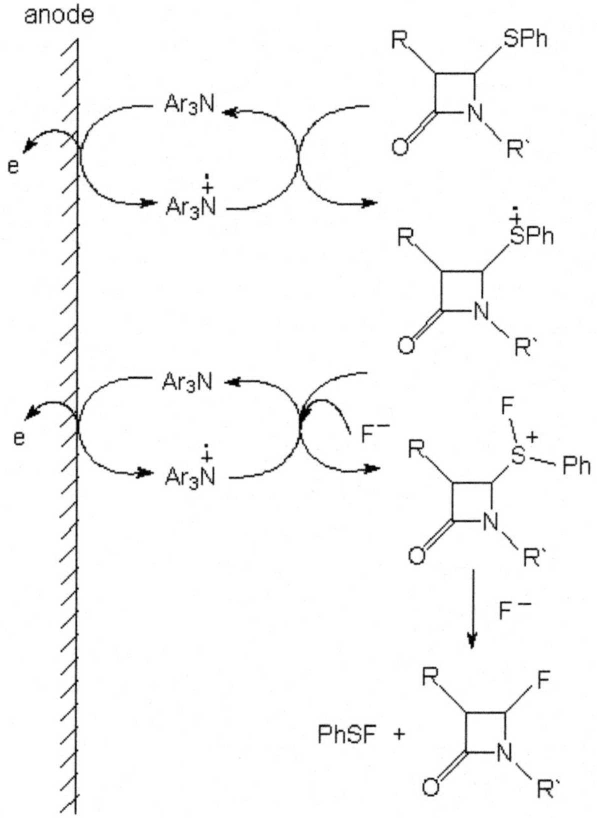

n	R	R¹	Electrolyte	F/mol	Yield,%
0	H	Ph	$Me_4NF*4HF$	3,5	64
0	Me	$4-MeC_6H_4$	$Me_4NF*4HF$	3	50
1	H	Ph	$Et_4NF*3HF$	2,6	71
1	H	$PhCH_2$	$Et_4NF*3HF$	2	70

The nature of electrodes matters within certain limits. The increase of anhydrous hydrogen fluoride results in increment of fluorination product. As example, it was shown by anodic fluorination of oxyindole.

Table 5 : *Electro-chemical obtaining of 4-fluoroazetidine-2-one derivative.*

Substrate		Electrolyte	F/mol	Product	Yield,%
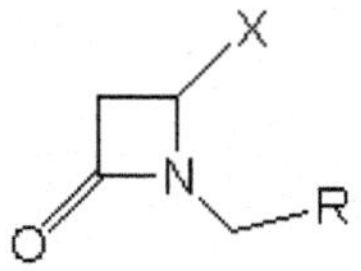				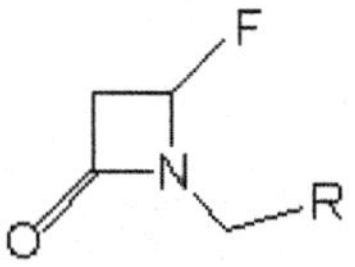	
X	R				
SiMe$_3$	Ph	Et$_3$N 3HF	8.0		80
SiMe$_3$	Ph	Et$_3$N·2HF	20.0		49
SiMe$_3$	Ph	Py· nHF	4.0		66
SiMe$_3$	Ph	Bu$_4$N·BF$_4$	4.0		смола
SiMe$_3$	Pr	Et$_3$N·3HF	18.0		78
		Et$_3$N·3HF			
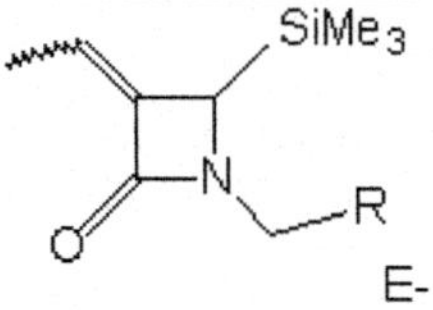 E-			10	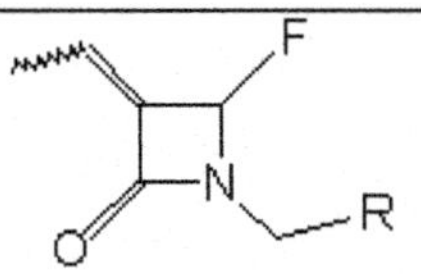 E-	68
Z-			6	Z-	80
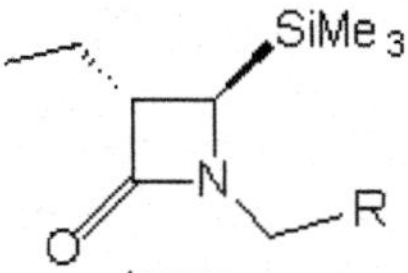 trans-		Et$_3$N ̄3HF	14	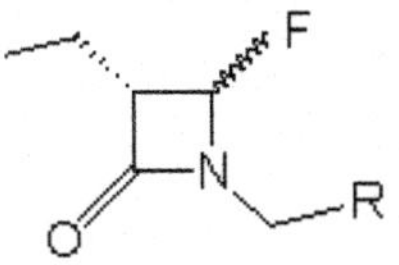 cis/trans = 1/1	88
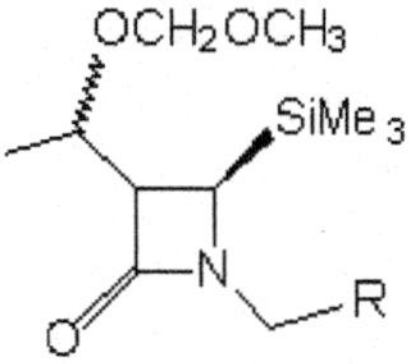 trans-		Et$_3$N·3HF	10	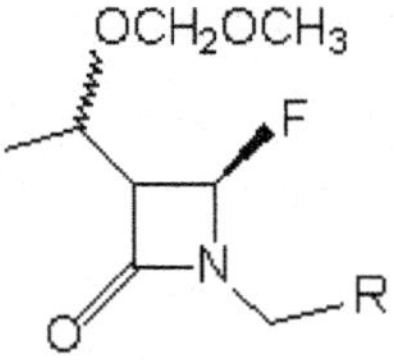 trans-	75

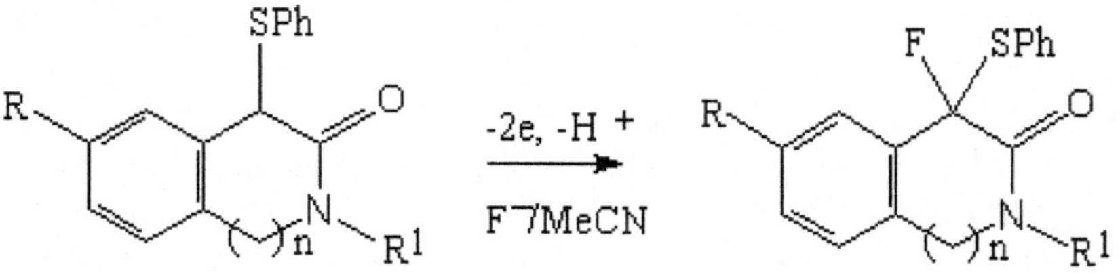

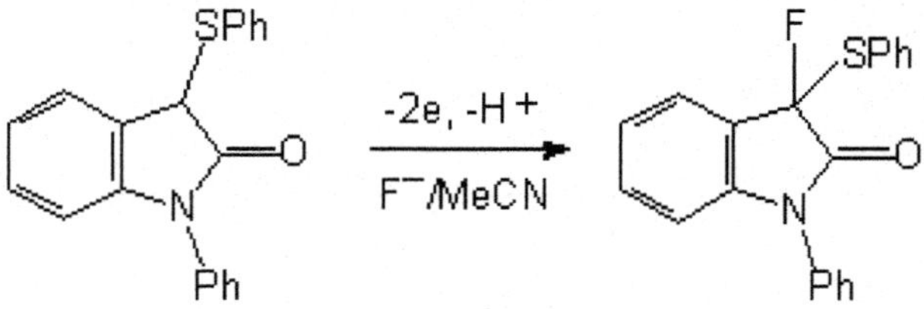

Anode	Yield, %	
	Et$_4$NF*3HF	Me$_4$NF*4HF
Pt	58	64
Carbon	28	60
C-steel	36	59
C-felt	21	42

At the same time anodic fluorination of oxyindole in Et$_4$NF*3HF/MeCN system with absence of SPh group results in fluorination of benzene ring without affecting of heterocyclic system.

This example shows the importance of SPh group in anodic fluorination of different derivatives.

During anodic fluorination of ethyl isonicotinate, pyrazole, pyridine, 1,10-phenanthroline the fluorination products containing fluorine in heterocycle and in methyl group were formed. However the yield of fluoro-derivatives is not very high.

It was shown, that during electro-chemical fluorination of 1-methylpyrazole-4-carboxylic acid ethyl ester **15** the using of 70% HF/Py/Et$_3$N complex on platinum anode results in formation of 1-methyl-5-fluoropyrazole-4-carboxylic acid ethyl ester **16** and 1-fluoromethyl-5-fluoropyrazole-4-carboxylic acid ethyl ester **17**. The highest yield was 40% and selectivity of fluorine attack *in position 5* was 83%.

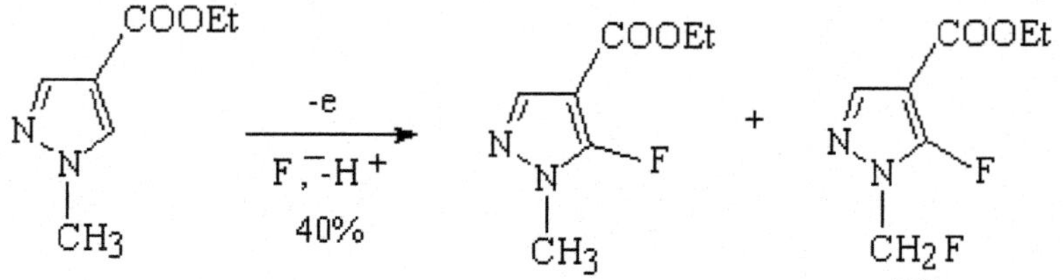

Anodic fluorination of pyridine in Et$_3$N*3HF system gives 2-fluoropyridine with 22% yield. Heterocyclic ring containing electron-seeking substituents complicate the direct fluorination of cycle and the yield of 2-fluoro-substituted derivative is low. For example, 2-fluoro-iso-nicotinic acid ethyl ester with high conversion is obtained using electrolysis of iso-nicotinic acid ester in Et$_3$N*3HF/MeCN system with 30% yield and 76% conversion.

Table : *Anodic fluorination of 2-methylpyrazole-4-carboxylic acid ethyl ester [61].*

Anode	System	Solvent	Voltage, V	Products 16 :17 ratio	Yield,%
Pt	Pt HF/Py	MeCN	2,5	25 :75	16
	HF/Py/Et3N(0.3)			57 :43	23
	HF/Py/Et3N(0.6)			83 :17	40
	HF/Py/Et3N(1.0)		3	100 :0	18
	HF/Py/Et3N(0.6)	THF	2,7	100 :0	2
	HF/Py/Et2NPh	MeCN	2,5	92 :8	39
C	HF/Py/Et3N	MeCN	2,1	97 :3	31

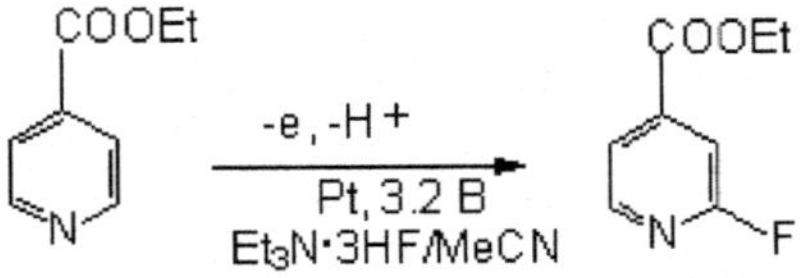

Nitrogen-containing heterocyclic compounds in electrolytes mentioned are subject to fluorination and as a rule hydrogen atoms are displaced in α-position to nitrogen atom of the cycle. For example, anodic fluorination of pyridazininone-3-ona derivative **18** in the medium 70% HF/Py/MeCN at Pt electrodes at room temperature gives mono-fluoro-derivative **19**.

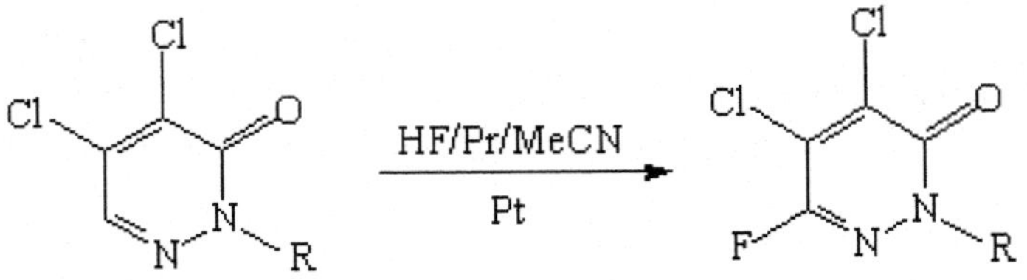

R = 3-CF$_3$C$_6$H$_4$, t-Bu

Anodic fluorination of 2H-1,4-benzothiazine-3(4)-ones derivatives **20** in acetonitrile in the presence of Et$_3$N*3HF results in formation of mono-fluoro-derivatives **21**, moreover fluorine displaces hydrogen in α-position to sulfur atom of ring.

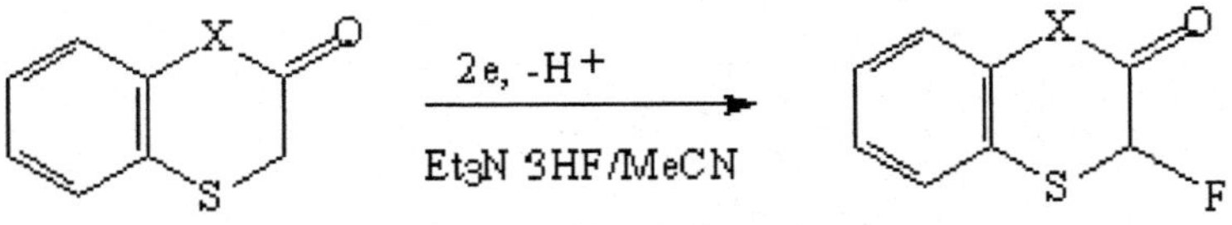

X	Potential of anode, V	F/mol	Yield of 21, %
NH	3.0	15	0
NHCOPh	2.0	2.2	77
N-Pr^i	1.5	2.5	88
n-Me	1.6	2.1	68

Anodic fluorination of biologically active flavone derivatives, having first potential E_p^{ox} within 2.36-2.52 V, results in formation of difluoro-and trifluoro-derivatives.

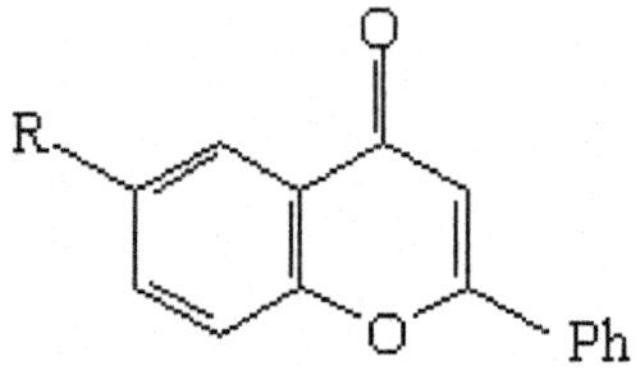

22 a-c

R   =   H(a) Me(b) Cl(c)

E_p^{ox} = 2.50   2.36   2.52

B vsSSCE

Anodic fluorination of flavone and 6-chloroflavone in Et$_3$N *3HF and Et$_4$NF*4HF systems mainly produced 3-mono-fluoro-derivative with small quantity of 2,3-difluoro-derivative. The process was carried out at 30°C and the ratio of products in reaction mixture depended on electrolyte used.

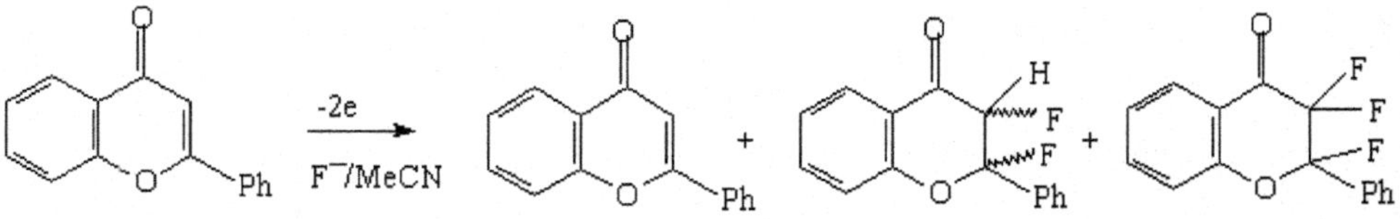

22a   22a   23   24

Electrolyte	Yield,%		
	22a	23 (cis :trans)	24
Et$_3$N.3HF	43	0	6
Et$_3$N.5HF	9	4	
Et$_4$NF.3HF	3	54(2 :1)	6
Et$_4$NF.4HF	7	68(2 :1)	9

At that the temperature has a key influence :

Temperature,°C	−10	0	10	20	30
Yield **22a**,%	9	27	31	43	58
Yield **24**,%	traces	3	4	6	19

Electro-chemical oxidization of caffeine, guanosine tetra-acetate and uridine tri-acetate, proceeding in electrolyte Et_3N. 3HF, results in formation of mono-fluoro-derivatives with small yield.

It should be noted that there is the high regioselectivity during anodic fluorination of α-phenylsulpho phenyl substituted lactams **25** in Et_3N.3HF/MeCN medium (Pt electrodes) when hydrogen is substituted by fluorine in α-position to Sph-group independently of ring size **26**.

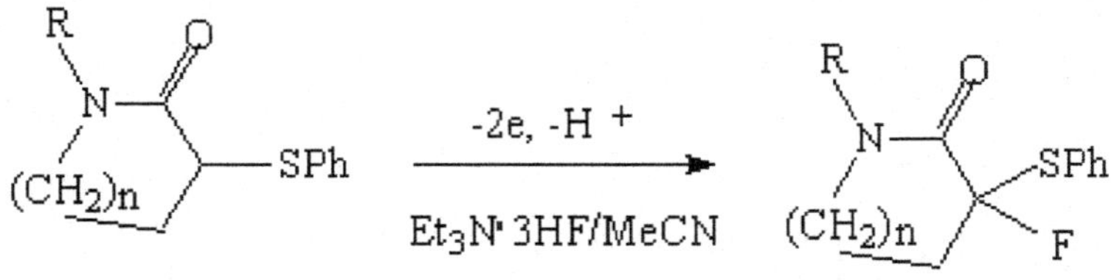

n	R	Potential of anode (V), relative to SSCE	F/mol	Yield of 26,%,
1	Me	1,8	2,5	85
1	cyclohexyl	1,8	2,5	84
2	Me	2,0	2,2	69
3	Me	1,9	4,0	compounds mixture

ANODIC FLUORINATION OF THE ORGANIC COMPOUND WITH CH-FRAGMENT CONTAINING ELECTRON-SEEKING SUBSTITUTES

High yield of fluoro-derivatives is achieved by anodic fluorination of substituted derivatives both aromatic and heterocyclic compounds, having protons movable enough in α-position to the main cycle. As a rule, most effective $Et_3N* 3HF$ system and platinum electrode are used.

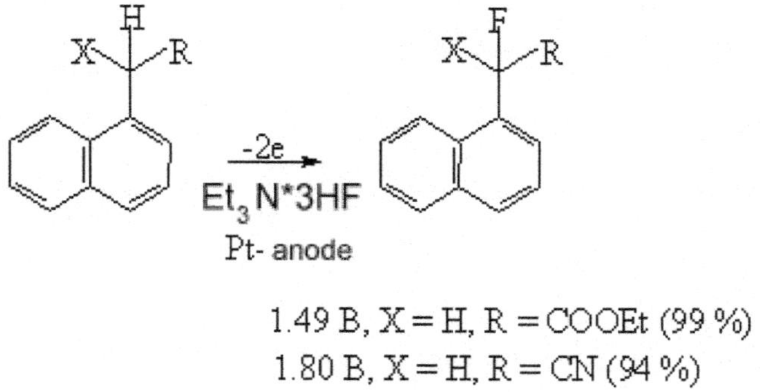

1.49 B, X = H, R = COOEt (99 %)
1.80 B, X = H, R = CN (94 %)

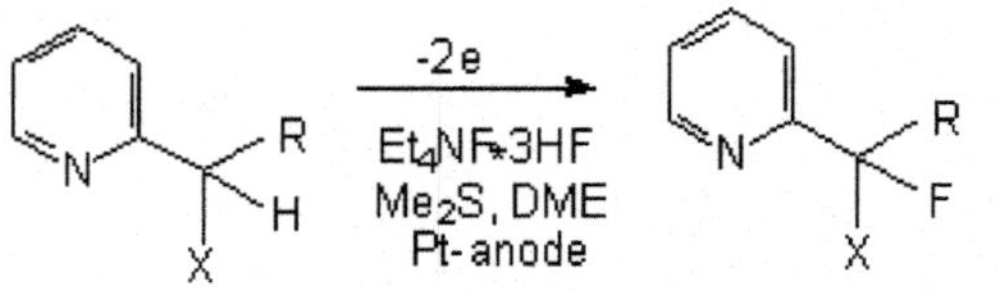

1.70 B, X = PhS, R = COOEt (55 %)
2.00 B, X = PhS, R = CN (56 %)

Table: Electro-chemical oxidization of caffeine and similar compounds (MeCN, Et$_3$N * 3HF).

Substrate	Oxidation potential B v SSCE	Time, h	Product	Yield, %
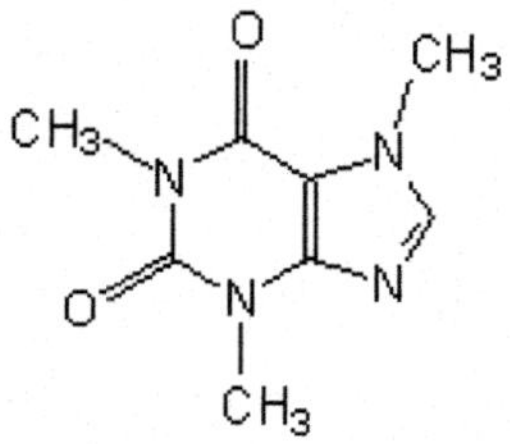	5	17	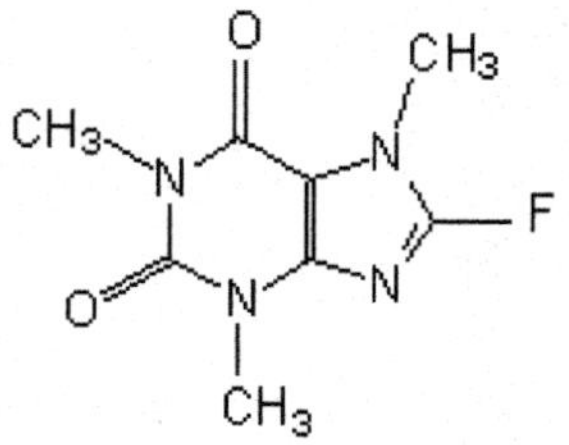	40.3
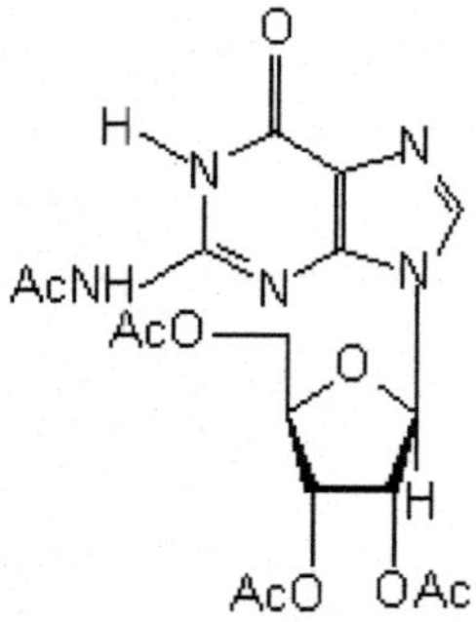	3	10	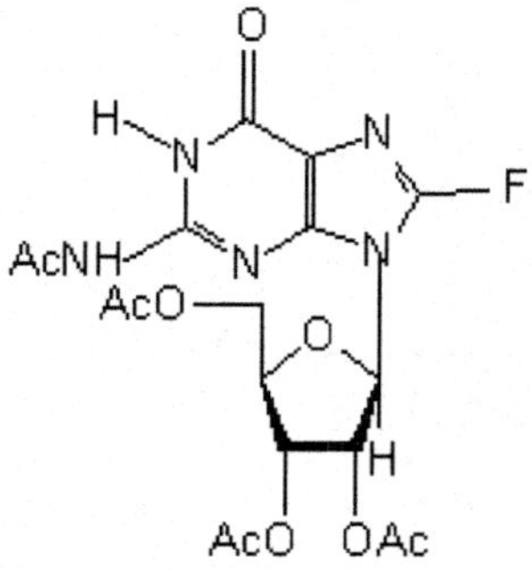	6.3
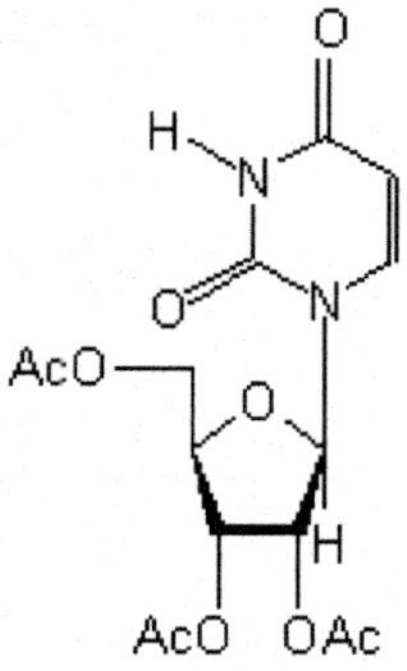	C.C.E. 100 мА	5.5	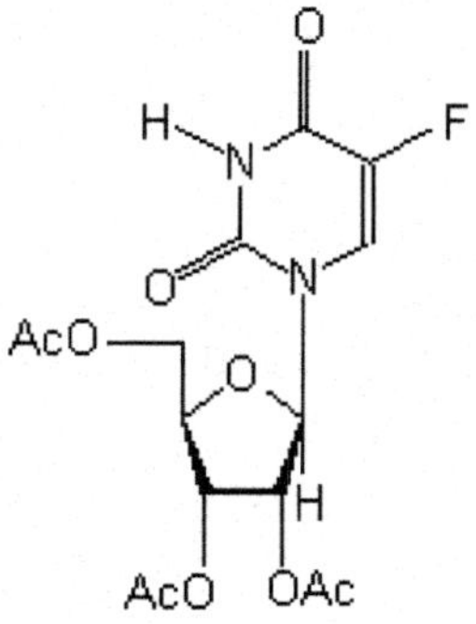	4.6

If in benzene ring there are electron-donating substituents, then during anodic fluorination active methene group of substituent will be affected mostly.

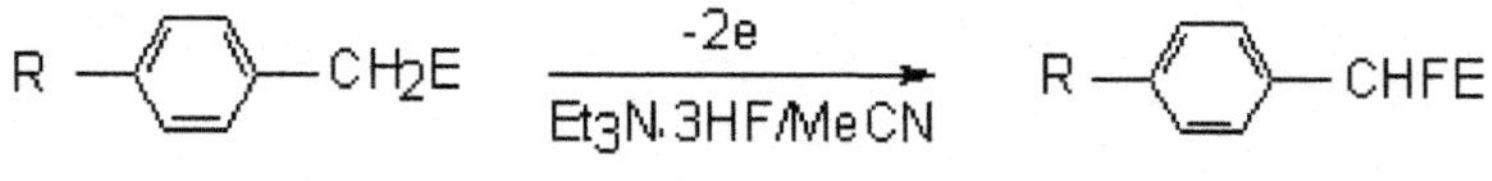

$$R \longrightarrow \hspace{-0.5em}\bigcirc\hspace{-0.5em} -CH_2E \xrightarrow[\text{Et}_3\text{N}\cdot 3\text{HF/MeCN}]{-2e} R \longrightarrow \hspace{-0.5em}\bigcirc\hspace{-0.5em} -CHFE$$

R	E	Potential, V	Conversion	Yield, %
MeO	COMe	1,2	100	72
MeO	COC_6H_4OMe	1,2	99	72
Cl	COOEt	1,7	75	36
MeO	COOEt	1,28	100	69
3,4(MeO)$_2$	COOEt	0,8	92	73
CH$_2$=CHCH$_2$O	COOEt	1,49	100	51
H	CN	2,33	44	22
MeO	SO$_2$Et	1,37	96	71
Cl	CN	1,88	65	64
MeO	CN	1,37	97	67

It is shown, that the reaction goes *via* initial generation of cation-radical of substituent atom neighbouring to-CH$_2$-fragment with following fluoride-ion attack of cationoid centre of intermediate carbcation, which is in α-position to carbonyl group.

Anodic fluorination of sulfides in Et$_3$N*3HF medium results in formation of mono-fluoro-and hem-difluoro-derivatives. At that elimination of SPh group occurs.

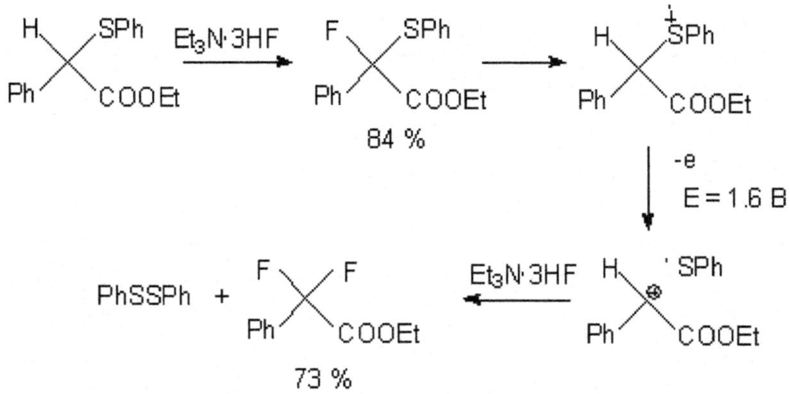

84 %

-e

E = 1.6 B

PhSSPh +

73 %

Anodic oxidization of benzylalkyl ketones, carboxylic acids esters, nitriles, benzyl derivatives in the system Et$_3$N*3HF/MeCN at platinum anode enables to obtain mono-fluoro-derivatives, some of which are biologically active.

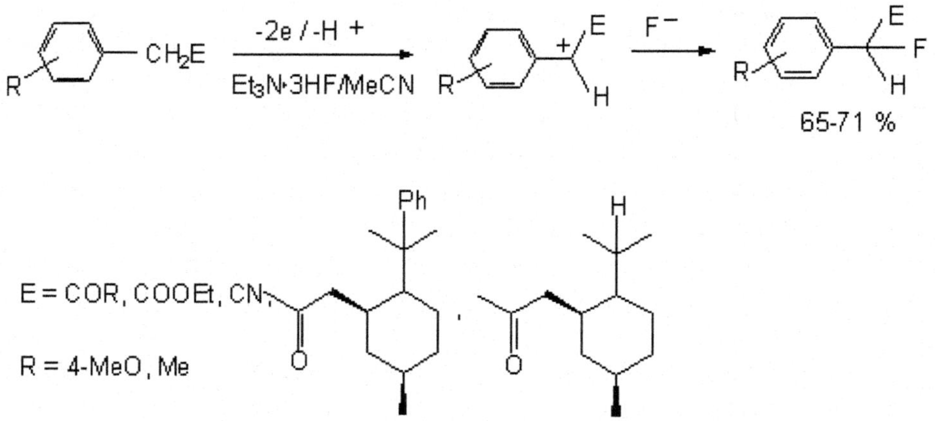

$$R-\text{C}_6\text{H}_4-\text{CH}_2\text{E} \xrightarrow[\text{Et}_3\text{N·3HF/MeCN}]{-2e/-\text{H}^+} R-\text{C}_6\text{H}_4-\overset{+}{\underset{H}{\text{C}}}(E) \xrightarrow{F^-} R-\text{C}_6\text{H}_4-\underset{H}{\overset{E}{\text{C}}}F$$

65-71 %

E = COR, COOEt, CN

R = 4-MeO, Me

Substituent in benzene ring influences greatly on the ratio of resulting products [91].

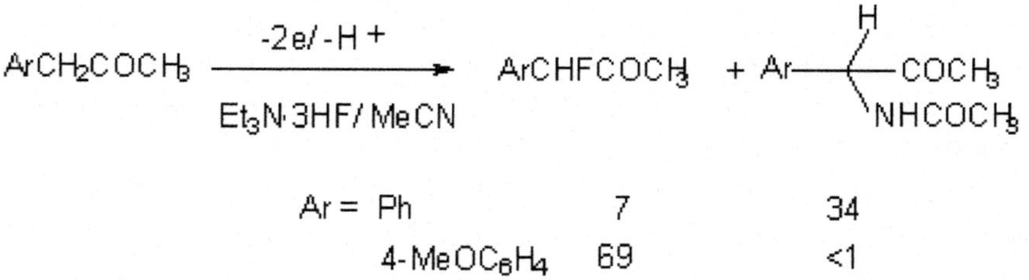

$$\text{ArCH}_2\text{COCH}_3 \xrightarrow[\text{Et}_3\text{N·3HF/MeCN}]{-2e/-\text{H}^+} \text{ArCHFCOCH}_3 + \text{Ar}-\underset{\text{NHCOCH}_3}{\overset{H}{\text{C}}}-\text{COCH}_3$$

| Ar = Ph | 7 | 34 |
| 4-MeOC₆H₄ | 69 | <1 |

At the same time electrophilic substituents in α-position to methene group have a light influence on the anodic fluorination process, while the increasing of voltage results in increasing of difluoro-derivative yield.

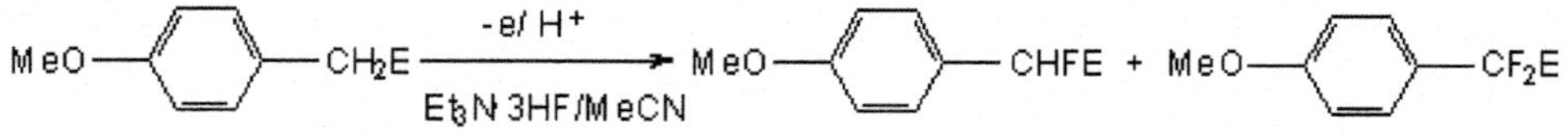

$$\text{MeO}-\text{C}_6\text{H}_4-\text{CH}_2\text{E} \xrightarrow[\text{Et}_3\text{N·3HF/MeCN}]{-e/H^+} \text{MeO}-\text{C}_6\text{H}_4-\text{CHFE} + \text{MeO}-\text{C}_6\text{H}_4-\text{CF}_2\text{E}$$

E	Voltage, V	Yield, %	Voltage, V	Yield, %
CH₃CO	1,2	69	1,6	61
4-MeOC₆H₄CO	1,2	71	1,55	50
COOEt	1,28	69	1,45	55
CN	1,35	65	1,60	50

The presence of methyl group in benzene ring results in its fluorination. Thus, p-methylbenzylsulfonate during anodic fluorination in Et₃N*3HF/MeCN system forms the mixture of electrolysis products.

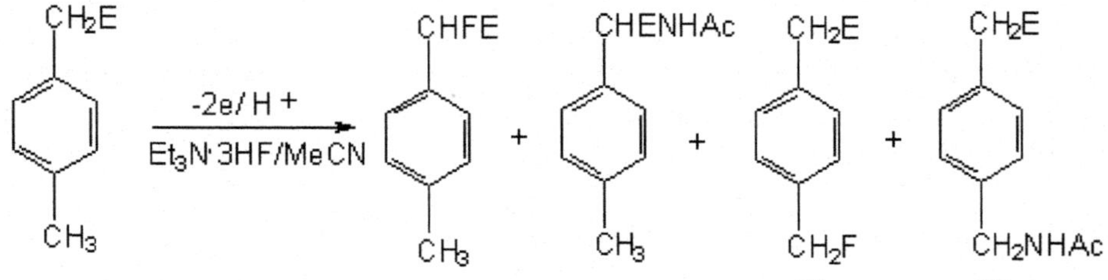

E	Voltage, V	Yield, %			
		27	28	29	30
COOEt	1,65	40	14	8	18
CN	1,73	58	<1	3	6
SO$_3$Et	1,78	15	0	21	30
H	1,80			10	25

Electro-chemical oxidization of aliphatic aldehydes in Et$_3$N*5HF or Py* nHF (n = 3-6) in acetonitrile or sulpholane results in formation of corresponding carboxylic acids fluoroanhydrides with high yield.

2,2-disubstituted cyclic ketones in Et$_4$N*5H-F at anodic fluorination produce fluoro-anhydrides of appropriate fluoro-containing carboxylic acids due to selective splitting of Ñ-Ñ bond between carbonyl atom of carbon and α-substituted carbon atom (table). Thus, electrolysis of 2,2-dimethylcyclohexanone in electrolyte Et$_4$N*5HF at 0°C with subsequent addition of methyl alcohol produces methyl 6-fluoro-6-methylheptanoic acid ester.

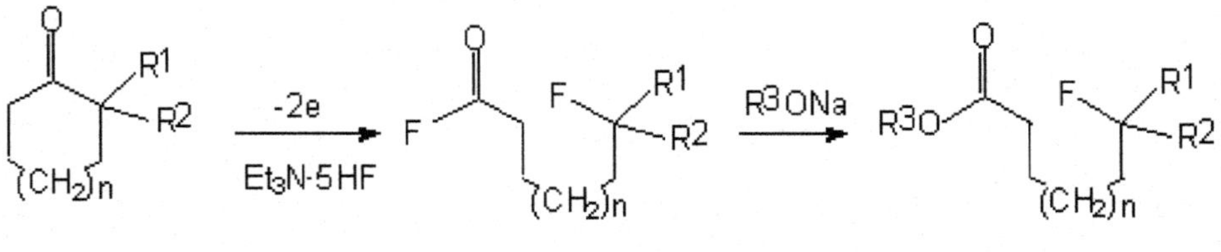

Probably, the reaction path is the following :

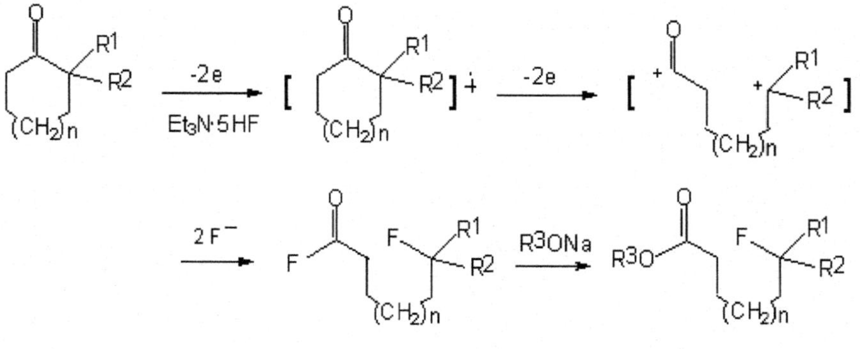

At the same time anodic fluorination of unsaturated carboxylic acids esters having cylcopentane and cyclohexane fragments in Et$_3$N*5HF system results in forming of expanded by one carbon atom cycle. 2-2-Difluorocycloalkanecarboxylic acids esters with high selectivity and good yield are produced.

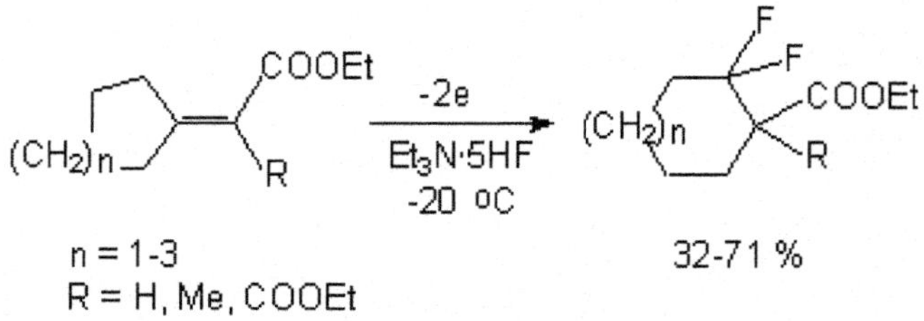

n = 1-3
R = H, Me, COOEt

32-71 %

Anodic fluorination of different N-substituted lactams using electrolyte Et$_3$N*3HF results in formation of corresponding mono-fluoro-containing products, in which fluorine atom is in α-position to nitrogen atom.

During electrolysis of α, β-unsaturated esters the hem-β,β-difluoro-derivatives of esters were formed in electrolyte due to re-arrangement of alkyl group from α to β-position. If alkenyl group is in this position, for example in case of carboxylic acids esters of dienes, then the fluorination with formation of vicinal difluoroderivatives will occur.

Table : Anodic fluorination of dienes esters.

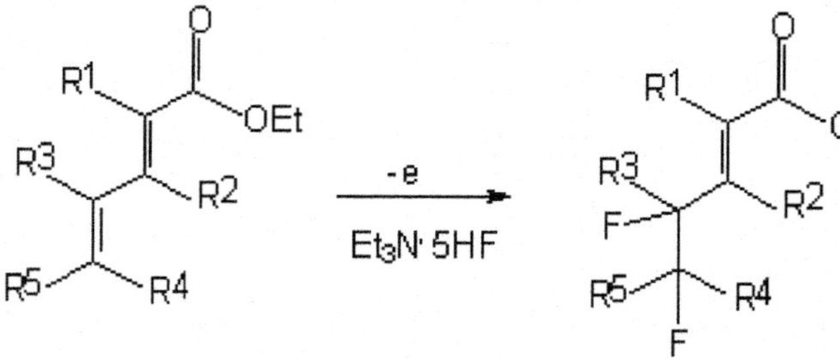

R^1	R^2	R^3	R^4	R^5	Electrolyte	F/mol	Potential of anode, V	Yield,%	Isomer ratio
H	H	H	Me	Me	Et$_3$N*5HF	2,5	1,6	73	–
H	H	H	H	Ph	Et$_4$NF*2HF	2,5	1,4	35	2,5 :1
H	H	H	H	Me	Et$_3$N*5HF	2,5	1,8	69	2 :1
H	H	H	H	n-C$_6$H$_{13}$	Et$_3$N*5HF	2,5	1,8	53	5 :1
H	H	Me	H	Me	Et$_3$N*5HF	2,5	1,6	61	1,5 :1
H	Me	H	Me	Me	Et$_3$N*5HF	2,5	1,6	16	–
H	Me	H	Me	Me	Et$_3$N*3HF	8	1,6	45	–
Me	H	H	Me	Me	Et$_3$N*5HF	2,5	1,25	86	
COOEt	H	H	Me	Me	Et$_3$N*5HF	2,5	1,6	85	3 :1
COOEt	H	H	Me	Me	Et$_3$N*2HF	2,5	1,6	4	3 :1

Table : Synthesis of carboxylic acids esters using electro-synthesis (electrolyte Et₃N*5HF, 0°C) of cyclic ketones.

Ketone	V,vs Ag/Ag+	F /mol	Product	Yield,%
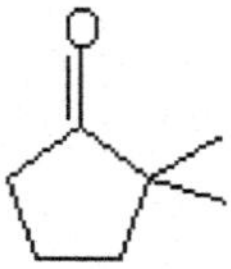	2.0	2.5		80
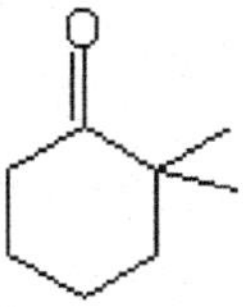	2.0	3.1	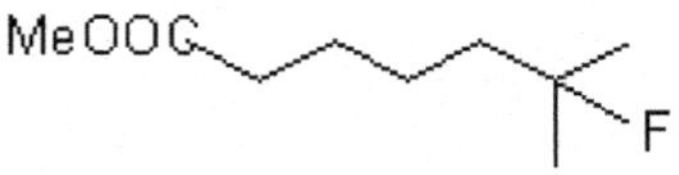	81
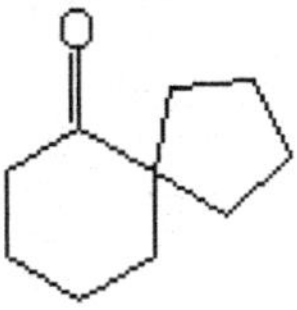	2.2	3.0	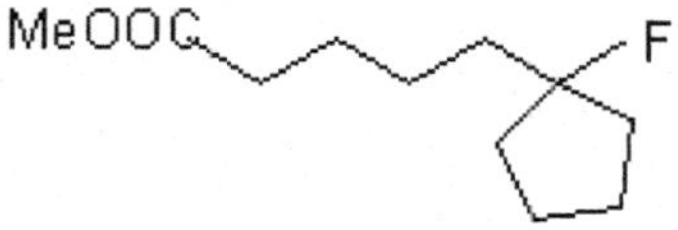	82
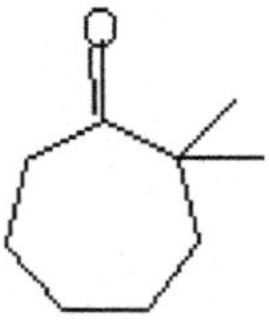	2.3	3.5	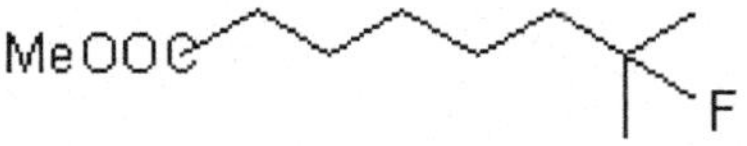	81
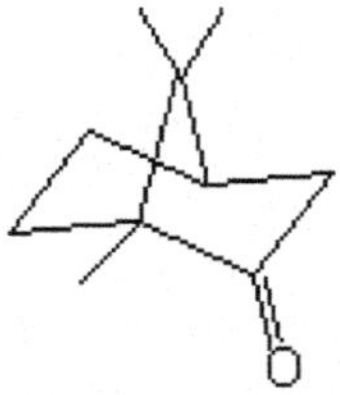	2.2	4.0	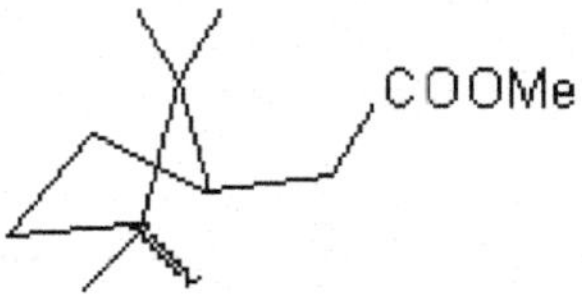	60
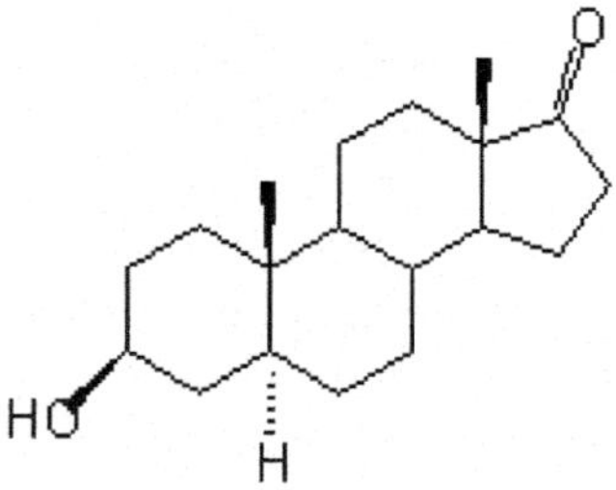	2.0	8.0	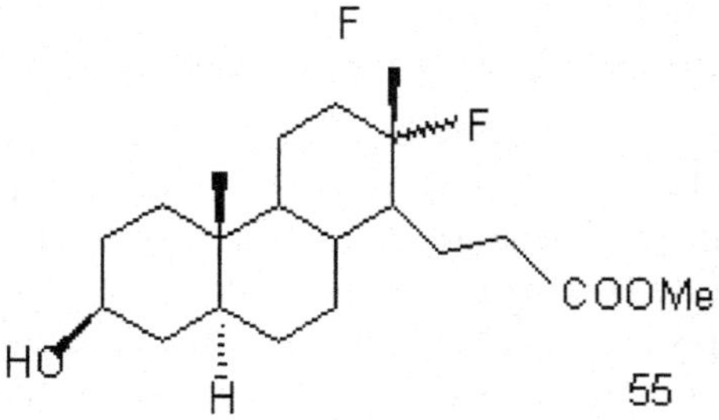	55

Table : Electro-chemical oxidization of unsaturated cyclic esters.

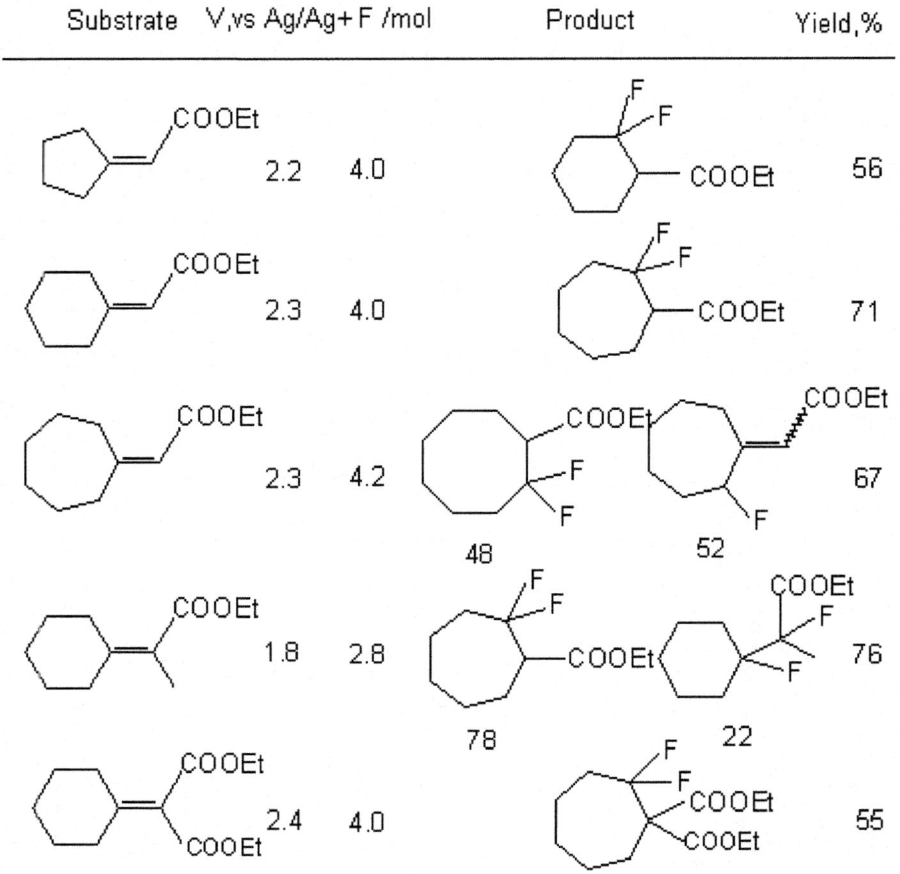

Substrate	V,vs Ag/Ag+	F /mol	Product	Yield,%

In case of compounds, containing two carb-ethoxy groups at multiple bonds of diene and electron-donating substituents (for example, methyl groups), anodic fluorination produces mixture of hem-difluoro-and vicinal difluoro-derivatives (ratio 3 : 1). This shows the realization of two fluorination processes pathways.

Olah's reagent (HF/Py) is most widely used as electrolyte, and methylene chloride is used as solvent. Another path is connected with use of Et₃N*3HF as electrolyte, and acetonitrile as solvent. Carbonyl compounds, containing enol form of-CH₂-fragment in α-position are fluorinated with high selectivity into α-position, producing as a rule monofluoroderivatives. It is important, that phenyl substituent should be at carbonyl group or this fragment should be in cyclic system. This circumstance results in the fact that anodic fluorination is widely used for selective introduction of fluorine atom into compounds, having electron-seeking groups.

Chapter 13

CONDUCTIVE POLYMER

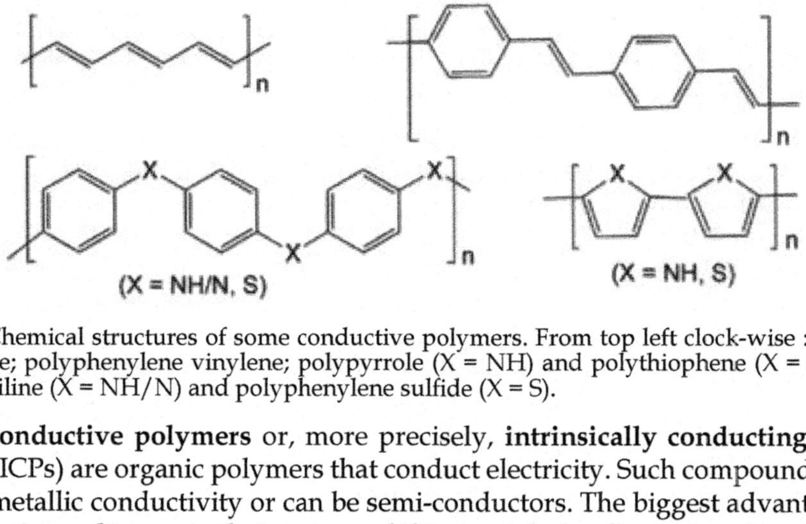

Fig. : Chemical structures of some conductive polymers. From top left clock-wise : polya-cetylene; polyphenylene vinylene; polypyrrole (X = NH) and polythiophene (X = S); and polyaniline (X = NH/N) and polyphenylene sulfide (X = S).

Conductive polymers or, more precisely, **intrinsically conducting polymers** (ICPs) are organic polymers that conduct electricity. Such compounds may have metallic conductivity or can be semi-conductors. The biggest advantage of conductive polymers is their processability, mainly by dispersion. Conductive polymers are generally not thermoplastics, *i.e.*, they are not thermoformable. But, like insulating polymers, they are organic materials. They can offer high electrical conductivity but do not show similar mechanical properties to other commercially available polymers. The electrical properties can be fine-tuned using the methods of organic synthesis and by advanced dispersion techniques.

HISTORY

Polyaniline was first described in the mid-19th century by Henry Letheby, who investigated the electro-chemical and chemical oxidation products of aniline in acidic media. He noted that reduced form was colourless but the oxidized forms were deep blue.

The first highly-conductive organic compounds were the charge transfer complexes. In the 1950s, researchers reported that polycyclic aromatic compounds formed semi-conducting charge-transfer complex salts with halogens. In 1954, researchers at Bell Labs and elsewhere reported organic charge transfer complexes with resistivities as low as 8 ohms-cm. In the early 1970s, researchers demonstrated salts of tetra-thiafulvalene show almost metallic conductivity, while super-conductivity was demonstrated in 1980. Broad research on charge transfer salts continues today. While these compounds were technically not polymers, this indicated that organic compounds can carry current. While organic conductors were previously intermittently discussed, the field was particularly energized by the prediction of super-conductivity following the discovery of BCS theory.

In 1963 Australians B.A. Bolto, D.E. Weiss, and co-workers reported derivatives of polypyrrole with resistivities as low as 1 ohm cm. cites multiple reports of similar high-conductivity oxidized polyacetylenes. With the notable exception of charge transfer complexes (some of which are even super-conductors), organic molecules were previously considered insulators or at best weakly conducting semi-conductors. Subsequently, DeSurville and co-workers reported high conductivity in a polyaniline. Likewise, in 1980, Diaz and Logan reported films of polyaniline that can serve as electrodes.

While mostly operating in the quantum realm of less than 100 nanometers, "molecular" electronic processes can collectively manifest on a macro scale. Examples include quantum tunneling, negative resistance, phonon-assisted hopping and polarons. In 1977, Alan J. Heeger, Alan MacDiarmid and Hideki Shirakawa reported similar high conductivity in oxidized iodine-doped polyacetylene. For this research, they were awarded the 2000 Nobel Prize in Chemistry *"for the discovery and development of conductive polymers."* Polyacetylene itself did not find practical applications, but drew the attention of scientists and encouraged the rapid growth of the field. Since the late 1980s, organic light-emitting diodes (OLEDs) have emerged as an important application of conducting polymers.

TYPES

The linear-backbone "polymer blacks" (polyacetylene, polypyrrole, and polyaniline) and their copolymers are the main class of conductive polymers. Poly(p-phenylene vinylene) (PPV) and its soluble derivatives have emerged as the prototypical electro-luminescent semi-conducting polymers. Today, poly(3-alkylthiophenes) are the archetypical materials for solar cells and transistors.

The following table presents some organic conductive polymers according to their composition. **The well-studied classes are written in bold** and *the less well studied ones are in italic.*

The main chain contains	Heteroatoms present		
	No heteroatom	Nitrogen-containing	Sulfur-containing
Aromatic cycles	• Poly(fluorene)s • polyphenylenes • polypyrenes • polyazulenes • polynaphthalenes	The N is in the aromatic cycle : • Poly(pyrrole)s (PPY) • Polycarbazoles • Polyindoles • Polyazepines The N is outside the aromatic cycle : • polyanilines (PANI)	The S is in the aromatic cycle : • Poly(thiophene)s (PT) • Poly(3,4-ethylenedioxythioPhene) (PEDOT) The S is outside the aromatic cycle : • Poly(p-phenylene sulfide) (PPS)
Double bonds	• Poly(acetylene)s (PAC)		
Aromatic cycles and double bonds	• Poly(p-phenylene vinylene) (PPV)		

SYNTHESIS

Conductive polymers are prepared by many methods. Most conductive polymers are prepared by oxidative coupling of monocyclic precursors. Such reactions entail dehydrogenation :

$$n \, H\text{-}[X]\text{-}H \rightarrow H\text{-}[X]_n\text{-}H + 2(n\text{-}1) \, H^+ + 2(n\text{-}1) \, e^-$$

The low solubility of most polymers presents challenges. Some researchers have addressed this through the formation of nanostructures and surfactant-stabilized conducting polymer dispersions in water. These include polyaniline nanofibers and PEDOT : PSS. These materials have lower molecular weights than that of some materials previously explored in the literature. However, in some cases, the molecular weight need not be high to achieve the desired properties.

MOLECULAR BASIS OF ELECTRICAL CONDUCTIVITY

The conductivity of such polymers is the result of several processes. For example, in traditional polymers such as polyethylenes, the valence electrons are bound in sp^3 hybridized covalent bonds. Such "sigma-bonding electrons" have low mobility and do not contribute to the electrical conductivity of the material. However, in conjugated materials, the situation is completely different. Conducting polymers have backbones of contiguous sp^2 hybridized carbon centers. One valence electron on each center resides in a p_z orbital, which is orthogonal to the other three sigma-bonds. All the p_z orbitals combine with each other to a molecule wide delocalized set of orbitals. The electrons in these delocalized orbitals have high mobility when the material is "doped" by oxidation, which removes some of these delocalized electrons. Thus, the conjugated p-orbitals form a one-dimensional electronic band, and the electrons within this band become mobile when it is partially emptied.

The band structures of conductive polymers can easily be calculated with a tight binding model. In principle, these same materials can be doped by reduction, which adds electrons to an otherwise unfilled band. In practice, most organic conductors are doped oxidatively to give p-type materials. The redox doping of organic conductors is analogous to the doping of silicon semi-conductors, whereby a small fraction silicon atoms are replaced by electron-rich, *e.g.*, phosphorus, or electron-poor, *e.g.*, boron, atoms to create n-type and p-type semi-conductors, respectively.

Although typically "doping" conductive polymers involves oxidizing or reducing the material, conductive organic polymers associated with a protic solvent may also be "self-doped."

Undoped conjugated polymers state are semi-conductors or insulators. In such compounds, the energy gap can be > 2 eV, which is too great for thermally activated conduction. Therefore, undoped conjugated polymers, such as polythiophenes, polyacetylenes only have a low electrical conductivity of around 10^{-10} to 10^{-8} S/cm. Even at a very low level of doping (< 1%), electrical conductivity increases several orders of magnitude up to values of around 0.1 S/cm. Subsequent doping of the conducting polymers will result in a saturation of the conductivity at values around 0.1–10 kS/cm for different polymers. Highest values reported up to now are for the conductivity of stretch oriented polyacetylene with confirmed values of about 80 kS/cm. Although the pi-electrons in polyactetylene are delocalized along the chain, pristine polyacetylene is not a metal. Polyacetylene has alternating single and double bonds which have lengths of 1.44 and 1.36 Å, respectively. Upon doping, the bond alteration is diminished in conductivity increases. Non-doping increases in conductivity can also be accomplished in a field effect transistor (organic FET or OFET) and by irradiation. Some materials also exhibit negative differential resistance and voltage-controlled "switching" analogous to that seen in inorganic amorphous semi-conductors.

Despite intensive research, the relationship between morphology, chain structure and conductivity is still poorly understood. Generally, it is assumed that conductivity should be higher for the higher degree of crystallinity and better alignment of the chains, however this could not be confirmed for polyaniline and was only recently confirmed for PEDOT, which are largely amorphous.

PROPERTIES AND APPLICATIONS

Due to their poor processability, conductive polymers have few large-scale applications. They have promise in antistatic materials and they have been incorporated into commercial displays and batteries, but there have had limitations due to the manufacturing costs, material inconsistencies, toxicity, poor solubility in solvents, and inability to directly melt process. Literature suggests they are also promising in organic solar cells, printing electronic circuits, organic light-emitting diodes, actuators, electrochromism, super-capacitors, chemical sensors and biosensors, flexible transparent displays, electromagnetic shielding and possibly replacement for the popular transparent conductor indium tin oxide. Another use is for microwave-absorbent coatings, particularly radar-absorptive coatings on stealth aircraft. Conducting polymers are rapidly gaining attraction in new applications with increasingly processable materials with better electrical and physical properties and lower costs. The new nanostructured forms of conducting polymers particularly, augment this field with their higher surface area and better dispersability.

With the availability of stable and reproducible dispersions, PEDOT and polyaniline have gained some large scale applications. While PEDOT (poly(3,4-ethylenedioxythiophene)) is mainly used in antistatic applications and as a transparent conductive layer in form of PEDOT :PSS dispersions (PSS=polystyrene sulfonic acid), polyaniline is widely used for printed circuit board manufacturing–in the final finish, for protecting copper from corrosion and preventing its solderability.

ELECTRO-LUMINESCENCE

Electro-luminescence is light emission stimulated by electrical current. In organic compounds, electro-luminescence has been known since the early 1950s, when Bernanose and co-workers first produced electro-luminescence in crystalline thin films of acridine orange and quinacrine. In 1960, researchers at Dow Chemical developed AC-driven electroluminescent cells using doping. In some cases, similar light emission is observed when a voltage is applied to a thin layer of a conductive organic polymer film. While electro-luminescence was originally mostly of academic interest, the increased conductivity of modern conductive polymers means enough power can be put through the device at low voltages to generate practical amounts of light. This property has led to the development of flat panel displays using organic LEDs, solar panels, and optical amplifiers.

BARRIERS TO APPLICATIONS

Since most conductive polymers require oxidative doping, the properties of the resulting state are crucial. Such materials are salt-like (polymer salt), which diminishes their solubility in organic solvents and water and hence their processability. Furthermore, the charged organic backbone is often unstable towards atmospheric moisture. The poor processability for many polymers requires the introduction of solubilizing or substituents, which can further complicate the synthesis.

Experimental and theoretical thermodynamical evidence suggests that conductive polymers may even be completely and principally insoluble so that they can only be processed by dispersion.

TRENDS

Most recent emphasis is on organic light emitting diodes and organic polymer solar cells. The Organic Electronics Association is an international platform to promote applications of organic semi-conductors. Conductive polymer products with embedded and improved electromagnetic interference (EMI) and electrostatic discharge (ESD) protection have led to both prototypes and products. For example,

Polymer Electronics Research Center at University of Auckland is developing a range of novel DNA sensor technologies based on conducting polymers, photoluminescent polymers and inorganic nanocrystals (quantum dots) for simple, rapid and sensitive gene detection. Typical conductive polymers must be "doped" to produce high conductivity. As of 2001, there remains to be discovered an organic polymer that is *intrinsically* electrically conducting.

POLYANILINE

Polyaniline (PANI) is a conducting polymer of the semi-flexible rod polymer family. Although the compound itself was discovered over 150 years ago, only since the early 1980s has polyaniline captured the intense attention of the scientific community. This interest is due to the rediscovery of high electrical conductivity. Amongst the family of conducting polymers and organic semi-conductors, polyaniline has many attractive processing properties. Because of its rich chemistry, polyaniline is one of the most studied conducting polymers of the past 50 years.

History

As described by Alan MacDiarmid, the first definitive report of polyaniline did not occur until 1862, which included an electro-chemical method for the determination of small quantities of aniline.

From the early 20th century on, occasional reports about the structure of PANI were published. Subsequent to his investigation of other highly-conductive organic materials, MacDiarmid demonstrated the conductive states of polyaniline which arose upon protonic doping of the emeradine form of polyaniline. Conductive polymers such as polyaniline remain of widespread interest, providing an opportunity to address fundamental issues of importance to condensed matter physics, including, for example, the metal-insulator transition, the Peierls Instability and quantum decoherence.

Synthesis and Properties

Fig. : Main polyaniline structures n+m = 1, x = half degree of polymerization.

Polymerized from the inexpensive aniline monomer, polyaniline can be found in one of three idealized oxidation states :

- **Leucoemeraldine**–white/clear & colourless $(C_6H_4NH)_n$
- **Emeraldine**–green for the emeraldine salt, blue for the emeraldine base $([C_6H_4NH]_2[C_6H_4N]_2)_n$
- **(per)nigraniline**–blue/violet $(C_6H_4N)_n$

In figure, x equals half the degree of polymerization (DP). Leucoemeraldine with n = 1, m = 0 is the fully reduced state. Pernigraniline is the fully oxidized state (n = 0, m = 1) with imine links instead of amine links. Studies have shown that most forms of polyaniline are one of the three states or physical mixtures of these components. The emeraldine (n = m = 0.5) form of polyaniline, often referred to as emeraldine base (EB), is neutral, if doped (protonated) it is called emeraldine salt (ES), with the imine nitrogens protonated by an acid. Protonation helps to delocalize the otherwise trapped diiminoquinone-diaminobenzene state. Emeraldine base is regarded as the most useful form of polyaniline due to its high stability at room temperature and the fact that, upon doping with acid, the resulting emeraldine salt form of polyaniline is highly electrically conducting. Leucoemeraldine and pernigraniline are poor conductors, even when doped with an acid.

The colour change associated with polyaniline in different oxidation states can be used in sensors and electro-chromic devices. Although colour is useful, the best method for making a polyaniline sensor is arguably to take advantage of the dramatic changes in electrical conductivity between the different oxidation states or doping levels. Treatment of emeraldine with acids increases the electrical conductivity by ten orders of magnitude. Undoped polyaniline has a conductivity of 6.28×10^{-9} S/m, while conductivities of 4.60×10^{-5} S/m can be achieved by doping to 4% HBr. The same material can be prepared by oxidation of leucoemeraldine.

Polyaniline is more noble than copper and slightly less noble than silver which is the basis for its broad use in printed circuit board manufacturing (as a final finish) and in corrosion protection.

Synthesis

Although the synthetic methods to produce polyaniline are quite simple, the mechanism of polymerization is probably complex. The formation of leucoemeraldine can be described as follows, where [O] is a generic oxidant :

$$n\ C_6H_5NH_2 + [O] \rightarrow [C_6H_4NH]_n + H_2O$$

The most common oxidant is ammonium persulfate. The components are each dissolved in 1 M hydrochloric acid (other acids can be used), and the two solutions slowly combined. The reaction is very exothermic. The polymer precipitates as an unstable dispersion with micrometer-scale particulates is prepared by oxidation of the emeraldine base, one typical oxidant being meta-chloroperoxybenzoic acid :

$$\{[C_6H_4NH]_2[C_6H_4N]_2\}_n + RCO_3H \rightarrow [C_6H_4N]_n + H_2O + RCO_2H$$

Processing

The synthesis of polyaniline nanostructures is facile.

Using special polymerisation procedures and surfactant dopants, the obtained polyaniline powder can be made dispersible and hence useful for practical applications. Bulk synthesis of polyaniline nanofibers has led to a highly scalable and commercially applicable form of polyaniline that has been researched extensively since their discovery in 2002.

A multi-stage model for the formation of emeraldine base is proposed. In the first stage of the reaction the pernigraniline PS salt oxidation state is formed. In the second stage pernigraniline is reduced to the emeraldine salt as aniline monomer gets oxidized to the radical cation. In the third stage this radical cation couples with ES salt. This process can be followed by light scattering analysis which allows the determination of the absolute molar mass. According to one study in the first step a DP of 265 is reached with the DP of the final polymer at 319. Approximately 19% of the final polymer is made up of the aniline radical cation which is formed during the reaction.

Polyaniline is typically produced in the form of long-chain polymer aggregates, surfactant (or dopant) stabilized nanoparticle dispersions, or stabilizer-free nanofiber dispersions depending on the supplier and synthetic route. Surfactant or dopant stabilized polyaniline dispersions have been available for commercial sale since the late 1990s.

Applications

Polyaniline and the other conducting polymers such as polythiophene, polypyrrole, and PEDOT/PSS have potential for applications due to their light weight, conductivity, mechanical flexibility and low cost. Polyaniline is especially attractive because it is relatively inexpensive, has three distinct oxidation states with different colours and has an acid/base doping response. This latter property makes polyaniline an attractive for acid/base chemical vapour sensors, super-capacitors and bio-sensors. The different colours, charges and conformations of the multiple oxidation states also make the material promising for applications such as actuators, super-capacitors and electro-chromics. They are suitable for manufacture of electrically conducting yarns, antistatic coatings, electromagnetic shielding, and flexible electrodes.

Attractive fields for current and potential utilization of polyaniline is in antistatics, charge dissipation or electrostatic dispersive (ESD) coatings and blends, electromagnetic interference shielding (EMI), anti-corrosive coatings, hole injection layers, transparent conductors, indium tin oxide replacements, actuators, chemical vapour and solution based sensors, electro-chromic coatings (for colour change windows, mirrors etc.), PEDOT-PSS replacements, toxic metal recovery, catalysis, fuel cells and active electronic components such as for non-volatile memory.

Currently, the major applications are printed circuit board manufacturing (final finishes, used in millions of m^2 every year), antistatic and ESD coatings, and corrosion protection.

POLYDIACETYLENES

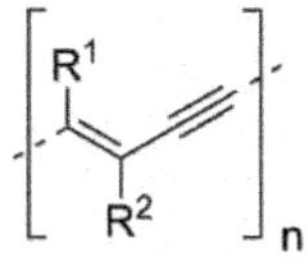

Fig. : General chemical structure of a polydiacetylene.

Polydiacetylenes (PDAs) are a family of conducting polymers. They are created by the photo-polymerization of diacetylene moieties. They have multiple applications from the development of organic films to immobilization of other molecules.

POLYDIOCTYLFLUORENE

Polydioctylfluorene (PDF) is an organic compound, a polymer of 9,9-dioctyl-fluorene, with formula $(C_{13}H_6(C_8H_{17})_2)_n$. It is an electro-luminescent conductive polymer that characteristically emits blue light. Like other polyfluorene polymers, it has been studied as a possible material for light-emitting diodes.

Structure

The monomer has an aromatic fluorene core-$C_{13}H_6$-with two aliphatic *n*-octyl-C_8H_{17} tails attached to the central carbon.

POLYPYRROLE

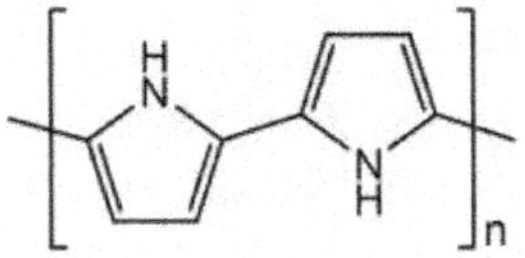

Fig. : Polypyrrole.

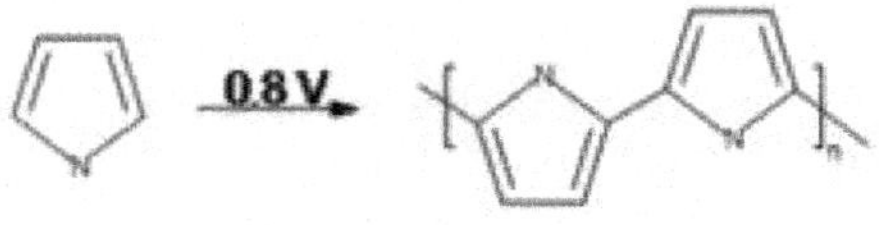

Fig. : Pyrrole can be polymerised electro-chemically.

Polypyrrole (PPy) is a type of organic polymer formed from by polymerization of pyrrole. Polypyrroles are conducting polymers, related members being polythiophene, polyaniline, and polyacetylene. The Nobel Prize in Chemistry was awarded in 2000 for work on conductive polymers including polypyrrole.

Synthesis

Some of the first examples of polypyrroles were reported in 1963 by Weiss and co-workers. These workers described the pyrolysis of tetra-iodopyrrole to produce highly conductive materials. Most commonly Ppy is prepared by oxidation of pyrrole, which can be achieved using ferric chloride in methanol :

$$n \; C_4H_4NH + 2 \; FeCl_3 \rightarrow (C_4H_2NH)_n + 2 \; FeCl_2 + 2 \; HCl$$

Polymerization is thought to occur *via* the formation of the pi-radical cation $C_4H_4NH^+$. This electrophile attacks the C-2 carbon of an unoxidized molecule of pyrrole to give a dimeric radical $(C_4H_4NH)_2]^+$. The process repeats itself many times.

Conductive forms of PPy are prepared by oxidation ("p-doping") of the polymer :

$$(C_4H_2NH)_n + x \; FeCl_3 \rightarrow (C_4H_2NH)_nCl_x + x \; FeCl_2$$

The polymerization and p-doping can also be affected electro-chemically. The resulting conductive polymer are peeled off of the anode.

Properties

Films of PPy are yellow but darken in air due to some oxidation. Doped films are blue or black depending on the degree of polymerization and film thickness. They are amorphous, showing only weak diffraction. PPy is described as "quasi-unidimensional" *vs* one-dimensional since there is some cross-linking and chain hopping. Undoped and doped films are insoluble in solvents but swellable. Doping makes the materials brittle. They are stable in air up to 150°C at which temperature the dopant starts to evolve (*e.g.*, as HCl).

PPy is an insulator, but its oxidized derivatives are good electrical conductors. The conductivity of the material depends on the conditions and reagents used in the oxidation. Conductivities range from 2 to 100 S/cm. Higher conductivities are associated with larger anions, such as tosylate. Doping the polymer requires that the material swell to accommodate the charge-compensating anions. The physical changes associated with this charging and discharging has been discussed as a form of artificial muscle.

Applications

PPy and related conductive polymers have two main application in electronic devices and for chemical sensors. PPy is also potential vehicle for drug delivery. The polymer matrix serves as a container for proteins.

Research Trends

Polypyrrole is also being investigated in low temperature fuel cell technology to increase the catalyst dispersion in the carbon support layers and to sensitize cathode electro-catalysts, as it has been inferred that the metal electro-catalysts (Pt, Co, etc.) when co-ordinated with the nitrogen in the pyrrole monomers show enhanced oxygen reduction activity.

Polypyrrole (together with other conjugated polymers such as polyaniline, poly (ethylenedioxythiophene) etc.) has been actively studied as a material for "artificial muscles", a technology that would offer numerous advantages over traditional motor actuating elements.

Polypyrrole was used to coat silica and reverse phase silica to yield a material capable of anion exchange and exhibiting hydrophobic interactions.

Polypyrrole was used in the microwave fabrication of multi-walled carbon nanotubes, a new method that allows to obtain CNTs in a matter of seconds.

Chemical and Engineering News reported in June 2013 that Chinese research has produced a water-resistant polyurethane sponge coated with a thin layer of polypyrrole that absorbs 20 times its weight in oil and is reusable.

Chapter 14

ELECTRO-CHEMICAL TECHNIQUES

CYCLIC VOLTAMMETRY

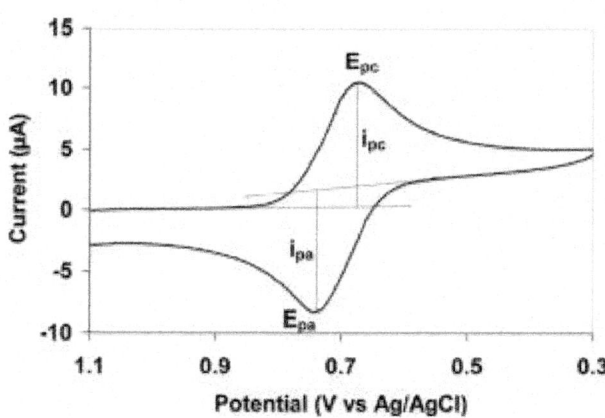

Fig. : Typical cyclic voltammogram where i_{pc} and i_{pa} show the peak cathodic and anodic current respectively for a reversible reaction.

Cyclic voltammetry or CV is a type of potentiodynamic electro-chemical measurement. In a cyclic voltammetry experiment the working electrode potential is ramped linearly *versus* time like linear sweep voltammetry. Cyclic voltammetry takes the experiment a step further than linear sweep voltammetry which ends when it reaches a set potential. When cyclic voltammetry reaches a set potential, the working electrode's potential ramp is inverted. This inversion can happen multiple times during a single experiment. The current at the working electrode is plotted versus the applied voltage to give the cyclic voltammogram trace. Cyclic voltammetry is generally used to study the electro-chemical properties of an analyte in solution.

Experimental Method

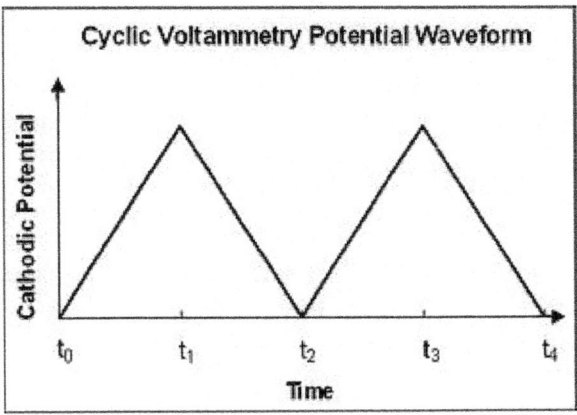

Fig. : Cyclic voltammetry waveform.

In cyclic voltammetry, the electrode potential ramps linearly *versus* time as shown. This ramping is known as the experiment's scan rate (V/s). The potential is applied between the reference electrode and the working electrode and the current is measured between the working electrode and the counter electrode. These data are then plotted as current (i) *vs.* potential (E). As the waveform shows, the forward scan produces a current peak for any analytes that can be reduced (or oxidized depending on the initial scan direction) through the range of the potential scanned. The current will increase as the potential reaches the reduction potential of the analyte, but then falls off as the concentration of the analyte is depleted close to the electrode surface. If the redox couple is reversible then when the applied potential is reversed, it will reach the potential that will reoxidize the product formed in the first reduction reaction, and produce a current of reverse polarity from the forward scan. This oxidation peak will usually have a similar shape to the reduction peak. As a result, information about the redox potential and electro-chemical reaction rates of the compounds are obtained.

For instance if the electronic transfer at the surface is fast and the current is limited by the diffusion of species to the electrode surface, then the current peak will be proportional to the square root of the scan rate. This relationship is described by the Cottrell equation. The CV experiment then samples only a small portion of the solution, the material within the diffusion layer.

Characterization

The utility of cyclic voltammetry is highly dependent on the analyte being studied. The analyte has to be redox active within the experimental potential window. It is also highly desirable for the analyte to display a reversible wave. A reversible wave is when an analyte is reduced or oxidized on a forward scan and is then reoxidized or rereduced in a predictable way on the return scan as shown in the first figure.

Even reversible couples contain polarization overpotential and thus display a hysteresis between absolute potential between the reduction (E_{pc}) and oxidation peak (E_{pa}). This overpotential emerges from a combination of analyte diffusion rates and the intrinsic activation barrier of transferring electrons from an electrode to analyte. A theoretical description of polarization overpotential is in part described by the Butler-Volmer equation and Cottrell equation. Conveniently in an ideal system the relationships reduces to, $|E_{pc} - E_{pa}| = \dfrac{57 \text{ mV}}{n}$, for an n electron process.

Reversible couples will display ratio of the peak currents passed at reduction (i_{pc}) and oxidation (i_{pa}) that is near unity ($1 = i_{pa}/i_{pc}$). This ratio can be perturbed for reversible couples in the presence of a following chemical reaction, stripping wave, or nucleation event.

When such reversible peaks are observed thermodynamic information in the form of half cell potential $E^0_{1/2}$ can be determined. When waves are semi-reversible such as when i_{pa}/i_{pc} is less than or greater than 1, it can be possible to determine even more information especially kinetic processes like following chemical reaction.

When waves are non-reversible it is impossible to determine what their thermodynamic $E^0_{1/2}$ is with cyclic voltammetry. This $E^0_{1/2}$ can be determined, however it often requires equal quantities of the analyte in both oxidation states. When a wave is non-reversible cyclic voltammetry can not determine if the wave is at its thermodynamic potential or shifted to a more extreme potential by some form of overpotential. The couple could be irreversible because of a following chemical process, a common example for transition metals is a shift in the geometry of the co-ordination sphere. If this is the case, then higher scan rates may show a reversible wave. It is also possible that the wave is irreversible due to a physical process most commonly some form of precipitation. Some speculation can be made in regards to irreversible waves however they are generally outside the scope of cyclic voltammetry.

Experimental Setup

The method uses reference electrode, working electrode, and counter electrode which in combination are sometimes referred to as a three-electrode setup. Electrolyte is usually added to the test solution to ensure sufficient conductivity. The combination of the solvent, electrolyte and specific working electrode material determines the range of the potential.

Electrodes are static and sit in unstirred solutions during cyclic voltammetry. This "still" solution method results in cyclic voltammetry's characteristic diffusion controlled peaks. This method also allows a portion of the analyte to remain after reduction or oxidation where it may display further redox activity. Stirring the solution between cyclic voltammetry traces is important as to supply the electrode surface with fresh analyte for each new experiment. The solubility of an analyte can change drastically with its overall charge. Since cyclic voltammetry usually alters the charge of the analyte it is common for reduced or oxidized analyte to precipitate out onto the electrode. This layering of analyte can insulate the electrode surface, display its own redox activity in subsequent scans, or at the very least alter the electrode surface. For this and other reasons it is often necessary to clean electrodes between scans.

Common materials for working electrodes include glassy carbon, platinum, and gold. These electrodes are generally encased in a rod of inert insulator with a disk exposed at one end. A regular working electrode has a radius within an order of magnitude of 1 mm. Having a controlled surface area with a defined shape is important for interpreting cyclic voltammetry results.

To run cyclic voltammetry experiments at high scan rates a regular working electrode is insufficient. High scan rates create peaks with large currents and increased resistances which result in distortions. Ultra-micro-electrodes can be used to minimize the current and resistance.

The counter electrode, also known as the auxiliary or second electrode, can be any material which conducts easily and won't react with the bulk solution. Reactions occurring at the counter electrode surface are un-important as long as it continues to conduct current well. To maintain the observed current the counter electrode will often oxidize or reduce the solvent or bulk electrolyte.

Reference electrodes are a complex subject and worth investigating elsewhere.

Variations

In some experiments an electro-active species is fixed to the surface of the electrode, for instance in microparticle voltammetry.

Potentiodynamic techniques also exist that add low-amplitude ac perturbation to a potential ramp and measure variable response in a single frequency (ac voltammetry) or in many frequencies simultaneously (potentiodynamic electrochemical impedance spectroscopy). The response in alternating current is two-dimensional–it is characterised by amplitude and phase. The amplitude and phase depend differently on frequency for constituents of ac response attributed to different processes (charge transfer, diffusion, double layer charging, etc.). Frequency response analysis enables simultaneous monitoring of the various processes that contribute to the potentiodynamic ac response of electro-chemical system.

Distinctions

Cyclic voltammetry is not a hydrodynamic technique. In a hydrodynamic technique flow is achieved at the electrode surface by stirring the solution, pumping the solution, or rotating the electrode as is the case with rotating disk electrodes and rotating ring-disk electrodes. These techniques target steady state conditions which appear the same scanned from the positive or the negative, thus limiting them to linear sweep voltammetry.

Applications

Cyclic voltammetry (CV) has become an important and widely used electro-analytical technique in many areas of chemistry. It is widely used to study a variety of redox processes, for obtaining stability of reaction products, the presence of intermediates in oxidation-reduction reactions, reaction and electron transfer kinetics, and the reversibility of a reaction. CV can also be used to determine the electron stoichiometry of a system, the diffusion coefficient of an analyte, and the formal reduction potential, which can be used as an identification tool. In addition, because concentration is proportional to current in a reversible, Nernstian system, concentration of an unknown solution can be determined by generating a calibration curve of current *vs.* concentration.

This latter application is gaining interest in the field of cellular biology where it is used to measure concentration of various chemicals in the cells of organisms, including living ones.

BULK ELECTROLYSIS

Bulk electrolysis is also known as *potentiostatic coulometry* or *controlled potential coulometry*. The experiment is a form of coulometry which generally employs a three electrode system controlled by a potentiostat. In the experiment the working electrode is held at a constant potential (volts) and current (amps) is monitored over time (seconds). In a properly run experiment an analyte is quantitatively converted from its original oxidation state to a new oxidation state, either reduced or oxidized. As the substrate is consumed, the current also decreases, approaching zero when the conversion nears completion.

The results of a *bulk electrolysis* are visually displayed as the total coulombs passed (total electric charge) plotted against time in seconds, even through the experiment measures electric current (amps) over time. This is done to show that the experiment is approaching an expected total number of coulombs.

Fundamental Relationships and Applications

The sample mass, molecular mass, number of electrons in the electrode reaction, and number of electrons passed during the experiment are all related by Faraday's laws of electrolysis. It follows that, if three of the values are known, then the fourth can be calculated. The bulk electrolysis can also be useful for synthetic purposes if the product of the electrolysis can be isolated. This is most convenient when the product is neutral and can be isolated from the electrolyte solution through extraction or when the product plates out on the electrode or precipitates in another fashion. Even if the product can not be isolated, other analytical techniques can be performed on the solution including NMR, EPR, UV-Vis, FTIR, among others techniques depending on the specific situation. In specially designed cells the solution can be actively monitored during the experiment.

Cell Design

In most three electrode experiments there are two isolated cells. One contains the auxiliary and working electrode, while the other contains the reference electrode. Strictly speaking, the reference electrode does not require a separate compartment. A Quasi-Reference Electrode such as a silver/silver chloride wire electrode can be exposed directly to the analyte solution. In such situations there is concern that the analyte and trace redox products may interact with the reference electrode and either render it useless or increase drift. As a result even these simple references are commonly sequestered in their own cells. The more complex references such as standard hydrogen electrode, saturated calomel electrode, or silver chloride electrode (specific concentration) can not directly mix the analyte solution for fear the electrode will fall apart or interact/react with the analyte.

A bulk electrolysis is best performed in a three part cell in which both the auxiliary electrode and reference electrode have their own cell which connects to the cell containing the working electrode. This isolates the undesired redox events taking place at the auxiliary electrode. During bulk electrolysis, the analyte undergoes a redox event at the working electrode. If the system was open, then it would be possible for the product of that reaction to diffuse back to the auxiliary electrode and undergo the inverse redox reaction. In addition to maintaining the proper current at the working electrode, the auxiliary electrode will experience extreme potentials often oxidizing or reducing the solvent or electrolyte to balance the current. In voltammetry experiments, the currents (amps) are so small and it is not a problem to decompose a small amount of solvent or electrolyte. In contrast, a bulk electrolysis involves currents greater by several orders of magnitude. At the auxiliary electrode, this greater current would decompose a significant amount of the solution/electrolyte and probably boiling the solution in the process all in an effort to balance the current. To mitigate this challenge the auxiliary cell will often contain a stoichiometric or greater amount of *sacrificial reductant* (ferrocene) or *sacrificial oxidant* (ferrocenium) to balance the overall redox reaction.

For ideal performance the auxiliary electrode should be similar in surface area, as close as possible, and evenly spaced with the working electrode. This is in an effort to prevent "hot spots". Hot spots are the result of current following the path of least resistance. This means much of the redox chemistry will occur at the points at either end of the shortest path between the working and auxiliary electrode. Heating associated with the capacitances resistance of the solution can occur at the area around these points, actually boiling the solution. The bubbling resulting from this isolated boiling of the solution can be confused with gas evolution.

Rates and Kinetics

The rate of such reactions/experiments is not determined by the concentration of the solution, but rather the mass transfer of the substrate in the solution to the electrode surface. Rates will increase when the volume of the solution is decreased, the solution is stirred more rapidly, or the area of the working electrode is increased. Since mass transfer is so important the solution is stirred during a bulk electrolysis. However, this technique is generally not considered a hydrodynamic technique, since a laminar flow of solution against the electrode is neither the objective or outcome of the siring.

Bulk electrolysis is occasionally cited in the literature as means to study electro-chemical reaction rates. However, bulk electrolysis is generally a poor method to study electro-chemical reaction rates since the rate of bulk electrolysis is generally governed by the specific cells ability to perform mass transfer. Rates slower than this mass transfer bottleneck are rarely of interest.

Efficiency and Thermodynamics

Electro-catalytic analyzes will often mention the *current efficiency* or faradaic efficiency of a given process determined by a bulk electrolysis experiment. For example if one molecule of hydrogen results from every two electrons inserted into an acidic solution then the faradaic efficiency would be 100%. This indicates that the electrons did not ended up performing some other reaction. For example the oxidation of water will often produce oxygen as well as hydrogen peroxide at the anode. Each of these products is related to its own faradaic efficiency which is tied to the experimental arrangement.

Nor is current efficiency the same as thermodynamic efficiency, since it never address the how much energy (potential in volts) is in the electrons added or removed. The voltage efficiency determined by the reactions overpotential is more directly related to the thermodynamics of the electro-chemical reaction. In fact the extent to which a reaction goes to completion is related to how much greater the applied potential is than the reduction potential of interest. In the case where multiple reduction potentials are of interest, it is often difficult to set an electrolysis potential a "safe" distance (such as 200 mV) past a redox event. The result is incomplete conversion of the substrate, or else conversion of some of the substrate to the more reduced form. This factor must be considered when analyzing the current passed and when attempting to do further analysis/isolation/experiments with the substrate solution.

REACTION MECHANISM

In chemistry, a **reaction mechanism** is the step by step sequence of elementary reactions by which overall chemical change occurs.

Although only the net chemical change is directly observable for most chemical reactions, experiments that suggest the possible sequence of steps in a reaction mechanism can often be designed. Recently, electrospray ionization mass spectrometry has been used to corroborate the mechanism of several organic reaction proposals.

Description

A chemical mechanism describes in detail exactly what takes place at each stage of an overall chemical reaction (transformation). It also describes each reactive intermediate, activated complex, and transition state, and which bonds are broken (and in what order), and which bonds are formed (and in what order). A complete mechanism must also account for all reactants used, the function of a catalyst, stereo-chemistry, all products formed and the amount of each. It must also describe the relative rates of the reaction steps and the rate equation for the overall reaction. Reaction intermediates are chemical species, often unstable and short-lived, which are not reactants or products of the overall chemical reaction, but are temporary products and reactants in the mechanism's reaction steps. Reaction intermediates are often free radicals or ions. Transition states can be unstable intermediate molecular states even in the elementary reactions. Transition states are commonly molecular entities involving an unstable number of bonds and/or unstable geometry. They correspond to maxima on the reaction co-ordinate, and to saddle points on the potential energy surface for the reaction.

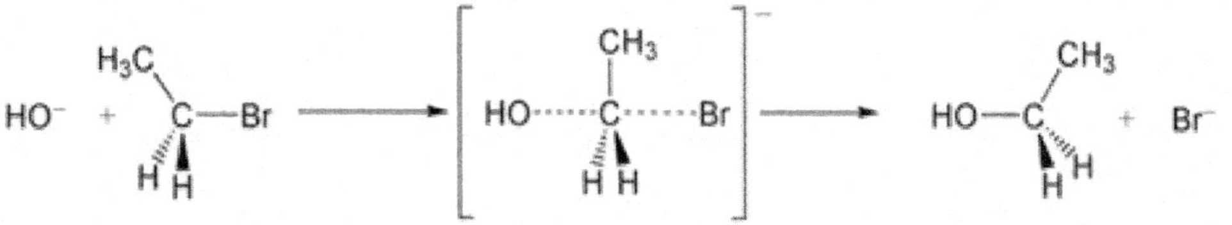

Fig. : S_N2 reaction mechanism. Note the negatively charged transition state in brackets in which the central carbon atom in question shows five bonds, an unstable condition.

The electron or arrow pushing method is often used in illustrating a reaction mechanism; for example.

A reaction mechanism must also account for the order in which molecules react. Often what appears to be a single-step conversion is in fact a multistep reaction.

Examples

Consider the following reaction :

$$CO + NO_2 \rightarrow CO_2 + NO$$

In this case, experiments have determined that this reaction takes place according to the rate law $r = k[NO_2]^2$. This form suggests that the rate-determining step is a reaction between two molecules of NO_2. A possible mechanism for the overall reaction that explains the rate law is :

$$2 NO_2 \rightarrow NO_3 + NO \text{ (slow)}$$
$$NO_3 + CO \rightarrow NO_2 + CO_2 \text{ (fast)}$$

Each step is called an elementary step, and each has its own rate law and molecularity. The elementary steps should add up to the original reaction.

When determining the overall rate law for a reaction, the slowest step is the step that determines the reaction rate. Because the first step is the slowest step, it is the rate-determining step. Because it involves the collision of two NO_2 molecules, it is a bimolecular reaction with a rate law of $r = k[NO_2]^2$. If we were to cancel out all the molecules that appear on both sides of the reaction, we would be left with the original reaction.

Other reactions may have mechanisms of several consecutive steps. In organic chemistry, one of the first reaction mechanisms proposed was that for the benzoin condensation, put forward in 1903 by A. J. Lapworth.

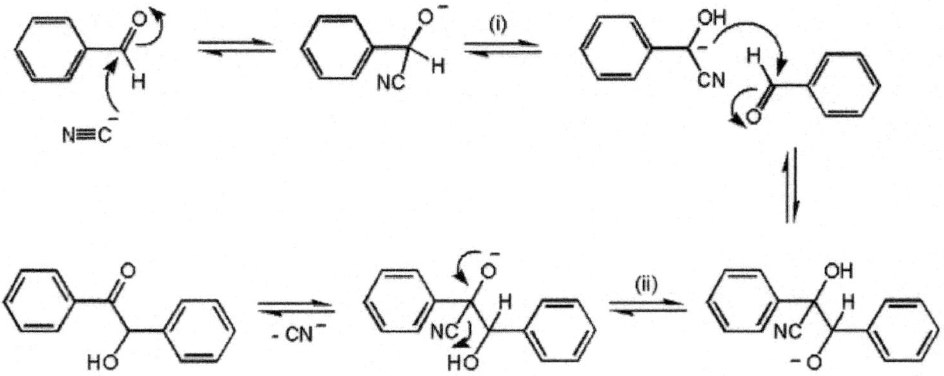

Fig. : Benzoin condensation reaction mechanism. Cyanide ion (CN⁻) acts as a catalyst here, entering at the first step and leaving in the last step. Proton (H⁺) transfers occur at (i) and (ii). The arrow pushing method is used in some of the steps to show where electron pairs go.

There are also more complex mechanisms such as chain reactions, in which the *propagation* steps of the chain form a closed cycle.

Modelling

A correct reaction mechanism is an important part of accurate predictive modeling. For many combustion and plasma systems, detailed mechanisms are not available or require development.

Even when information is available, identifying and assembling the relevant data from a variety of sources, reconciling discrepant values and extra-polating to different conditions can be a difficult process without expert help. Rate constants or thermochemical data are often not available in the literature, so computational chemistry techniques or group-additivity methods must be used to obtain the required parameters.

At the different stages of a reaction mechanism's elaboration, appropriate methods must be used. One approach can involve the use of cross-over experiments.

MOLECULARITY

Molecularity in chemistry is the number of colliding molecular entities that are involved in a single reaction step. While the order of a reaction is derived experimentally, the molecularity is a theoretical concept and can only be applied to elementary reactions. In elementary reactions, the reaction order, the molecularity and the stoichiometric coefficient are the same, although only numerically, because they are different concepts.

- A reaction involving one molecular entity is called unimolecular.
- A reaction involving two molecular entities is called bimolecular.
- A reaction involving three molecular entities is called termolecular or trimolecular. Termolecular reactions in solutions or gas mixtures are very rare, because of the improbability of three molecular entities simultaneously colliding. However the term *termolecular* is also used to refer to three body association reactions of the type :

$$A + B \xrightarrow{M} C$$

Where the M over the arrow denotes that to conserve energy and momentum a second reaction with a third body is required. After the initial bimolecular collision of A and B an energetically excited reaction intermediate is formed, then, it collides with a M body, in a second bimolecular reaction, transferring the excess energy to it.

The reaction can be explained as two consecutive reactions :

$A + B \rightarrow AB^*$

$AB^* + M \rightarrow C + M$

These reactions frequently have a pressure and temperature dependence region of transition between second and third order kinetics.

Catalytic reactions are often three-component, but in practice a complex of the starting materials is first formed and the rate-determining step is the reaction of this complex into products, not an adventitious collision between the two species and the catalyst. For example, in hydrogenation with a metal catalyst, molecular dihydrogen first dissociates onto the metal surface into hydrogen atoms bound to the surface, and it is these monatomic hydrogens that react with the starting material, also previously adsorbed onto the surface.

CHEMICAL KINETICS

Chemical kinetics, also known as **reaction kinetics**, is the study of rates of chemical processes. Chemical kinetics includes investigations of how different experimental conditions can influence the speed of a chemical reaction and yield information about the reaction's mechanism and transition states, as well as the construction of mathematical models that can describe the characteristics of a chemical reaction. In 1864, Peter Waage and Cato Guldberg pioneered the development of chemical kinetics by formulating the law of mass action, which states that the speed of a chemical reaction is proportional to the quantity of the reacting substances.

Chemical kinetics deals with the experimental determination of reaction rates from which rate laws and rate constants are derived. Relatively simple rate laws exist for zero-order reactions (for which reaction rates are independent of concentration), first-order reactions, and second-order reactions, and can be derived for others. In consecutive reactions, the rate-determining step often determines the kinetics. In consecutive first-order reactions, a steady state approximation can simplify the rate law. The activation energy for a reaction is experimentally determined through the Arrhenius equation and the Eyring equation. The main factors that influence the reaction rate include : the physical state of the reactants, the concentrations of the reactants, the temperature at which the reaction occurs, and whether or not any catalysts are present in the reaction.

Factors Affecting Reaction Rate

Nature of the Reactants

Depending upon what substances are reacting, the reaction rate varies. Acid/base reactions, the formation of salts, and ion exchange are fast reactions. When covalent bond formation takes place between the molecules and when large molecules are formed, the reactions tend to be very slow. Nature and strength of bonds in reactant molecules greatly influence the rate of its transformation into products.

Physical State

The physical state (solid, liquid, or gas) of a reactant is also an important factor of the rate of change. When reactants are in the same phase, as in aqueous solution, thermal motion brings them into contact. However, when they are in different phases, the reaction is limited to the interface between the reactants. Reaction can occur only at their area of contact; in the case of a liquid and a gas, at the surface of the liquid. Vigorous shaking and stirring may be needed to bring the reaction to completion. This means that the more finely divided a solid or liquid reactant the greater its surface area per unit volume and the more contact it makes with the other reactant, thus the faster the reaction. To make an analogy, for example, when one starts a fire, one uses wood chips and small branches — one does not start with large logs right away. In organic chemistry, on water reactions are the exception to the rule that homogeneous reactions take place faster than heterogeneous reactions.

Concentration

The reactions are due to collisions of reactant species. The frequency with which the molecules or ions collide depends upon their concentrations. The more crowded the molecules are, the more likely they are to collide and react with one another. Thus, an increase in the concentrations of the reactants will result in the corresponding increase in the reaction rate, while a decrease in the concentrations will have a reverse effect. For example, combustion that occurs in air (21% oxygen) will occur more rapidly in pure oxygen.

Temperature

Temperature usually has a major effect on the rate of a chemical reaction. Molecules at a higher temperature have more thermal energy. Although collision frequency is greater at higher temperatures, this alone contributes only a very small proportion to the increase in rate of reaction. Much more important is the fact that the proportion of reactant molecules with sufficient energy to react (energy greater than activation energy : $E > E_a$) is significantly higher and is explained in detail by the Maxwell–Boltzmann distribution of molecular energies.

The 'rule of thumb' that the rate of chemical reactions doubles for every 10°C temperature rise is a common misconception. This may have been generalized from the special case of biological systems, where the a (temperature coefficient) is often between 1.5 and 2.5.

A reaction's kinetics can also be studied with a temperature jump approach. This involves using a sharp rise in temperature and observing the relaxation time of the return to equilibrium. A particularly useful form of temperature jump apparatus is a shock tube, which can rapidly jump a gas's temperature by more than 1000 degrees.

Catalysts

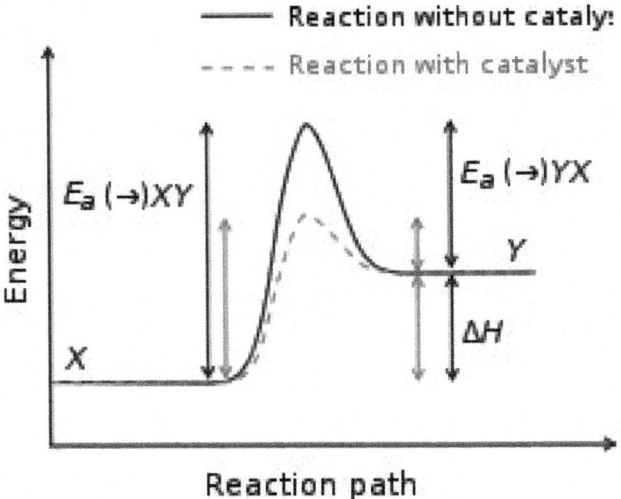

Fig. : Generic potential energy diagram showing the effect of a catalyst in a hypothetical endothermic chemical reaction. The presence of the catalyst opens a different reaction pathway (shown in red) with a lower activation energy. The final result and the overall thermodynamics are the same.

A catalyst is a substance that accelerates the rate of a chemical reaction but remains chemically unchanged afterwards. The catalyst increases rate reaction by providing a different reaction mechanism to occur with a lower activation energy. In autocatalysis a reaction product is itself a catalyst for that reaction leading to positive feedback. Proteins that act as catalysts in bio-chemical reactions are called enzymes. Michaelis–Menten kinetics describe the rate of enzyme mediated reactions. A catalyst does not affect the position of the equilibria, as the catalyst speeds up the backward and forward reactions equally.

In certain organic molecules, specific substituents can have an influence on reaction rate in neighbouring group participation.

Agitating or mixing a solution will also accelerate the rate of a chemical reaction, as this gives the particles greater kinetic energy, increasing the number of collisions between reactants and, therefore, the possibility of successful collisions.

Pressure

Increasing the pressure in a gaseous reaction will increase the number of collisions between reactants, increasing the rate of reaction. This is because the activity of a gas is directly proportional to the partial pressure of the gas. This is similar to the effect of increasing the concentration of a solution.

In addition to this straightforward mass-action effect, the rate coefficients themselves can change due to pressure. The rate coefficients and products of many high-temperature gas-phase reactions change if an inert gas is added to the mixture; variations on this effect are called fall-off and chemical activation. These phenomena are due to exothermic or endothermic reactions occurring faster than heat transfer, causing the reacting molecules to have non-thermal energy distributions (non-Boltzmann distribution). Increasing the pressure increases the heat transfer rate between the reacting molecules and the rest of the system, reducing this effect.

Condensed-phase rate coefficients can also be affected by (very high) pressure; this is a completely different effect than fall-off or chemical-activation. It is often studied using diamond anvils.

A reaction's kinetics can also be studied with a pressure jump approach. This involves making fast changes in pressure and observing the relaxation time of the return to equilibrium.

Equilibrium

While a chemical kinetics is concerned with the rate of a chemical reaction, thermodynamics determines the extent to which reactions occur. In a reversible reaction, chemical equilibrium is reached when the rates of the forward and reverse reactions are equal (the principle of detailed balance) and the concentrations of the reactants and products no longer change. This is demonstrated by, for example, the Haber–Bosch process for combining nitrogen and hydrogen to produce ammonia. Chemical clock reactions such as the Belousov–Zhabotinsky reaction demonstrate that component concentrations can oscillate for a long time before finally attaining the equilibrium.

Free Energy

In general terms, the free energy change (ΔG) of a reaction determines whether a chemical change will take place, but kinetics describes how fast the reaction is. A reaction can be very exothermic and have a very positive entropy change but will not happen in practice if the reaction is too slow. If a reactant can produce two different products, the thermodynamically most stable one will in general form, except in special circumstances when the reaction is said to be under kinetic reaction control. The Curtin–Hammett principle applies when determining the product ratio for two reactants interconverting rapidly, each going to a different product. It is possible to make predictions about reaction rate constants for a reaction from free-energy relationships.

The kinetic isotope effect is the difference in the rate of a chemical reaction when an atom in one of the reactants is replaced by one of its isotopes.

Chemical kinetics provides information on residence time and heat transfer in a chemical reactor in chemical engineering and the molar mass distribution in polymer chemistry.

Applications

The mathematical models that describe chemical reaction kinetics provide chemists and chemical engineers with tools to better understand and describe chemical processes such as food decomposition, micro-organism growth, stratospheric ozone decomposition, and the complex chemistry of biological systems. These models can also be used in the design or modification of chemical reactors to optimize product yield, more efficiently separate products, and eliminate environmentally harmful by-products. When performing catalytic cracking of heavy hydrocarbons into gasoline and light gas, for example, kinetic models can be used to find the temperature and pressure at which the highest yield of heavy hydrocarbons into gasoline will occur. Kinetics is also a basic aspect of chemistry.

CHEMICAL THERMODYNAMICS

Chemical thermodynamics is the study of the interrelation of heat and work with chemical reactions or with physical changes of state within the confines of the laws of thermodynamics. Chemical thermodynamics involves not only laboratory measurements of various thermodynamic properties, but also the application of mathematical methods to the study of chemical questions and the *spontaneity* of processes.

The structure of chemical thermodynamics is based on the first two laws of thermodynamics. Starting from the first and second laws of thermodynamics, four equations called the "fundamental equations of Gibbs" can be derived. From these four, a multitude of equations, relating the thermodynamic properties of the thermodynamic system can be derived using relatively simple mathematics. This outlines the mathematical framework of chemical thermodynamics.

History

In 1865, the German physicist Rudolf Clausius, in his *Mechanical Theory of Heat*, suggested that the principles of thermochemistry, *e.g.* the heat evolved in combustion reactions, could be applied to the principles of thermodynamics. Building on the work of Clausius, between the years 1873-76 the American mathematical physicist Willard Gibbs published a series of three papers, the most famous one being the paper *On the Equilibrium of Heterogeneous Substances*. In these papers, Gibbs showed how the first two laws of thermodynamics could be measured graphically and mathematically to determine both the thermodynamic equilibrium of chemical reactions as well as their tendencies to occur or proceed. Gibbs' collection of papers provided the first unified body of thermodynamic theorems from the principles developed by others, such as Clausius and Sadi Carnot.

During the early 20th century, two major publications successfully applied the principles developed by Gibbs to chemical processes, and thus established the foundation of the science of chemical thermodynamics. The first was the 1923 textbook *Thermodynamics and the Free Energy of Chemical Substances* by Gilbert N. Lewis and Merle Randall. This book was responsible for supplanting the chemical affinity for the term free energy in the English-speaking world. The second was the 1933 book *Modern Thermodynamics by the methods of Willard Gibbs* written by E. A. Guggenheim. In this manner, Lewis, Randall, and Guggenheim are considered as the founders of modern chemical thermodynamics because of the major contribution of these two books in unifying the application of thermodynamics to chemistry.

Overview

The primary objective of chemical thermodynamics is the establishment of a criterion for the determination of the feasibility or spontaneity of a given transformation. In this manner, chemical thermodynamics is typically used to predict the energy exchanges that occur in the following processes :

1. Chemical reactions
2. Phase changes
3. The formation of solutions

The following state functions are of primary concern in chemical thermodynamics :

*   Internal energy (U)

- Enthalpy (H)
- Entropy (S)
- Gibbs free energy (G)

Most identities in chemical thermodynamics arise from application of the first and second laws of thermodynamics, particularly the law of conservation of energy, to these state functions.

The 3 Laws of Thermodynamics

1. The energy of the universe is constant.
2. In any spontaneous process, there is always an increase in entropy of the universe
3. The entropy of a perfect crystal at 0 Kelvin is zero.

Chemical Energy

Chemical energy is the potential of a chemical substance to undergo a transformation through a chemical reaction or to transform other chemical substances. Breaking or making of chemical bonds involves energy, which may be either absorbed or evolved from a chemical system.

Energy that can be released (or absorbed) because of a reaction between a set of chemical substances is equal to the difference between the energy content of the products and the reactants. This change in energy is called the change in internal energy of a chemical reaction. Where $\Delta U°_{f\,reactants}$ is the internal energy of formation of the reactant molecules that can be calculated from the bond energies of the various chemical bonds of the molecules under consideration and $\Delta U°_{f\,products}$ is the internal energy of formation of the product molecules. The internal energy change of a process is equal to the heat change if it is measured under conditions of constant volume, as in a closed rigid container such as a bomb calorimeter. However, under conditions of constant pressure, as in reactions in vessels open to the atmosphere, the measured heat change is not always equal to the internal energy change, because pressure-volume work also releases or absorbs energy. (The heat change at constant pressure is called the enthalpy change; in this case the enthalpy of formation).

Another useful term is the heat of combustion, which is the energy released due to a combustion reaction and often applied in the study of fuels. Food is similar to hydrocarbon fuel and carbohydrate fuels, and when it is oxidized, its caloric content is similar.

In chemical thermodynamics the term used for the chemical potential energy is chemical potential, and for chemical transformation an equation most often used is the Gibbs-Duhem equation.

Chemical Reactions

In most cases of interest in chemical thermodynamics there are internal degrees of freedom and processes, such as chemical reactions and phase transitions, which always create entropy unless they are at equilibrium, or are maintained at a "running equilibrium" through "quasi-static" changes by being coupled to constraining devices, such as pistons or electrodes, to deliver and receive external work. Even for homogeneous "bulk" materials, the free energy functions depend on the composition, as do all the extensive thermodynamic potentials, including the internal energy. If the quantities { N_i }, the number of chemical species, are omitted from the formulae, it is impossible to describe compositional changes.

Gibbs Function

For a "bulk" (unstructured) system they are the last remaining extensive variables. For an unstructured, homogeneous "bulk" system, there are still various *extensive* compositional variables { N_i } that G depends on, which specify the composition, the amounts of each chemical substance, expressed as the numbers of molecules present or (dividing by Avogadro's number), the numbers of moles :

$G = G(T, P, \{N_i\})$.

For the case where only PV work is possible

$$dG = -SdT + VdP + \sum_i \mu_i dN_i$$

in which μ_i is the chemical potential for the i-th component in the system :

$$\mu_i = \left(\frac{\partial G}{\partial N_i} \right)_{T,P,N_{j \neq i, etc.}} \cdot$$

The expression for dG is especially useful at constant T and P, conditions which are easy to achieve experimentally and which approximates the condition in living creatures :

$$(dG)_{T, P} = \sum_i \mu_i dN_i.$$

Chemical Affinity

While this formulation is mathematically defensible, it is not particularly transparent since one does not simply add or remove molecules from a system. There is always a *process* involved in changing the composition; *e.g.*, a chemical reaction (or many), or movement of molecules from one phase (liquid) to another (gas or solid). We should find a notation which does not seem to imply that the amounts of the components (N_i} can be changed independently. All real processes obey conservation of mass, and in addition, conservation of the numbers of atoms of each kind. Whatever molecules are transferred to or from should be considered part of the "system".

Consequently we introduce an explicit variable to represent the degree of advancement of a process, a progress variable ξ for the *extent of reaction* (Prigogine & Defay, p. 18; Prigogine, pp. 4–7; Guggenheim, p. 37.62), and to the use of the partial derivative $\partial G/\partial \xi$ (in place of the widely used "ΔG", since the quantity at issue is not a finite change). The result is an understandable expression for the dependence of dG on chemical reactions (or other processes). If there is just one reaction :

$$(dG)_{T,P} = \left(\frac{\partial G}{\partial \xi} \right)_{T,P} d\xi.$$

If we introduce the *stoichiometric coefficient* for the *i-th* component in the reaction :

$$v_i = \partial N_i / \partial \xi$$

which tells how many molecules of i are produced or consumed, we obtain an algebraic expression for the partial derivative :

$$\left(\frac{\partial G}{\partial \xi} \right)_{T,P} = \sum_i \mu_i v_i = -A$$

where, (De Donder; Progoine & Defay, p. 69; Guggenheim, pp. 37,240), we introduce a concise and historical name for this quantity, the "affinity", symbolized by **A**, as introduced by Théophile de Donder in 1923. The minus sign comes from the fact the affinity was defined to represent the rule that spontaneous changes will ensue only when the change in the Gibbs free energy of the process is negative, meaning that the chemical species have a positive affinity for each other. The differential for G takes on a simple form which displays its dependence on compositional change :

$$(dG)_{T,P} = -A \, d\xi.$$

If there are a number of chemical reactions going on simultaneously, as is usually the case

$$(dG)_{T,P} = - \sum_k A_k d\xi_k.$$

a set of reaction coordinates $\{\xi_j\}$, avoiding the notion that the amounts of the components $\{N_i\}$ can be changed independently. The expressions above are equal to zero at thermodynamic equilibrium, while in the general case for real systems, they are negative because all chemical reactions proceeding at a finite rate produce entropy. This can be made even more explicit by introducing the reaction *rates* $d\xi_k / dt$. For each and every *physically independent process* (Prigogine & Defay, p. 38; Prigogine, p. 24)

$$A \, \dot{\xi} \le 0.$$

This is a remarkable result since the chemical potentials are intensive system variables, depending only on the local molecular milieu. They cannot "know" whether the temperature and pressure (or any other system variables) are going to be held constant over time. It is a purely local criterion and must hold regardless of any such constraints. Of course, it could have been obtained by taking partial derivatives of any of the other fundamental state functions, but nonetheless is a general criterion for ($-T$ times) the entropy production from that spontaneous process; or at least any part of it that is not captured as external work.

We now relax the requirement of a homogeneous "bulk" system by letting the chemical potentials and the affinity apply to any locality in which a chemical reaction (or any other process) is occurring. By accounting for the entropy production due to irreversible processes, the inequality for dG is now replaced by an equality

$$dG = -SdT + VdP - \sum_k A_k d\xi_k + W'$$

or

$$dG_{T,P} = -\sum_k A_k d\xi_k + W'.$$

Any decrease in the Gibbs function of a system is the upper limit for any iso-thermal, isobaric work that can be captured in the surroundings, or it may simply be dissipated, appearing as T times a corresponding increase in the entropy of the system and/or its surrounding. Or it may go partly toward doing external work and partly toward creating entropy. The important point is that the *extent of reaction* for a chemical reaction may be coupled to the displacement of some external mechanical or electrical quantity in such a way that one can advance only if the other one also does. The coupling may occasionally be *rigid*, but it is often flexible and variable.

Solutions

In solution chemistry and biochemistry, the Gibbs free energy decrease ($\partial G/\partial \xi$, in molar units, denoted cryptically by ΔG) is commonly used as a surrogate for ($-T$ times) the entropy produced by spontaneous chemical reactions in situations where there is no work being done; or at least no "useful" work; *i.e.*, other than perhaps some $\pm PdV$. The assertion that all *spontaneous reactions have a negative ΔG* is merely a restatement of the fundamental thermodynamic relation, giving it the physical dimensions of energy and somewhat obscuring its significance in terms of entropy. When there is no useful work being done, it would be less misleading to use the Legendre transforms of the entropy appropriate for constant T, or for constant T and P, the Massieu functions $-F/T$ and $-G/T$ respectively.

Non Equilibrium

Generally the systems treated with the conventional chemical thermodynamics are either at equilibrium or near equilibrium. Ilya Prigogine developed the ther-modynamic treatment of open systems that are far from equilibrium. In doing so he has discovered phenomena and structures of completely new and completely unexpected types. His generalized, non-linear and irreversible thermodynamics has found surprising applications in a wide variety of fields.

The non-equilibrium thermodynamics has been applied for explaining how ordered structures *e.g.* the biological systems, can develop from disorder. Even if Onsager's relations are utilized, the classical principles of equilibrium in thermodynamics still show that linear systems close to equilibrium always develop into states of disorder which are stable to perturbations and cannot explain the occurrence of ordered structures.

Prigogine called these systems dissipative systems, because they are formed and maintained by the dissipative processes which take place because of the exchange of energy between the system and its environment and because they disappear if that exchange ceases. They may be said to live in symbiosis with their environment.

The method which Prigogine used to study the stability of the dissipative structures to perturbations is of very great general interest. It makes it possible to study the most varied problems, such as city traffic problems, the stability of insect communities, the development of ordered biological structures and the growth of cancer cells to mention but a few examples.

System Constraints

In this regard, it is crucial to understand the role of walls and other *constraints*, and the distinction between *independent* processes and *coupling*. Contrary to the clear implications of many reference sources, the previous analysis is not restricted to homogeneous, isotropic bulk systems which can deliver only PdV work to the outside world, but applies even to the most structured systems. There are complex systems with many chemical "reactions" going on at the same time, some of which are really only parts of the same, overall process. An *independent* process is one that *could* proceed even if all others were unaccountably stopped in their tracks. Understanding this is perhaps a "thought experiment" in chemical kinetics, but actual examples exist.

A gas reaction which results in an increase in the number of molecules will lead to an increase in volume at constant external pressure. If it occurs inside a cylinder closed with a piston, the equilibrated reaction can proceed only by doing work against an external force on the piston. The extent variable for the reaction can increase only if the piston moves, and conversely, if the piston is pushed inward, the reaction is driven backwards.

Similarly, a redox reaction might occur in an electro-chemical cell with the passage of current in wires connecting the electrodes. The half-cell reactions at the electrodes are constrained if no current is allowed to flow. The current might be dissipated as joule heating, or it might in turn run an electrical device like a motor doing mechanical work. An automobile lead-acid battery can be recharged, driving the chemical reaction backwards. In this case as well, the reaction is not an independent process. Some, perhaps most, of the Gibbs free energy of reaction may be delivered as external work.

The hydrolysis of ATP to ADP and phosphate can drive the force times distance work delivered by living muscles, and synthesis of ATP is in turn driven by a redox chain in mitochondria and chloroplasts, which involves the transport of ions across the membranes of these cellular organelles. The coupling of processes here, and in the previous examples, is often not complete. Gas can leak slowly past a piston, just as it can slowly leak out of a rubber balloon. Some reaction may occur in a battery even if no external current is flowing. There is usually a coupling coefficient, which may depend on relative rates, which determines what percentage of the driving free energy is turned into external work, or captured as "chemical work"; a misnomer for the free energy of another chemical process.

ELECTRON-TRANSFER DISSOCIATION

Electron-transfer dissociation (ETD) is a method of fragmenting ions in a mass spectrometer. Similar to electron-capture dissociation, ETD induces fragmentation of cations (*e.g.* peptides or proteins) by transferring electrons to them. It was invented by Donald F. Hunt, Joshua Coon, John E. P. Syka and Jarrod Marto at the University of Virginia.

ETD Fragmentation

ETD does not use free electrons but employs radical anions (*e.g.* anthracene or azobenzene) for this purpose :

$$[M + nH]^{n+} + A^- \rightarrow [[M + nH]^{(n-1)+}]* + A \rightarrow \text{fragments}.$$

where A is the anion. ETD cleaves randomly along the peptide backbone (so called c and z ions) while side chains and modifications such as phosphorylation are left intact. The technique only works well for higher charge state ions ($z>2$), however relative to collision-induced dissociation (CID), ETD is advantageous for the fragmentation of longer peptides or even entire proteins. This makes the technique important for top-down proteomics.

Much like ECD, ETD is believed to be particularly effective for peptides with modifications such as phosphorylation.

NEGATIVE ELECTRON-TRANSFER DISSOCIATION

Negative electron-transfer dissociation (NETD) is an ion/ion reaction in which an electron from an anionic specie is transferred to cationic reagent. Following this transfer event, the electron deficient anion undergoes internal re-arrangement and fragments. NETD is the ion/ion analogue of electron-detachment dissociation (EDD).

Peptide Fragmentation Mechanism

NETD is compatible with fragmenting peptide and proteins along the backbone at the C_α-C bond. The resulting fragments are usually a·-and x-type product ions.

Chapter 15

ALKANE

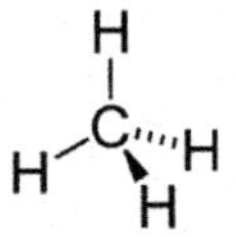

Fig. : Chemical structure of methane, the simplest alkane.

In organic chemistry, an **alkane**, or **paraffin** (a still-used historical name that also has other meanings), is a saturated hydrocarbon. Alkanes consist only of hydrogen and carbon atoms, all bonds are single bonds, and the carbon atoms are not joined in cyclic structures but instead form an open chain. They have the general chemical formula C_nH_{2n+2}. Alkanes belong to a homologous series of organic compounds in which the members differ by a molecular mass of 14.03u (mass of a methanediyl group, $-CH_2-$, one carbon atom of mass 12.01u, and two hydrogen atoms of mass ≈1.01u each). There are two main commercial sources : crude oil and natural gas.

Each carbon atom has 4 bonds (either C-H or C-C bonds), and each hydrogen atom is joined to a carbon atom (H-C bonds). A series of linked carbon atoms is known as the carbon skeleton or carbon backbone. The number of carbon atoms is used to define the size of the alkane (*e.g.*, C_2-alkane).

An **alkyl group**, generally abbreviated with the symbol R, is a functional group or side-chain that, like an alkane, consists solely of single-bonded carbon and hydrogen atoms, for example a methyl or ethyl group.

The simplest possible alkane (the parent molecule) is methane, CH_4. There is no limit to the number of carbon atoms that can be linked together, the only limitation being that the molecule is acyclic, is saturated, and is a hydrocarbon. Saturated oils and waxes are examples of larger alkanes where the number of carbons in the carbon backbone is greater than 10.

Alkanes are not very reactive and have little biological activity. All alkanes are colourless and odourless. Alkanes can be viewed as a molecular tree upon which can be hung the more biologically active/reactive portions (functional groups) of the molecule.

STRUCTURE CLASSIFICATION

Saturated hydrocarbons can be :

- Linear (general formula $C_{nH_{2n+2}}$) wherein the carbon atoms are joined in a snake-like structure
- Branched (general formula $C_{nH_{2n+2}}$, $n > 3$) wherein the carbon backbone splits off in one or more directions
- Cyclic (general formula $C_{nH_{2n}}$, $n > 2$) wherein the carbon backbone is linked so as to form a loop.

According to the definition by IUPAC, the former two are alkanes, whereas the third group is called cycloalkanes. Saturated hydrocarbons can also combine any of the linear, cyclic (*e.g.*, polycyclic) and branching structures, and they are still alkanes (no general formula) as long as they are acyclic (*i.e.*, having no loops). They also have single covalent bonds between their carbons.

ISOMERISM

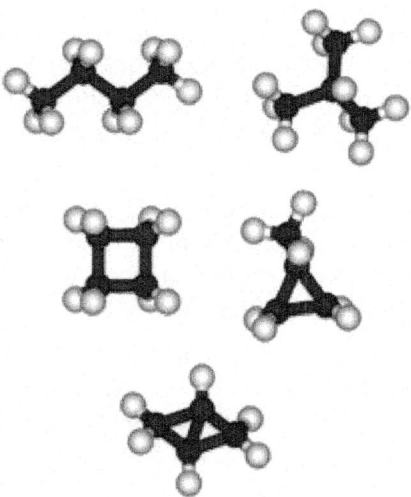

Fig. : Different C_4-alkanes and-cycloalkanes (left to right) : *n*-butane and isobutane are the two C_4H_{10} isomers; cyclobutane and methylcyclopropane are the two C_4H_8 isomers.

Bicyclo[1.1.0] butane is the only C_4H_6 compound and has no isomer; tetrahedrane (not shown) is the only C_4H_4 compound and has also no isomer.

Alkanes with more than three carbon atoms can be arranged in various different ways, forming structural isomers. The simplest isomer of an alkane is the one in which the carbon atoms are arranged in a single chain with no branches. This isomer is sometimes called the *n*-isomer (*n* for "normal", although it is not necessarily the most common). However, the chain of carbon atoms may also be branched at one or more points. The number of possible isomers increases rapidly with the number of carbon atoms. For example :

- C_1 : methane only
- C_2 : ethane only
- C_3 : propane only
- C_4 : 2 isomers : *n*-butane and isobutane
- C_5 : 3 isomers : pentane, isopentane, and neopentane
- C_6 : 5 isomers : hexane, 2-methylpentane, 3-methylpentane, 2,2-dimethylbutane, and 2,3-dimethylbutane
- C_{12} : 355 isomers
- C_{32} : 27,711,253,769 isomers
- C_{60} : 22,158,734,535,770,411,074,184 isomers, many of which are not stable.

Branched alkanes can be chiral. For example 3-methylhexane and its higher homologues are chiral due to their stereogenic center at carbon atom number 3. In addition to these isomers, the chain of carbon atoms may form one or more loops. Such compounds are called cycloalkanes.

NOMENCLATURE

The IUPAC nomenclature (systematic way of naming compounds) for alkanes is based on identifying hydrocarbon chains. Unbranched, saturated hydrocarbon chains are named systematically with a Greek numerical prefix denoting the number of carbons and the suffix "-ane".

In 1866, August Wilhelm von Hofmann suggested systematizing nomenclature by using the whole sequence of vowels a, e, i, o and u to create suffixes-ane,-ene,-ine (or-yne),-one,-une, for the hydrocarbons C_nH_{2n+2}, C_nH_{2n}, C_nH_{2n-2}, C_nH_{2n-4}, C_nH_{2n-6}. Now, the first three name hydrocarbons with single, double and triple bonds; "-one" represents a ketone; "-ol" represents an alcohol or OH group; "-oxy-" means an ether and refers to oxygen between two carbons, so that methoxymethane is the IUPAC name for dimethyl ether.

It is difficult or impossible to find compounds with more than one IUPAC name. This is because shorter chains attached to longer chains are prefixes and the convention includes brackets. Numbers in the name, referring to which carbon a group is attached to, should be as low as possible, so that 1-is implied and usually omitted from names of organic compounds with only one side-group. Symmetric compounds will have two ways of arriving at the same name.

LINEAR ALKANES

Straight-chain alkanes are sometimes indicated by the prefix n-(for *normal*) where a non-linear isomer exists. Although this is not strictly necessary, the usage is still common in cases where there is an important difference in properties between the straight-chain and branched-chain isomers, *e.g.*, n-hexane or 2-or 3-methylpentane.

The members of the series (in terms of number of carbon atoms) are named as follows :

Methane, CH_4-one carbon and four hydrogen

Ethane, C_2H_6-two carbon and six hydrogen

Propane, C_3H_8-three carbon and 8 hydrogen

Butane, C_4H_{10}-four carbon and 10 hydrogen

Pentane, C_5H_{12}-five carbon and 12 hydrogen

Hexane, C_6H_{14}-six carbon and 14 hydrogen.

The first four names were derived from methanol, ether, propionic acid and butyric acid, respectively. Alkanes with five or more carbon atoms are named by adding the suffix-**ane** to the appropriate numerical multiplier prefix with elision of any terminal vowel (-a or-o) from the basic numerical term. Hence, pentane, C_5H_{12}; hexane, C_6H_{14}; heptane, C_7H_{16}; octane, C_8H_{18}; etc. The prefix is generally Greek, however alkanes with a carbon atom count ending in nine, for example nonane, use the Latin prefix **non-**.

BRANCHED ALKANES

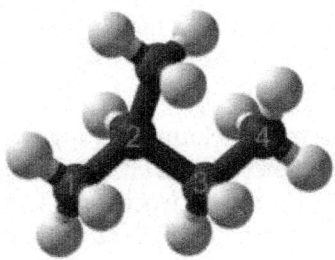

Fig. : Ball-and-stick model of isopentane (common name) or 2-methylbutane (IUPAC systematic name).

Simple branched alkanes often have a common name using a prefix to distinguish them from linear alkanes, for example n-pentane, isopentane, and neopentane.

IUPAC naming conventions can be used to produce a systematic name.

The key steps in the naming of more complicated branched alkanes are as follows :

• Identify the longest continuous chain of carbon atoms

- Name this longest root chain using standard naming rules
- Name each side chain by changing the suffix of the name of the alkane from "-ane" to "-yl"
- Number the root chain so that sum of the numbers assigned to each side group will be as low as possible
- Number and name the side chains before the name of the root chain
- If there are multiple side chains of the same type, use prefixes such as "di-" and "tri-" to indicate it as such, and number each one.
- Add side chain names in alphabetical (disregarding "di-" etc. prefixes) order in front of the name of the root chain

Table : Comparison of nomenclatures for three isomers of C_5H_{12}.

Common name	n-pentane	isopentane	neopentane
IUPAC name	pentane	2-methylbutane	2,2-dimethylpropane
Structure			

CYCLIC ALKANES

So-called cyclic alkanes are, in the technical sense, *not* alkanes, but cycloalkanes. They are hydrocarbons just like alkanes, but contain one or more rings.

Simple cycloalkanes have a prefix "cyclo-" to distinguish them from alkanes. Cycloalkanes are named as per their acyclic counterparts with respect to the number of carbon atoms, *e.g.*, cyclopentane (C_5H_{10}) is a cycloalkane with 5 carbon atoms just like pentane (C_5H_{12}), but they are joined up in a five-membered ring. In a similar manner, propane and cyclopropane, butane and cyclobutane, etc.

Substituted cycloalkanes are named similar to substituted alkanes — the cycloalkane ring is stated, and the substituents are according to their position on the ring, with the numbering decided by Cahn-Ingold-Prelog rules.

TRIVIAL NAMES

The trivial (non-systematic) name for alkanes is *paraffins*. Together, alkanes are known as the *paraffin series*. Trivial names for compounds are usually historical artifacts. They were coined before the development of systematic names, and have been retained due to familiar usage in industry. Cycloalkanes are also called naphthenes.

It is almost certain that the term *paraffin* stems from the petro-chemical industry. Branched-chain alkanes are called *isoparaffins*. The use of the term "paraffin" is a general term and often does not distinguish between pure compounds and mixtures of isomers, *i.e.*, compounds with the same chemical formula, *e.g.*, pentane and isopentane.

Examples

The following trivial names are retained in the IUPAC system :

* Isobutane for 2-methylpropane
* Isopentane for 2-methylbutane
* Neopentane for 2,2-dimethylpropane.

PHYSICAL PROPERTIES

All alkanes are colourless and odourless.

Table of alkanes

Alkane	Formula	Boiling point [°C]	Melting point [°C]	Density [g·cm⁻³] (at 20°C)
Methane	CH_4	–162	–182	gas
Ethane	C_2H_6	–89	–183	gas
Propane	C_3H_8	–42	–188	gas
Butane	C_4H_{10}	0	–138	gas
Pentane	C_5H_{12}	36	–130	0.626 (liquid)
Hexane	C_6H_{14}	69	–95	0.659 (liquid)
Heptane	C_7H_{16}	98	–91	0.684 (liquid)
Octane	C_8H_{18}	126	–57	0.703 (liquid)
Nonane	C_9H_{20}	151	–54	0.718 (liquid)
Decane	$C_{10}H_{22}$	174	–30	0.730 (liquid)
Undecane	$C_{11}H_{24}$	196	–26	0.740 (liquid)
Dodecane	$C_{12}H_{26}$	216	–10	0.749 (liquid)
Hexadecane	$C_{16}H_{34}$	287	18	0.773 (liquid)
Icosane	$C_{20}H_{42}$	343	37	solid
Triacontane	$C_{30}H_{62}$	450	66	solid
Tetracontane	$C_{40}H_{82}$	525	82	solid
Pentacontane	$C_{50}H_{102}$	575	91	solid
Hexacontane	$C_{60}H_{122}$	625	100	solid

Boiling point

Alkanes experience inter-molecular van der Waals forces. Stronger inter-molecular van der Waals forces give rise to greater boiling points of alkanes.

There are two determinants for the strength of the van der Waals forces :

* The number of electrons surrounding the molecule, which increases with the alkane's molecular weight
* The surface area of the molecule

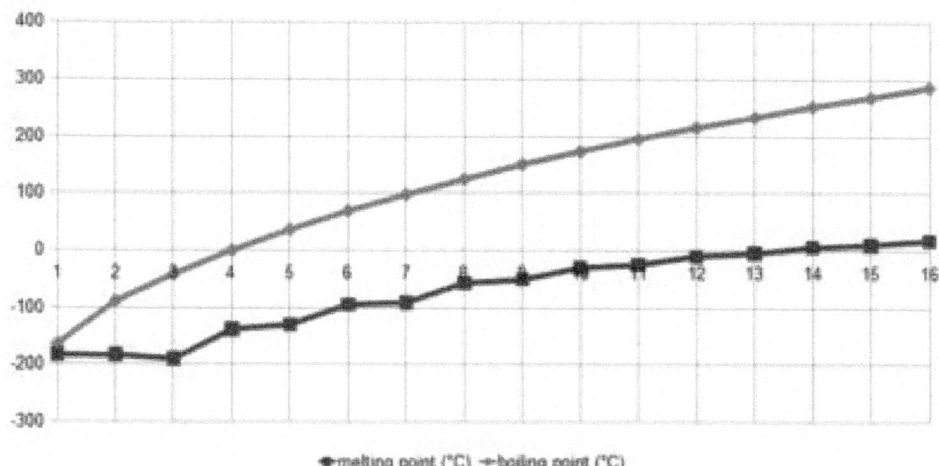

—melting point (°C) —boiling point (°C)

Fig. : Melting (blue) and boiling (orange) points of the first 16 n-alkanes in°C.

Under standard conditions, from CH_4 to C_4H_{10} alkanes are gaseous; from C_5H_{12} to $C_{17}H_{36}$ they are liquids; and after $C_{18}H_{38}$ they are solids. As the boiling point of alkanes is primarily determined by weight, it should not be a surprise that the boiling point has almost a linear relationship with the size (molecular weight) of the molecule. As a rule of thumb, the boiling point rises 20–30°C for each carbon added to the chain; this rule applies to other homologous series.

A straight-chain alkane will have a boiling point higher than a branched-chain alkane due to the greater surface area in contact, thus the greater van der Waals forces, between adjacent molecules. For example, compare isobutane (2-methylpropane) and n-butane (butane), which boil at −12 and 0°C, and 2,2-dimethylbutane and 2,3-dimethylbutane which boil at 50 and 58°C, respectively. For the latter case, two molecules 2,3-dimethylbutane can "lock" into each other better than the cross-shaped 2,2-dimethylbutane, hence the greater van der Waals forces.

On the other hand, cycloalkanes tend to have higher boiling points than their linear counterparts due to the locked conformations of the molecules, which give a plane of inter-molecular contact.

Melting Point

The melting points of the alkanes follow a similar trend to boiling points for the same reason. That is, (all other things being equal) the larger molecule the higher the melting point. There is one significant difference between boiling points and melting points. Solids have more rigid and fixed structure than liquids. This rigid structure requires energy to break down. Thus the better put together solid structures will require more energy to break apart. The odd-numbered alkanes have a lower trend in melting points than even numbered alkanes. This is because even numbered alkanes pack well in the solid phase, forming a well-organized structure, which requires more energy to break apart. The odd-number alkanes pack less well and so the "looser" organized solid packing structure requires less energy to break apart.

The melting points of branched-chain alkanes can be either higher or lower than those of the corresponding straight-chain alkanes, again depending on the ability of the alkane in question to pack well in the solid phase : This is particularly true for isoalkanes (2-methyl isomers), which often have melting points higher than those of the linear analogues.

Conductivity and Solubility

Alkanes do not conduct electricity, nor are they substantially polarized by an electric field. For this reason they do not form hydrogen bonds and are insoluble in polar solvents such as water. Since the hydrogen bonds between individual water molecules are aligned away from an alkane molecule, the co-existence of an alkane and water leads to an increase in molecular order (a reduction in entropy). As there is no significant bonding between water molecules and alkane molecules, the second law of thermodynamics suggests that this reduction in entropy should be minimized by minimizing the contact between alkane and water : Alkanes are said to be hydrophobic in that they repel water.

Their solubility in non-polar solvents is relatively good, a property that is called lipophilicity. Different alkanes are, for example, miscible in all proportions among themselves.

The density of the alkanes usually increases with increasing number of carbon atoms, but remains less than that of water. Hence, alkanes form the upper layer in an alkane-water mixture.

Molecular Geometry

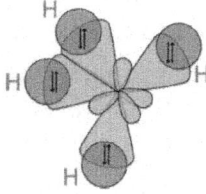

Fig. : sp^3-hybridization in methane.

The molecular structure of the alkanes directly affects their physical and chemical characteristics. It is derived from the electron configuration of carbon, which has four valence electrons. The carbon atoms in alkanes are always sp^3 hybridized, that is to say that the valence electrons are said to be in four equivalent orbitals derived from the combination of the 2s orbital and the three 2p orbitals. These orbitals, which have identical energies, are arranged spatially in the form of a tetrahedron, the angle of $\cos^{-1}(-\frac{1}{3}) \approx 109.47°$ between them.

Bond Lengths and Bond Angles

An alkane molecule has only C–H and C–C single bonds. The former result from the overlap of a sp³-orbital of carbon with the 1s-orbital of a hydrogen; the latter by the overlap of two sp³-orbitals on different carbon atoms. The bond lengths amount to 1.09×10^{-10} m for a C–H bond and 1.54×10^{-10} m for a C–C bond.

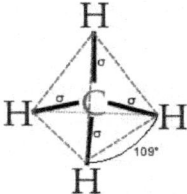

Fig. : The tetrahedral structure of methane.

The spatial arrangement of the bonds is similar to that of the four sp³-orbitals — they are tetrahedrally arranged, with an angle of 109.47° between them. Structural formulae that represent the bonds as being at right angles to one another, while both common and useful, do not correspond with the reality.

Conformation

The structural formula and the bond angles are not usually sufficient to completely describe the geometry of a molecule. There is a further degree of freedom for each carbon–carbon bond : the torsion angle between the atoms or groups bound to the atoms at each end of the bond. The spatial arrangement described by the torsion angles of the molecule is known as its conformation.

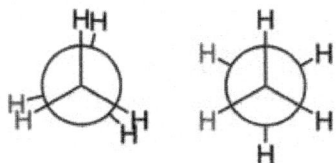

Fig. : Newman projections of the two conformations of ethane : eclipsed on the left, staggered on the right.

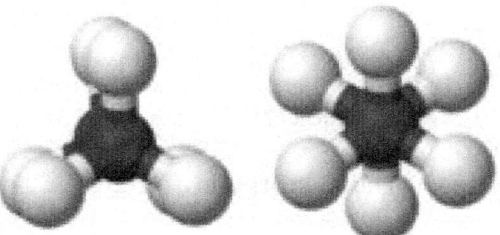

Fig. : Ball-and-stick models of the two rotamers of ethane.

Ethane forms the simplest case for studying the conformation of alkanes, as there is only one C–C bond. If one looks down the axis of the C–C bond, one will see the so-called Newman projection. The hydrogen atoms on both the front and rear carbon atoms have an angle of 120° between them, resulting from the projection of the base of the tetrahedron onto a flat plane. However, the torsion angle between a given hydrogen atom attached to the front carbon and a given hydrogen atom attached to the rear carbon can vary freely between 0° and 360°. This is a consequence of the free rotation about a carbon–carbon single bond. Despite this apparent freedom, only two limiting conformations are important : eclipsed conformation and staggered conformation.

The two conformations, also known as rotamers, differ in energy : The staggered conformation is 12.6 kJ/mol lower in energy (more stable) than the eclipsed conformation (the least stable).

This difference in energy between the two conformations, known as the torsion energy, is low compared to the thermal energy of an ethane molecule at ambient temperature. There is constant rotation about the C-C bond. The time taken for an ethane molecule to pass from one staggered conformation to the next, equivalent to the rotation of one CH_3-group by 120° relative to the other, is of the order of 10^{-11} seconds.

The case of higher alkanes is more complex but based on similar principles, with the anti-periplanar conformation always being the most favoured around each carbon-carbon bond. For this reason, alkanes are usually shown in a zigzag arrangement in diagrams or in models. The actual structure will always differ somewhat from these idealized forms, as the differences in energy between the conformations are small compared to the thermal energy of the molecules : Alkane molecules have no fixed structural form, whatever the models may suggest.

Spectroscopic Properties

Virtually all organic compounds contain carbon–carbon and carbon–hydrogen bonds, and so show some of the features of alkanes in their spectra. Alkanes are notable for having no other groups, and therefore for the *absence* of other characteristic spectroscopic features of different functional group like-OH,-CHO,-COOH etc.

Infrared Spectroscopy

The carbon–hydrogen stretching mode gives a strong absorption between 2850 and 2960 cm^{-1}, while the carbon–carbon stretching mode absorbs between 800 and 1300 cm^{-1}. The carbon–hydrogen bending modes depend on the nature of the group : methyl groups show bands at 1450 cm^{-1} and 1375 cm^{-1}, while methylene groups show bands at 1465 cm^{-1} and 1450 cm^{-1}. Carbon chains with more than four carbon atoms show a weak absorption at around 725 cm^{-1}.

NMR Spectroscopy

The proton resonances of alkanes are usually found at $\delta_H = 0.5$–1.5. The carbon-13 resonances depend on the number of hydrogen atoms attached to the carbon :

δ_C = 8–30 (primary, methyl,-CH_3), 15–55 (secondary, methylene,-CH_2-), 20–60 (tertiary, methyne, C-H) and quaternary. The carbon-13 resonance of quaternary carbon atoms is characteristically weak, due to the lack of Nuclear Overhauser effect and the long relaxation time, and can be missed in weak samples, or samples that have not been run for a sufficiently long time.

Mass Spectrometry

Alkanes have a high ionization energy, and the molecular ion is usually weak. The fragmentation pattern can be difficult to interpret, but, in the case of branched chain alkanes, the carbon chain is preferentially cleaved at tertiary or quaternary carbons due to the relative stability of the resulting free radicals. The fragment resulting from the loss of a single methyl group (M−15) is often absent, and other fragment are often spaced by intervals of fourteen mass units, corresponding to sequential loss of CH_2-groups.

CHEMICAL PROPERTIES

Alkanes are only weakly reactive with ionic and other polar substances. The acid dissociation constant (pK_a) values of all alkanes are above 60, hence they are practically inert to acids and bases. This inertness is the source of the term *paraffins* (with the meaning here of "lacking affinity"). In crude oil the alkane molecules have remained chemically unchanged for millions of years.

However redox reactions of alkanes, in particular with oxygen and the halogens, are possible as the carbon atoms are in a strongly reduced condition; in the case of methane, the lowest possible oxidation state for carbon (−4) is reached. Reaction with oxygen (*if* present in sufficient quantity to satisfy the reaction stoichiometry) leads to combustion without any smoke, producing carbon dioxide and water. Free radical halogenation reactions occur with halogens, leading to the production of haloalkanes. In addition, alkanes have been shown to interact with, and bind to, certain transition metal complexes in.

Free radicals, molecules with unpaired electrons, play a large role in most reactions of alkanes, such as cracking and reformation where long-chain alkanes are converted into shorter-chain alkanes and straight-chain alkanes into branched-chain isomers.

In highly branched alkanes, the bond angle may differ significantly from the optimal value (109.5°) in order to allow the different groups sufficient space. This causes a tension in the molecule, known as steric hindrance, and can substantially increase the reactivity.

Reactions with Oxygen (Combustion Reaction)

All alkanes react with oxygen in a combustion reaction, although they become increasingly difficult to ignite as the number of carbon atoms increases. The general equation for complete combustion is :

$$C_nH_{2n+2} + (1.5n+0.5)O_2 \rightarrow (n+1)H_2O + nCO_2$$

or $C_nH_{2n+2} + ((3n+1)/2)O_2 \rightarrow (n+1)H_2O + nCO_2$

In the absence of sufficient oxygen, carbon monoxide or even soot can be formed, as shown below :

$$C_nH_{(2n+2)} + (n+0.5)O_2 \rightarrow (n+1)H_2O + nCO$$

$$C_nH_{(2n+2)} + (0.5n+0.5)O_2 \rightarrow (n+1)H_2O + nC$$

For example methane :

$$2CH_4 + 3O_2 \rightarrow 2CO + 4H_2O$$

$$CH_4 + 1.5O_2 \rightarrow CO + 2H_2O$$

The standard enthalpy change of combustion, $\Delta_c H^\circ$, for alkanes increases by about 650 kJ/mol per CH_2 group. Branched-chain alkanes have lower values of $\Delta_c H^\circ$ than straight-chain alkanes of the same number of carbon atoms, and so can be seen to be somewhat more stable.

Reactions with Halogens

Alkanes react with halogens in a so-called *free radical halogenation* reaction. The hydrogen atoms of the alkane are progressively replaced by halogen atoms. Free-radicals are the reactive species that participate in the reaction, which usually leads to a mixture of products. The reaction is highly exothermic, and can lead to an explosion.

These reactions are an important industrial route to halogenated hydrocarbons. There are three steps :

- **Initiation** the halogen radicals form by homolysis. Usually, energy in the form of heat or light is required.
- **Chain reaction** or **Propagation** then takes place — the halogen radical abstracts a hydrogen from the alkane to give an alkyl radical. This reacts further.
- **Chain termination** where step the radicals recombine.

Experiments have shown that all halogenation produces a mixture of all possible isomers, indicating that all hydrogen atoms are susceptible to reaction. The mixture produced, however, is not a statistical mixture : Secondary and tertiary hydrogen atoms are preferentially replaced due to the greater stability of secondary and tertiary free-radicals. An example can be seen in the monobromination of propane.

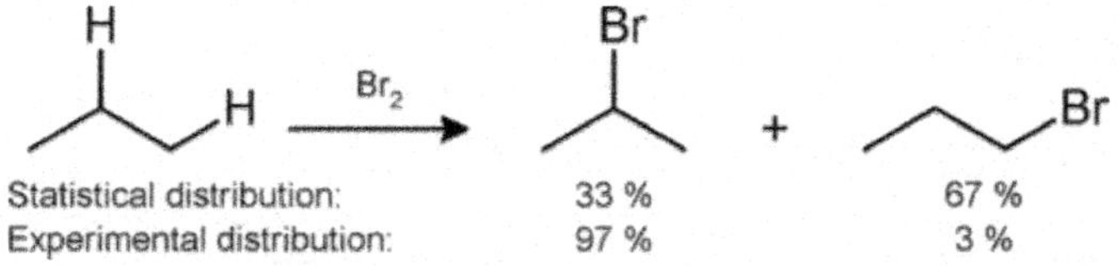

Statistical distribution:		33 %		67 %
Experimental distribution:		97 %		3 %

Cracking

Cracking breaks larger molecules into smaller ones. This can be done with a thermal or catalytic method. The thermal cracking process follows a homolytic mechanism with formation of free-radicals. The catalytic cracking process involves the presence of acid catalysts (usually solid acids such as silica-alumina and zeolites), which promote a heterolytic (asymmetric) breakage of bonds yielding pairs of ions of opposite charges, usually a carbocation and the very unstable hydride anion. Carbon-localized free-radicals and cations are both highly unstable and undergo processes of chain re-arrangement, C-C scission in position beta (*i.e.*, cracking) and intra-and inter-molecular hydrogen transfer or hydride transfer. In both types of processes, the corresponding reactive intermediates (radicals, ions) are permanently regenerated, and thus they proceed by a self-propagating chain mechanism. The chain of reactions is eventually terminated by radical or ion recombination.

Isomerization and Reformation

Dragan and his colleague were the first to report about isomerization in alkanes. Isomerization and reformation are processes in which straight-chain alkanes are heated in the presence of a platinum catalyst. In isomerization, the alkanes become branched-chain isomers. In other words, it does not lose any carbons or hydrogens, keeping the same molecular weight. In reformation, the alkanes become cycloalkanes or aromatic hydrocarbons, giving off hydrogen as a by-product. Both of these processes raise the octane number of the substance. Butane is the most common alkane that is put under the process of isomerization, as it makes many branched alkanes with high octane numbers.

Other Reactions

Alkanes will react with steam in the presence of a nickel catalyst to give hydrogen. Alkanes can be chlorosulfonated and nitrated, although both reactions require special conditions. The fermentation of alkanes to carboxylic acids is of some technical importance. In the Reed reaction, sulfur dioxide, chlorine and light convert hydrocarbons to sulfonyl chlorides. Nucleophilic Abstraction can be used to separate an alkane from a metal. Alkyl groups can be transferred from one compound to another by transmetalation reactions.

OCCURRENCE

Occurrence of Alkanes in the Universe

Alkanes form a small portion of the atmospheres of the outer gas planets such as Jupiter (0.1% methane, 0.0002% ethane), Saturn (0.2% methane, 0.0005% ethane), Uranus (1.99% methane, 0.00025% ethane) and Neptune (1.5% methane, 1.5 ppm ethane). Titan (1.6% methane), a satellite of Saturn, was examined by the *Huygens* probe, which indicated that Titan's atmosphere periodically rains liquid methane onto the moon's surface. Also on Titan the Cassini mission has imaged seasonal Methane/Ethane lakes near the polar regions of Titan. Methane and ethane have also been detected in the tail of the comet Hyakutake. Chemical analysis showed that the abundances of ethane and methane were roughly equal, which is thought to imply that its ices formed in inter-stellar space, away from the Sun, which would have evaporated these volatile molecules. Alkanes have also been detected in meteorites such as carbonaceous chondrites.

Occurrence of Alkanes on Earth

Traces of methane gas (about 0.0002% or 1745 ppb) occur in the Earth's atmosphere, produced primarily by methanogenic micro-organisms, such as Archaea in the gut of ruminants.

The most important commercial sources for alkanes are natural gas and oil. Natural gas contains primarily methane and ethane, with some propane and butane : oil is a mixture of liquid alkanes and other hydrocarbons. These hydrocarbons were formed when marine animals and plants (zooplankton and phytoplankton) died and sank to the bottom of ancient seas and were covered with sediments in an anoxic environment and converted over many millions of years at high temperatures and high pressure to their current form. Natural gas resulted thereby for example from the following reaction :

$$C_6H_{12}O_6 \rightarrow 3CH_4 + 3CO_2$$

These hydrocarbon deposits, collected in porous rocks trapped beneath impermeable cap rocks, comprise commercial oil fields. They have formed over millions of years and once exhausted cannot be readily replaced. The depletion of these hydrocarbons reserves is the basis for what is known as the energy crisis.

Methane is also present in what is called biogas, produced by animals and decaying matter, which is a possible renewable energy source.

Alkanes have a low solubility in water, so the content in the oceans is negligible; however, at high pressures and low temperatures (such as at the bottom of the oceans), methane can co-crystallize with water to form a solid methane clathrate (methane hydrate). Although this cannot be commercially exploited at the present time, the amount of combustible energy of the known methane clathrate fields exceeds the energy content of all the natural gas and oil deposits put together. Methane extracted from methane clathrate is therefore a candidate for future fuels.

Biological Occurrence

Acyclic alkanes occur in nature in various ways.

Bacteria and Archaea

Certain types of bacteria can metabolize alkanes : they prefer even-numbered carbon chains as they are easier to degrade than odd-numbered chains.

On the other hand, certain archaea, the methanogens, produce large quantities of methane by the metabolism of carbon dioxide or other oxidized organic compounds. The energy is released by the oxidation of hydrogen :

$$CO_2 + 4H_2 \rightarrow CH_4 + 2H_2O$$

Methanogens are also the producers of marsh gas in wetlands, and release about two billion tonnes of methane per year — the atmospheric content of this gas is produced nearly exclusively by them. The methane output of cattle and other herbivores, which can release up to 150 liters per day, and of termites, is also due to methanogens. They also produce this simplest of all alkanes in the intestines of humans. Methanogenic archaea are, hence, at the end of the carbon cycle, with carbon being released back into the atmosphere after having been fixed by photosynthesis. It is probable that our current deposits of natural gas were formed in a similar way.

Fungi and Plants

Alkanes also play a role, if a minor role, in the biology of the three eukaryotic groups of organisms : fungi, plants and animals. Some specialized yeasts, *e.g.*, *Candida tropicale*, *Pichia* sp., *Rhodotorula* sp., can use alkanes as a source of carbon and/or energy. The fungus *Amorphotheca resinae* prefers the longer-chain alkanes in aviation fuel, and can cause serious problems for aircraft in tropical regions.

In plants, the solid long-chain alkanes are found in the plant cuticle and epicuticular wax of many species, but are only rarely major constituents. They protect the plant against water loss, prevent the leaching of important minerals by the rain, and protect against bacteria, fungi, and harmful insects. The carbon chains in plant alkanes are usually odd-numbered, between twenty-seven and thirty-three carbon atoms in length and are made by the plants by decarboxylation of even-numbered fatty acids. The exact composition of the layer of wax is not only species-dependent, but changes also with the season and such environmental factors as lighting conditions, temperature or humidity.

More volatile short-chain alkanes are also produced by and found in plant tissues. The Jeffrey pine is noted for producing exceptionally high levels of *n*-heptane in its resin, for which reason its distillate was designated as the zero point for one octane rating. Floral scents have also long been known to contain volatile alkane components, and *n*-nonane is a significant component in the scent of some roses. Emission of gaseous and volatile alkanes such as ethane, pentane, and hexane by plants has also been documented at low levels, though they are not generally considered to be a major component of biogenic air pollution.

Edible vegetable oils also typically contain small fractions of biogenic alkanes with a wide spectrum of carbon numbers, mainly 8 to 35, usually peaking in the low to upper 20s, with concentrations up to dozens of milligrams per kilogram (parts per million by weight) and sometimes over a hundred for the total alkane fraction.

Animals

Alkanes are found in animal products, although they are less important than unsaturated hydrocarbons. One example is the shark liver oil, which is approximately 14% pristane (2,6,10,14-tetra-methylpentadecane, $C_{19}H_{40}$). They are important as pheromones, chemical messenger materials, on which insects depend for communication. In some species, *e.g.* the support beetle *Xylotrechus colonus*, pentacosane ($C_{25}H_{52}$), 3-methylpentaicosane ($C_{26}H_{54}$) and 9-methylpentaicosane ($C_{26}H_{54}$) are transferred by body contact. With others like the tsetse fly *Glossina morsitans morsitans*, the pheromone contains the four alkanes 2-methylheptadecane ($C_{18}H_{38}$), 17,21-dimethylheptatriacontane ($C_{39}H_{80}$), 15,19-dimethylheptatriacontane ($C_{39}H_{80}$) and 15,19,23-trimethylheptatriacontane ($C_{40}H_{82}$), and acts by smell over longer distances. Waggle-dancing honey bees produce and release two alkanes, tricosane and pentacosane.

Ecological Relations

One example, in which both plant and animal alkanes play a role, is the ecological relationship between the sand bee (*Andrena nigroaenea*) and the early spider orchid (*Ophrys sphegodes*); the latter is dependent for pollination on the former. Sand bees use pheromones in order to identify a mate; in the case of *A. nigroaenea*, the females emit a mixture of tricosane ($C_{23}H_{48}$), pentacosane ($C_{25}H_{52}$) and heptacosane ($C_{27}H_{56}$) in the ratio 3 : 3 : 1, and males are attracted by specifically this odour. The orchid takes advantage of this mating arrangement to get the male bee to collect and disseminate its pollen; parts of its flower not only resemble the appearance of sand bees, but also produce large quantities of the three alkanes in the same ratio as female sand bees. As a result numerous males are lured to the blooms and attempt to copulate with their imaginary partner : although this endeavour is not crowned with success for the bee, it allows the orchid to transfer its pollen, which will be dispersed after the departure of the frustrated male to different blooms.

PRODUCTION

Petroleum Refining

As stated earlier, the most important source of alkanes is natural gas and crude oil. Alkanes are separated in an oil refinery by fractional distillation and processed into many different products.

Fischer-Tropsch

The Fischer-Tropsch process is a method to synthesize liquid hydrocarbons, including alkanes, from carbon monoxide and hydrogen. This method is used to produce substitutes for petroleum distillates.

Laboratory Preparation

There is usually little need for alkanes to be synthesized in the laboratory, since they are usually commercially available. Also, alkanes are generally non-reactive chemically or biologically, and do not undergo functional group inter-conversions cleanly. When alkanes are produced in the laboratory, it is often a side-product of a reaction. For example, the use of n-butyllithium as a strong base gives the conjugate acid, n-butane as a side-product :

$$C_4H_9Li + H_2O \rightarrow C_4H_{10} + LiOH$$

However, at times it may be desirable to make a portion of a molecule into an alkane like functionality (alkyl group). For example, an ethyl group is an alkyl group; when this is attached to a hydroxy group, it gives ethanol, which is not an alkane. To do so, the best-known methods are hydrogenation of alkenes :

$$RCH=CH_2 + H_2 \rightarrow RCH_2CH_3 \ (R = alkyl)$$

Alkanes or alkyl groups can also be prepared directly from alkyl halides in the Corey-House-Posner-Whitesides reaction. The Barton-McCombie deoxygenation removes hydroxyl groups from alcohols *e.g.*

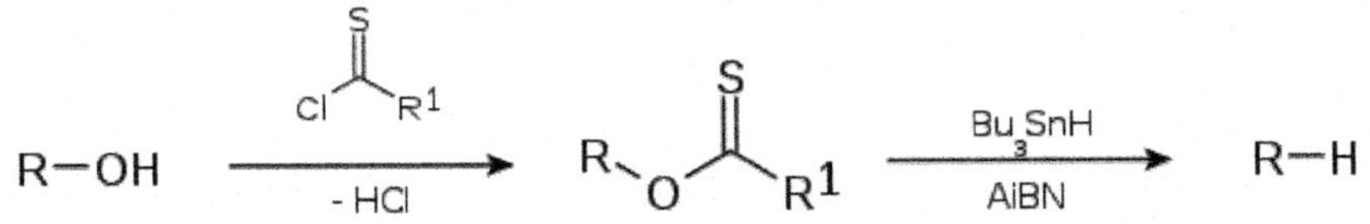

and the Clemmensen reduction removes carbonyl groups from aldehydes and ketones to form alkanes or alkyl-substituted compounds *e.g.* :

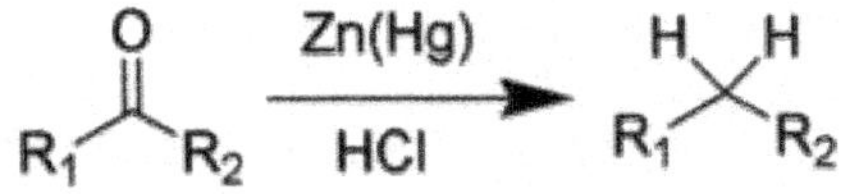

APPLICATIONS

The applications of a certain alkane can be determined quite well according to the number of carbon atoms. The first four alkanes are used mainly for heating and cooking purposes, and in some countries for electricity generation. Methane and ethane are the main components of natural gas; they are normally stored as gases under pressure. It is, however, easier to transport them as liquids : This requires both compression and cooling of the gas.

Propane and butane can be liquefied at fairly low pressures, and are well known as liquified petroleum gas (LPG). Propane, for example, is used in the propane gas burner and as a fuel for cars, butane in disposable cigarette lighters. The two alkanes are used as propellants in aerosol sprays.

From pentane to octane the alkanes are reasonably volatile liquids. They are used as fuels in internal combustion engines, as they vapourise easily on entry into the combustion chamber without forming droplets, which would impair the uniformity of the combustion. Branched-chain alkanes are preferred as they are much less prone to premature ignition, which causes knocking, than their straight-chain homologues. This propensity to premature ignition is measured by the octane rating of the fuel, where 2,2,4-trimethylpentane (*isooctane*) has an arbitrary value of 100, and heptane has a value of zero. Apart from their use as fuels, the middle alkanes are also good solvents for non-polar substances.

Alkanes from nonane to, for instance, hexadecane (an alkane with sixteen carbon atoms) are liquids of higher viscosity, less and less suitable for use in gasoline. They form instead the major part of diesel and aviation fuel. Diesel fuels are characterized by their cetane number, cetane being an old name for hexadecane. However, the higher melting points of these alkanes can cause problems at low temperatures and in polar regions, where the fuel becomes too thick to flow correctly.

Alkanes from hexadecane upwards form the most important components of fuel oil and lubricating oil. In the latter function, they work at the same time as anti-corrosive agents, as their hydrophobic nature means that water cannot reach the metal surface. Many solid alkanes find use as paraffin wax, for example, in candles. This should not be confused however with true wax, which consists primarily of esters.

Alkanes with a chain length of approximately 35 or more carbon atoms are found in bitumen, used, for example, in road surfacing. However, the higher alkanes have little value and are usually split into lower alkanes by cracking.

Some synthetic polymers such as polyethylene and polypropylene are alkanes with chains containing hundreds of thousands of carbon atoms. These materials are used in innumerable applications, and billions of kilograms of these materials are made and used each year.

ENVIRONMENTAL TRANSFORMATIONS

When released in the environment, alkanes don't undergo rapid biodegradation, because they have no functional groups (like hydroxyl or carbonyl) that are needed by most organisms in order to metabolize the compound.

However, some bacteria can metabolize some alkanes (especially those linear and short), by oxidizing the terminal carbon atom. The product is an alcohol, that could be next oxidized to an aldehyde, and finally to a carboxylic acid. The resulting fatty acid could be metabolized through the fatty acid degradation pathway.

Hazards

Methane is explosive when mixed with air (1–8% CH_4). Other lower alkanes can also form explosive mixtures with air. The lighter liquid alkanes are highly flammable, although this risk decreases with the length of the carbon chain. Pentane, hexane, heptane, and octane are classed as *dangerous for the environment* and *harmful*. The straight-chain isomer of hexane is a neurotoxin.

Considerations for Detection/Risk Control

- Methane is lighter than air (possibility of accumulation under roofs)
- Ethane is slightly heavier than air (possibility of pooling at ground levels/pits)
- Propane is heavier than air (possibility of pooling at ground levels/pits)
- Butane is heavier than air (possibility of pooling at ground levels/pits).

ALKENE

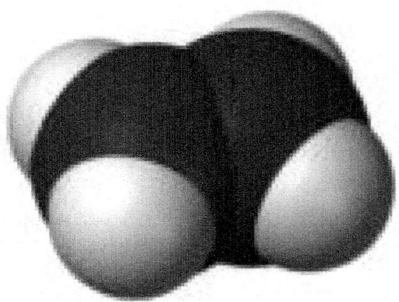

Fig. : A 3D model of ethylene, the simplest alkene.

In organic chemistry, an **alkene, olefin,** or **olefine** is an unsaturated chemical compound containing at least one carbon–carbon double bond. The simplest acyclic alkenes, with only one double bond and no other functional groups, known as mono-enes, form a homologous series of hydrocarbons with the general formula C_nH_{2n}. They have two hydrogen atoms less than the corresponding alkane (with the same number of carbon atoms).

The simplest alkene is ethylene (C_2H_4), which has the International Union of Pure and Applied Chemistry (IUPAC) name *ethene*. In the petro-chemical industry alkenes are often called *olefins*. For bridged alkenes, the Bredt's rule states that a double bond cannot be placed at the bridgehead of a bridged ring system, unless the rings are large enough (8 or more atoms). Aromatic compounds are often drawn as cyclic alkenes, but their structure and properties are different and they are not considered to be alkenes.

Structure

Bonding

Like single covalent bonds, double bonds can be described in terms of overlapping atomic orbitals, except that, unlike a single bond (which consists of a single sigma bond), a carbon–carbon double bond consists of one sigma bond and one pi bond. This double bond is stronger than a single covalent bond (611 kJ/mol for C=C *vs.* 347 kJ/mol for C–C) and also shorter with an average bond length of 1.33 Angstroms (133 pm).

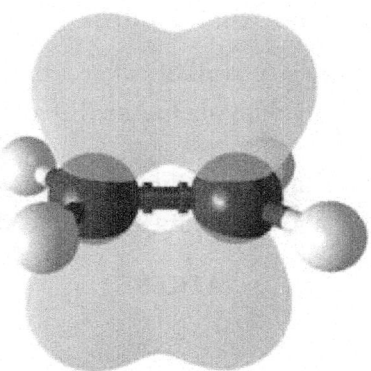

Fig. : Ethylene (ethene), showing the pi bond in green.

Each carbon of the double bond uses its three sp^2 hybrid orbitals to form sigma bonds to three atoms. The unhybridized $2p$ atomic orbitals, which lie perpendicular to the plane created by the axes of the three sp^2 hybrid orbitals, combine to form the pi bond. This bond lies outside the main C–C axis, with half of the bond on one side and half on the other.

Rotation about the carbon–carbon double bond is restricted because it involves breaking the pi bond, which requires a large amount of energy (264 kJ/mol in ethylene). As a consequence, substituted alkenes may exist as one of two isomers, called *cis* or *trans* isomers. More complex alkenes may be named using the E-Z notation, used to describe molecules having three or four different substituents (side groups). For example, of the isomers of butene, the two methyl groups of (Z)-but-2-ene (aka *cis*-2-butene) face the same side of the double bond, and in (E)-but2-ene (aka *trans*-2-butene) the methyl groups face the opposite side. These two isomers of butene are slightly different in their chemical and physical properties.

It is certainly not impossible to twist a double bond. In fact, a 90° twist requires an energy approximately equal to half the strength of a pi bond. The misalignment of the p orbitals is less than expected because pyramidalization takes place. *trans*-Cyclooctene is a stable strained alkene and the orbital misalignment is only 19° with a dihedral angle of 137° (normal 120°) and a degree of pyramidalization of 18°. This explains the dipole moment of 0.8 D for this compound (cis-isomer 0.4 D) where a value of zero is expected. The *trans* isomer of cycloheptene is only stable at low temperatures.

Shape

As predicted by the VSEPR model of electron pair repulsion, the molecular geometry of alkenes includes bond angles about each carbon in a double bond of about 120°. The angle may vary because of steric strain introduced by non-bonded interactions created by functional groups attached to the carbons of the double bond. For example, the C-C-C bond angle in propylene is 123.9°.

Physical Properties

The physical properties of alkenes are comparable with those of alkanes. The main differences between the two are that the acidity levels of alkenes are much higher than the ones in alkanes. The physical state depends on molecular mass. The simplest alkenes, ethene, propene and butene are gases. Linear alkenes of approximately five to sixteen carbons are liquids, and higher alkenes are waxy solids.

Reactions

Alkenes are relatively stable compounds, but are more reactive than alkanes due to the presence of a carbon–carbon pi-bond. It is also attributed to the presence of pi-electrons in the molecule. The majority of the reactions of alkenes involve the rupture of this pi bond, forming new single bonds.

Alkenes serve as a feedstock for the petro-chemical industry because they can participate in a wide variety of reactions.

Addition Reactions

Alkenes react in many addition reactions, which occur by opening up the double-bond. Most addition reactions to alkenes follow the mechanism of electrophilic addition. Examples of addition reactions are hydrohalogenation, halogenation, halohydrin formation, oxymercuration, hydroboration, dichlorocarbene addition, Simmons–Smith reaction, catalytic hydrogenation, epoxidation, radical polymerization and hydroxylation.

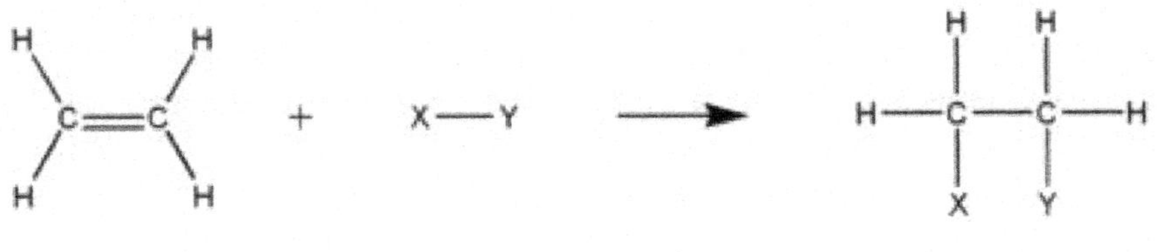

Hydrogenation

Hydrogenation of alkenes produces the corresponding alkanes. The reaction is carried out under pressure at a temperature of 200°C in the presence of a metallic catalyst. Common industrial catalysts are based on platinum, nickel or palladium. For laboratory syntheses, Raney nickel (an alloy of nickel and aluminium) is often employed. The simplest example of this reaction is the catalytic hydrogenation of ethylene to yield ethane :

$$CH_2=CH_2 + H_2 \rightarrow CH_3\text{-}CH_3$$

Hydration

Hydration, the addition of water across the double bond of alkenes, yields alcohols. The reaction is catalyzed by strong acids such as sulfuric acid. This reaction is carried out on an industrial scale to make ethanol.

$$CH_2=CH_2 + H_2O \rightarrow CH_3\text{-}CH_2OH$$

Alkenes can also be converted into alcohols via the oxymercuration–demercuration reaction or hydroboration–oxidation reaction.

Halogenation

In electrophilic halogenation the addition of elemental bromine or chlorine to alkenes yields vicinal dibromo-and dichloroalkanes (1,2-dihalides or ethylene dihalides), respectively. The decolouration of a solution of bromine in water with dichloromethylene as catalyst is an analytical test for the presence of alkenes :

$$CH_2=CH_2 + Br_2 \rightarrow BrCH_2\text{-}CH_2Br$$

It is also used as a quantitative test of unsaturation, expressed as the bromine number of a single compound or mixture. The reaction works because the high electron density at the double bond causes a temporary shift of electrons in the Br-Br bond causing a temporary induced dipole. This makes the Br closest to the double bond slightly positive and therefore an electrophile.

Hydrohalogenation

Hydrohalogenation is the addition of hydrohalic acids such as HCl or HI to alkenes to yield the corresponding haloalkanes.

$$CH_3\text{-}CH=CH_2 + HI \rightarrow CH_3\text{-}CHI\text{-}CH_2\text{-}\mathbf{H}$$

If the two carbon atoms at the double bond are linked to a different number of hydrogen atoms, the halogen is found preferentially at the carbon with fewer hydrogen substituents (Markovnikov's rule). The use of radical initiators or other chemicals can lead to the opposite product result, but the reactive-intermediate structures and mechanisms are different. Hydrobromic acid in particular is prone to forming radicals in the presence of various impurities or even atmospheric oxygen, leading to the reversal of the Markovnikov result :

$$CH_3\text{-}CH=CH_2 + HBr \rightarrow CH_3\text{-}CHH\text{-}CH_2\text{-}\mathbf{Br}$$

Halohydrin Formation

Alkenes react with water and halogens to form halohydrins by an addition reaction. Markovnikov regiochemistry and anti stereochemistry occur.

$$CH_2=CH_2 + X_2 + H_2O \rightarrow XCH_2\text{-}CH_2OH + HX$$

Oxidation

Alkenes are oxidized with a large number of oxidizing agents. In the presence of oxygen, alkenes burn with a bright flame to produce carbon dioxide and water. Catalytic oxidation with oxygen or the reaction with percarboxylic acids yields epoxides. Reaction with ozone in ozonolysis leads to the breaking of the double bond, yielding two aldehydes or ketones. Reaction with concentrated, hot $KMnO_4$ (or other oxidizing salts) in an acidic solution will yield ketones or carboxylic acids.

$$R_1\text{-CH=CH-R}_2 + O_3 \rightarrow R_1\text{-CHO} + R_2\text{-CHO} + H_2O$$

This reaction can be used to determine the position of a double bond in an unknown alkene.

The oxidation can be stopped at the vicinal diol rather than full cleavage of the alkene by using milder (dilute, lower temperature) $KMnO_4$ or with osmium tetroxide or other oxidants.

Photo-oxygenation

In the presence of an appropriate photo-sensitiser, such as methylene blue and light, alkenes can undergo reactions with reactive oxygen species generated by the photo-sensitiser, such as hydroxyl radicals, singlet oxygen or superoxide ion. These reactive photo-chemical intermediates are generated in what are known as Type I, Type II, and Type III processes, respectively. These various alternative processes and reactions can be controlled by choice of specific reaction conditions, leading to a wide range of different products. A common example is the [4+2] cycloaddition of singlet oxygen with a diene such as cyclopentadiene to yield an endoperoxide :

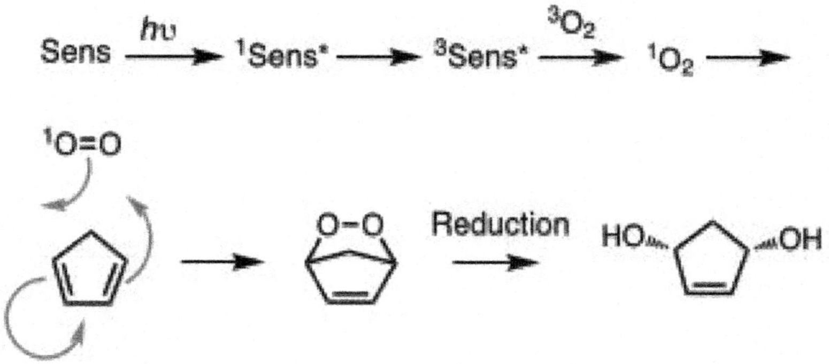

Another example is the Schenck ene reaction, in which singlet oxygen reacts with an allylic structure to give a transposed allyl peroxide :

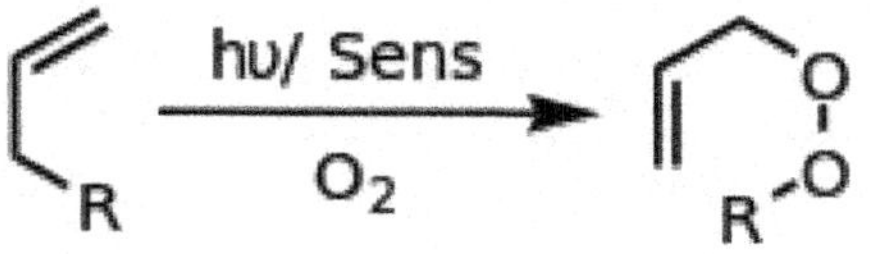

Polymerization

Polymerization of alkenes is a reaction that yields polymers of high industrial value at great economy, such as the plastics polyethylene and polypropylene. Polymers from alkene monomers are referred to in a general way as *polyolefins* or in rare instances as *polyalkenes*. A polymer from alpha-olefins is called a polyal-phaolefin (PAO). Polymerization can proceed via either a free-radical or an ionic mechanism, converting the double to a single bond and forming single bonds to join the other monomers. Polymerization of conjugated dienes such as buta-1,3-diene or isoprene (2-methylbuta-1,3-diene) results in largely 1,4-addition with possibly some 1,2-addition of the diene monomer to a growing polymer chain.

Metal Complexation

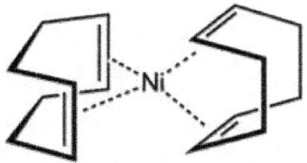

Fig. : Structure of bis(cyclooctadiene) nickel(0), a metal–alkene complex.

Alkenes can be ligands in metal complexes. The two carbon centres bond to the metal using the C-C pi-and pi* orbitals. Mono-and diolefins are often used as ligands in stable complexes. Cyclooctadiene and norbornadiene are popular chelating agents, and even ethylene itself is sometimes used as a ligand, for example, in Zeise's salt. In addition, metal–alkene complexes are intermediates in many metal-catalyzed reactions including hydrogenation, hydroformylation, and polymerization.

Reaction Overview

Reaction name	Product	Comment
Hydrogenation	alkanes	addition of hydrogen
Hydroalkenylation	alkenes	hydrometalation/insertion/beta elimination by metal catalyst
Halogen addition reaction	1,2-dihalide	electrophilic addition of halogens
Hydrohalogenation (Markovnikov)	haloalkanes	addition of hydrohalic acids
Antimarkovnikov hydrohalogenation	haloalkanes	free radicals mediated addition of hydrohalic acids
Hydroamination	amines	addition of N-H bond across C-C double bond
Hydroformylation	aldehydes	industrial process, addition of CO and H_2
Sharpless bishydroxylation	diols	oxidation, reagent : osmium tetroxide, chiral ligand
Woodward cis-hydroxylation	diols	oxidation, reagents : iodine, silver acetate
Ozonolysis	aldehydes or ketones	reagent : ozone
Olefin metathesis	alkenes	two alkenes rearrange to form two new alkenes

(Contd...)

(Contd...)

Reaction name	Product	Comment
Diels-Alder reaction	cyclohexenes	cycloaddition with a diene
Pauson-Khand reaction	cyclopentenones	cycloaddition with an alkyne and CO
Hydroboration–oxidation	alcohols	reagents : borane, then a peroxide
Oxymercuration-reduction	alcohols	electrophilic addition of mercuric acetate, then reduction
Prins reaction	1,3-diols	electrophilic addition with aldehyde or ketone
Paterno–Büchi reaction	oxetanes	photo-chemical reaction with aldehyde or ketone
Epoxidation	epoxide	electrophilic addition of a peroxide
Cyclopropanation	cyclopropanes	addition of carbenes or carbenoids
Hydroacylation	ketones	oxidative addition/reductive elimination by metal catalyst

SYNTHESIS

Industrial Methods

Alkenes are produced by hydrocarbon cracking. Raw materials are mostly natural gas condensate components (principally ethane and propane) in the US and Mideast and naphtha in Europe and Asia. Alkanes are broken apart at high temperatures, often in the presence of a zeolite catalyst, to produce a mixture of primarily aliphatic alkenes and lower molecular weight alkanes. The mixture is feedstock and temperature dependent, and separated by fractional distillation. This is mainly used for the manufacture of small alkenes (up to six carbons).

Related to this is catalytic dehydrogenation, where an alkane loses hydrogen at high temperatures to produce a corresponding alkene. This is the reverse of the catalytic hydrogenation of alkenes.

This process is also known as reforming. Both processes are endothermic and are driven towards the alkene at high temperatures by entropy.

Catalytic synthesis of higher α-alkenes (of the type $RCH=CH_2$) can also be achieved by a reaction of ethylene with the organo-metallic compound triethylaluminium in the presence of nickel, cobalt, or platinum.

Elimination Reactions

One of the principal methods for alkene synthesis in the laboratory is the elimination of alkyl halides, alcohols, and similar compounds. Most common is the β-elimination *via* the E2 or E1 mechanism, but α-eliminations are also known.

The E2 mechanism provides a more reliable β-elimination method than E1 for most alkene syntheses. Most E2 eliminations start with an alkyl halide or alkyl sulfonate ester (such as a tosylate or triflate). When an alkyl halide is used, the reaction is called a dehydrohalogenation. For unsymmetrical products, the more substituted alkenes (those with fewer hydrogens attached to the C=C) tend to predominate. Two common methods of elimination reactions are dehydrohalogenation of alkyl halides and dehydration of alcohols. A typical example is shown below; note that if possible, the H is *anti* to the leaving group, even though this leads to the less stable Z-isomer.

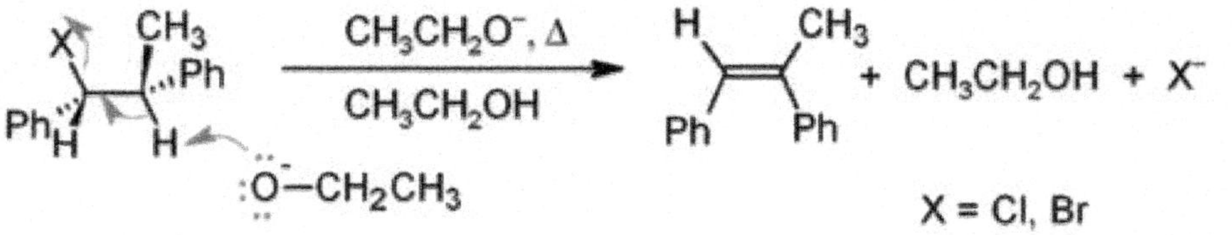

Alkenes can be synthesized from alcohols via dehydration, in which case water is lost via the E1 mechanism. For example, the dehydration of ethanol produces ethene :

$$CH_3CH_2OH + H_2SO_4 \rightarrow H_2C=CH_2 + H_3O^+ + HSO_4^-$$

An alcohol may also be converted to a better leaving group (*e.g.*, xanthate), so as to allow a milder *syn*-elimination such as the Chugaev elimination and the Grieco elimination. Related reactions include eliminations by β-haloethers (the Boord olefin synthesis) and esters (ester pyrolysis).

Alkenes can be prepared indirectly from alkyl amines. The amine or ammonia is not a suitable leaving group, so the amine is first either alkylated (as in the Hofmann elimination) or oxidized to an amine oxide (the Cope reaction) to render a smooth elimination possible. The Cope reaction is a *syn*-elimination that occurs at or below 150°C, for example :

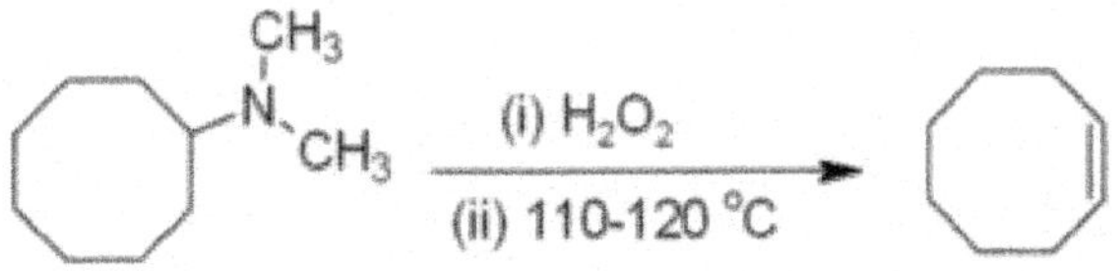

The Hofmann elimination is unusual in that the *less* substituted (non-Saytseff) alkene is usually the major product.

Alkenes are generated from α-halo sulfones in the Ramberg-Bäcklund reaction, *via* a three-membered ring sulfone intermediate.

Synthesis from Carbonyl Compounds

Another important method for alkene synthesis involves construction of a new carbon–carbon double bond by coupling of a carbonyl compound (such as an aldehyde or ketone) to a carbanion equivalent. Such reactions are sometimes called *olefinations*. The most well-known of these methods is the Wittig reaction, but other related methods are known.

The Wittig reaction involves reaction of an aldehyde or ketone with a Wittig reagent (or phosphorane) of the type $Ph_3P=CHR$ to produce an alkene and $Ph_3P=O$. The Wittig reagent is itself prepared easily from triphenylphosphine and an alkyl halide. The reaction is quite general and many functional groups are tolerated, even esters, as in this example :

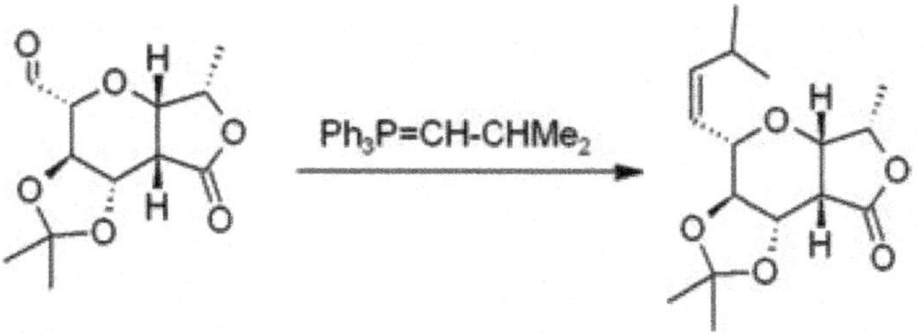

Related to the Wittig reaction is the Peterson olefination. This uses a less accessible silicon-based reagent in place of the phosphorane, but it allows for the selection of E or Z products. If an E-product is desired, another alternative is the Julia olefination, which uses the carbanion generated from a phenyl sulfone. The Takai olefination based on an organochromium intermediate also delivers E-products. A titanium compound, Tebbe's reagent, is useful for the synthesis of methylene compounds; in this case, even esters and amides react.

A pair of carbonyl compounds can also be reductively coupled together (with reduction) to generate an alkene. Symmetrical alkenes can be prepared from a single aldehyde or ketone coupling with itself, using Ti metal reduction (the McMurry reaction). If two different ketones are to be coupled, a more complex, indirect method such as the Barton–Kellogg reaction may be used.

A single ketone can also be converted to the corresponding alkene via its tosylhydrazone, using sodium methoxide (the Bamford–Stevens reaction) or an alkyllithium (the Shapiro reaction).

Synthesis from Alkenes : Olefin Metathesis and Hydrovinylation

Alkenes can be prepared by exchange with other alkenes, in a reaction known as olefin metathesis. Frequently, loss of ethene gas is used to drive the reaction towards a desired product. In many cases, a mixture of geometric isomers is obtained, but the reaction tolerates many functional groups. The method is particularly effective for the preparation of cyclic alkenes, as in this synthesis of muscone :

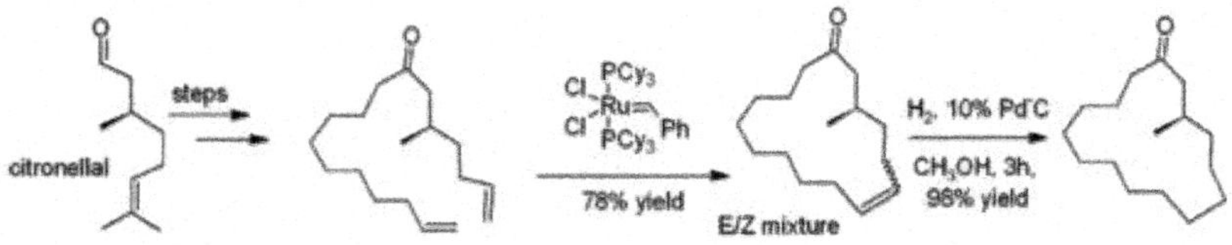

Transition metal catalyzed hydrovinylation is another important alkene synthesis process starting from alkene itself. In general, it involves the addition of a hydrogen and a vinyl group (or an alkenyl group) across a double bond. The hydrovinylation reaction was first reported by Alderson, Jenner, and Lindsey by using rhodium and ruthenium salts, other metal catalysts commonly employed nowadays included iron, cobalt, nickel, and palladium. The addition can be done highly regio-and stereo-selectively, the choices of metal centers, ligands, substrates and counterions often play very important role. Recent studies showed that the use of N-heterocyclic carbene with Ni can be useful for the selective preparations of functionalized geminal olefins or 1,1-disubstituted alkenes.

From Alkynes

Reduction of alkynes is a useful method for the stereoselective synthesis of disubstituted alkenes. If the *cis*-alkene is desired, hydrogenation in the presence of Lindlar's catalyst (a heterogeneous catalyst that consists of palladium deposited on calcium carbonate and treated with various forms of lead) is commonly used, though hydroboration followed by hydrolysis provides an alternative approach. Reduction of the alkyne by sodium metal in liquid ammonia gives the *trans*-alkene.

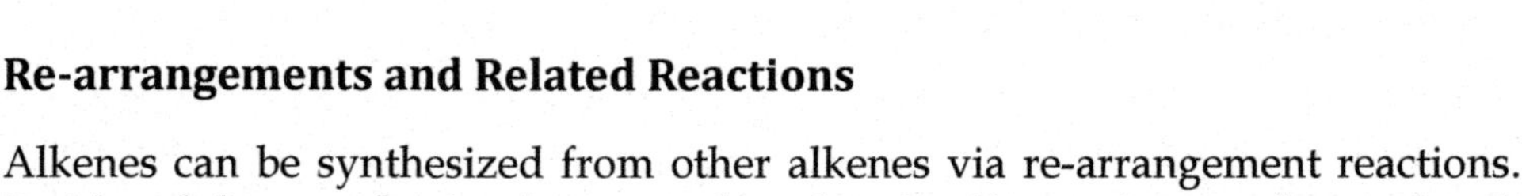

For the preparation multi-substituted alkenes, carbometalation of alkynes can give rise to a large variety of alkene derivatives.

Re-arrangements and Related Reactions

Alkenes can be synthesized from other alkenes via re-arrangement reactions. Besides olefin metathesis, a large number of pericyclic reactions can be used such as the ene reaction and the Cope re-arrangement.

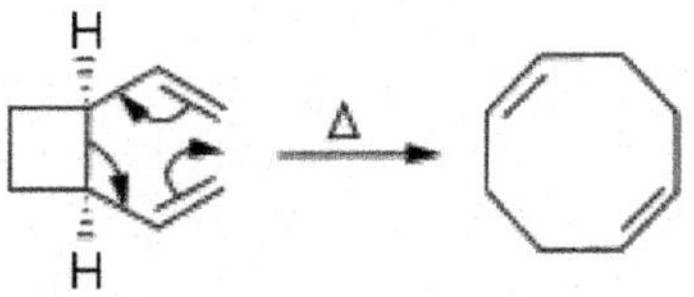

In the Diels-Alder reaction, a cyclohexene derivative is prepared from a diene and a reactive or electron-deficient alkene.

NOMENCLATURE

IUPAC Names

To form the root of the IUPAC names for alkenes, simply change the-an-infix of the parent to-en-. For example, CH_3-CH_3 is the alkane *ethANe*. The name of CH_2=CH_2 is therefore *ethENe*.

In higher alkenes, where isomers exist that differ in location of the double bond, the following numbering system is used :

1. Number the longest carbon chain that contains the double bond in the direction that gives the carbon atoms of the double bond the lowest possible numbers.
2. Indicate the location of the double bond by the location of its first carbon.
3. Name branched or substituted alkenes in a manner similar to alkanes.
4. Number the carbon atoms, locate and name substituent groups, locate the double bond, and name the main chain.

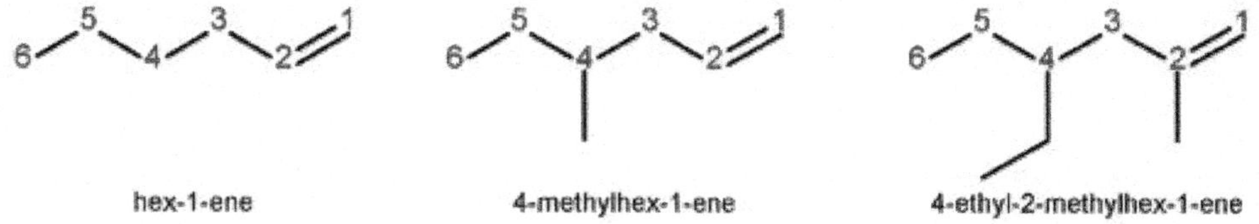

hex-1-ene   4-methylhex-1-ene   4-ethyl-2-methylhex-1-ene

Fig. : Naming substituted hex-1-enes.

Cis-trans Notation

In the specific case of disubstituted alkenes where the two carbons have one substituent each, Cis-trans notation may be used. If both substituents are on the same side of the bond, it is defined as (cis-). If the substituents are on either side of the bond, it is defined as (trans-).

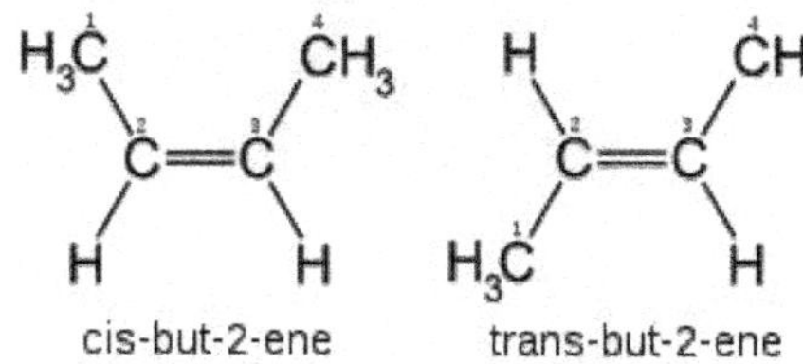

cis-but-2-ene   trans-but-2-ene

Fig. : The difference between *cis*-and *trans*-isomers.

E, Z Notation

When an alkene has more than one substituent (especially necessary with 3 or 4 substituents), the double bond geometry is described using the labels E and Z. These labels come from the German words "entgegen," meaning "opposite," and "zusammen," meaning "together." Alkenes with the higher priority groups (as determined by CIP rules) on the same side of the double bond have these groups together and are designated Z. Alkenes with the higher priority groups on opposite sides are designated E. A mnemonic to remember this : Z notation has the higher priority groups on "ze zame zide."

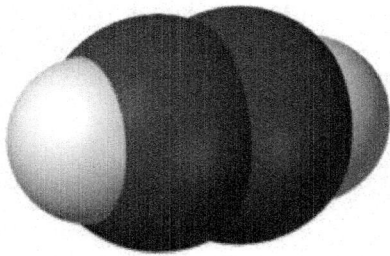

<table>
<tr><td align="center">(Z)-2-butene
zusammen (together)</td><td align="center">(E)-2-butene
entgegen (opposite)</td></tr>
<tr><td align="center">NB. the methyl groups are
on the same side</td><td align="center">NB. the methyl groups are
on opposite sides</td></tr>
</table>

Fig. : The difference between E and Z isomers.

Groups Containing C=C Double Bonds

IUPAC recognizes two names for hydrocarbon groups containing carbon–carbon double bonds, the vinyl group and the allyl group.

$$\{-CH=CH_2 \qquad \{-CH_2-CH=CH_2$$

vinyl   allyl

ALKYNE

Fig. : A 3D model of ethyne (acetylene), the simplest alkyne.

In organic chemistry, an **alkyne** is an unsaturated hydrocarbon which has at least one carbon – carbon triple bond between two carbon atoms. The simplest acyclic alkynes with only one triple bond and no other functional groups form a homologous series with the general chemical formula C_nH_{2n-2}. Alkynes are traditionally known as acetylenes, although the name *acetylene* also refers specifically to C_2H_2, known formally as ethyne using IUPAC nomenclature. Like other hydrocarbons, alkynes are generally hydrophobic but tend to be more reactive.

Chemical Properties

Alkynes are characteristically more unsaturated than alkenes. Thus they add two equivalents of bromine whereas an alkene adds only one equivalent. In some reactions, alkynes are less reactive than alkenes. For example, in a molecule with an-ene and an-yne group, addition occurs preferentially at the-ene. Possible explanations involve the two π-bonds in the alkyne delocalising, which would reduce the energy of the π-system or the stability of the intermediates during the reaction. They show greater tendency to polymerize or oligomerize than alkenes do. The resulting polymers, called polyacetylenes (which do not contain alkyne units) are conjugated and can exhibit semi-conducting properties.

Structure and Bonding

In acetylene, the H–C≡C bond angles are 180°. By virtue of this bond angle, alkynes tend to be rod-like. Correspondingly, cyclic alkynes are rare. Benzyne is highly unstable. The C≡C bond distance of 121 picometers is much shorter than the C=C distance in alkenes (134 pm) or the C-C bond in alkanes (153 pm).

Fig. : Illustrative alkynes : a, acetylene, b, two depictions of propyne, c, 1-butyne, d, 2-butyne, e, the naturally-occurring 1-phenylhepta-1,3,5-triyne, and f, the strained cycloheptyne. Triple bonds are highlighted blue.

The triple bond is very strong with a bond strength of 839 kJ/mol. The sigma bond contributes 369 kJ/mol, the first pi bond contributes 268 kJ/mol and the second pi-bond of 202 kJ/mol bond strength. Bonding usually discussed in the context of molecular orbital theory, which recognizes the triple bond as arising from overlap of s and p orbitals. In the language of valence bond theory, the carbon atoms in an alkyne bond are sp hybridized : they each have two unhybridized p orbitals and two sp hybrid orbitals. Overlap of an sp orbital from each atom forms one sp-sp sigma bond. Each p orbital on one atom overlaps one on the other atom, forming two pi bonds, giving a total of three bonds. The remaining sp orbital on each atom can form a sigma bond to another atom, for example to hydrogen atoms in the parent acetylene. The two sp orbitals project on opposite sides of the carbon atom.

Terminal and Internal Alkynes

Internal alkynes feature carbon substituents on each acetylenic carbon. Symmetrical examples include diphenylacetylene and 3-hexyne.

Terminal alkynes have the formula RC_2H. An example is methylacetylene (propyne using IUPAC nomenclature). Terminal alkynes, like acetylene itself, are mildly acidic, with pK_a values of around 25. They are far more acidic than alkenes and alkanes, which have pK_a values of around 40 and 50, respectively. The acidic hydrogen on terminal alkynes can be replaced by a variety of groups resulting in halo-, silyl-, and alkoxoalkynes. The carbanions generated by deprotonation of terminal alkynes are called acetylides.

Synthesis

Commercially, the dominant alkyne is acetylene itself, which is used as a fuel and a precursor to other compounds, *e.g.*, acrylates. Hundreds of millions of kilograms are produced annually by partial oxidation of natural gas :

$$2\,CH_4 + 3/2\,O_2 \rightarrow HC{\equiv}CH + 3\,H_2O$$

Propyne, also industrially useful, is also prepared by thermal cracking of hydrocarbons. Most other industrially useful alkyne derivatives are prepared from acetylene, *e.g. via* condensation with formaldehyde.

Specialty alkynes are prepared by dehydrohalogenation of vicinal alkyl dihalides or vinyl halides. Metal acetylides can be coupled with primary alkyl halides. *Via* the Fritsch-Buttenberg-Wiechell re-arrangement, alkynes are prepared from vinyl bromides. Alkynes can be prepared from aldehydes using the Corey-Fuchs reaction and from aldehydes or ketones by the Seyferth–Gilbert homologation. In the alkyne zipper reaction, alkynes are generated from other alkynes by treatment with a strong base.

Reactions

Featuring a reactive functional group, alkynes participate in many organic reactions.

Addition of Hydrogen, Halogens, and Related Reagents

Alkynes characteristically undergo reactions that show that they are "doubly unsaturated", meaning that each alkyne unit is capable of adding two equivalents of H_2, halogens or related HX reagents (X = halide, pseudohalide, etc.). Depending on catalysts and conditions, alkynes add one or two equivalents of hydrogen. Hydrogenation to the alkene is usually more desirable since alkanes are less useful :

$$RC{\equiv}CR' + H_2 \rightarrow \textit{cis}\text{-}RCH{=}CR'H$$

The largest scale application of this technology is the conversion of acetylene to ethylene in refineries. The steam cracking of alkanes affords a few per cent acetylene, which is selectively hydrogenated in the presence of a palladium/silver catalyst. For more complex alkynes, the Lindlar catalyst is widely recommended to avoid formation of the alkane, for example in the conversion of phenylacetylene to styrene.

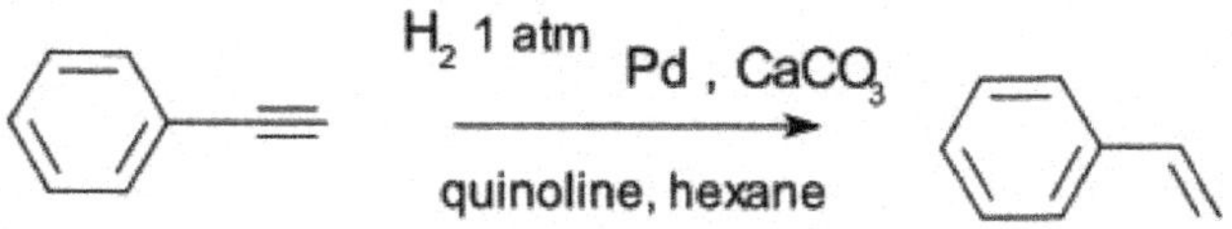

Similarly, halogenation of alkynes gives the vinyl dihalides or alkyl tetrahalides :

$$RC{\equiv}CR' + 2\ Br_2 \rightarrow RCBr_2CRBr_2$$

The addition of non-polar E-H bonds across $C{\equiv}C$ is general for silanes, boranes, and related hydrides. The hydroboration of alkynes gives vinylic boranes which oxidize to the corresponding aldehyde or ketone. In the thiolyne reaction the substrate is a thiol.

Hydrohalogenation gives the corresponding vinyl halides or alkyl dihalides, again depending on the number of equivalents of HX added. The addition of water to alkynes is a related reaction except the initial enol intermediate converts to the ketone or aldehyde. Illustrative is the hydration of phenylacetylene gives acetophenone, and the $(Ph_3P)AuCH_3$-catalyzed hydration of 1,8-nonadiyne to 2,8-non-anedione :

$$PhC{\equiv}CH + H_2O \rightarrow PhCOCH_3$$
$$HC{\equiv}CC_6H_{12}C{\equiv}CH + 2H_2O \rightarrow CH_3COC_6H_{12}COCH_3.$$

Cycloadditions and Oxidation

Alkynes undergo diverse cycloaddition reactions. Most notable is the Diels–Alder reaction with 1,3-dienes to give 1,4-cyclohexadienes. This general reaction has been extensively developed and electrophilic alkynes are especially effective dienophiles. The "cycloadduct" derived from the addition of alkynes to 2-pyrone eliminates carbon dioxide to give the aromatic compound. Other specialized cycloadditions include multi-component reactions such as alkyne trimerisation to give aromatic compounds and the [2+2+1]cycloaddition of an alkyne, alkene and carbon monoxide in the Pauson–Khand reaction. Non-carbon reagents also undergo cyclization, *e.g.* Azide alkyne Huisgen cycloaddition to give triazoles. Cycloaddition processes involving alkynes are often catalyzed by metals, *e.g.* enyne metathesis and alkyne metathesis, which allows the scrambling of carbyne (RC) centers :

$$RC{\equiv}CR + R'C{\equiv}CR' \rightleftarrows 2\ RC{\equiv}CR'$$

Oxidative cleavage of alkynes proceeds via cycloaddition to metal oxides. Most famously, potassium permanganate converts alkynes to a pair of carboxylic acids.

Reactions Specific for Terminal Alkynes

In addition to undergoing the reactions characteristic of internal alkynes, terminal alkynes are reactive as weak acids, with pK_a values (25) between that of ammonia

(35) and ethanol (16). The acetylide conjugate base is stabilized as a result of the high s character of the sp orbital, in which the electron pair resides. Electrons in an s orbital benefit from closer proximity to the positively charged atom nucleus, and are therefore, lower in energy. Treatment of terminal alkynes with a strong base gives the corresponding metal acetylides :

$$RC≡CH + MX → RC≡CM + HX (MX = NaNH_2, LiBu, RMgX).$$

The reactions of alkynes with certain metal cations, *e.g.* Ag^+ also gives acetylides. Thus, few drops of diamminesilver (I) hydroxide $(Ag(NH_3)_2OH)$ reacts with terminal alkynes signaled by formation of a white precipitate of the silver acetylide. Acetylide derivatives are synthetically useful nucleophiles that participate in C-C bond forming reactions, as illustrated in the area called "Reppe Chemistry".

In the Favorskii reaction, terminal alkynes add to carbonyl compounds to give the hydroxyalkyne. Coupling of terminal alkynes to give di-alkynes is effected in the Cadiot-Chodkiewicz coupling, Glaser coupling, and the Eglinton coupling reactions. Terminal alkynes can also be coupled to aryl or vinyl halides as in the Sonogashira coupling.

Alkynes in Nature and Medicine

According to Ferdinand Bohlmann, the first naturally occurring acetylenic compound, dehydromatricaria ester, was isolated from an *Artemisia* species in 1826. In the nearly two centuries that have followed, well over a thousand naturally occurring acetylenes have been discovered and reported. Polyynes, a sub-set of this class of natural products, have been isolated from a wide variety of plant species, cultures of higher fungi, bacteria, marine sponges, and corals. Some acids like tariric acid contains an alkyne group. Diynes and triynes, species with the linkage RC≡C-C≡CR' and RC≡C-C≡C-C≡CR' respectively, occur in certain plants (*Ichthyothere, Chrysanthemum, Cicuta, Oenanthe* and other members of the Asteraceae and Apiaceae families). Some examples are cicutoxin, oenanthotoxin, falcarinol and carotatoxin. These compounds are highly bioactive, *e.g.* as nematocides. 1-Phenylhepta-1,3,5-triyne is illustrative of a naturally occurring triyne.

Alkynes occur in some pharmaceuticals, including the contraceptive norethynodrel. A carbon–carbon triple bond is also present in marketed drugs such as the anti-retroviral Efavirenz and the anti-fungal Terbinafine. Molecules called ene-diynes feature a ring containing an alkene ("ene") between two alkyne groups ("diyne"). These compounds, *e.g.* calicheamicin, are some of the most aggressive antitumor drugs known, so much so that the ene-diyne sub-unit is sometimes referred to as a "warhead." Ene-diynes undergo re-arrangement *via* the Bergman cyclization, generating highly reactive radical intermediates that attack DNA within the tumor.

CYCLOALKANE

Cycloalkanes (also called naphthenes, but distinct from naphthalene) are types of hydrocarbon compounds that have one or more rings of carbon atoms in the chemical structure of their molecules. Alkanes are types of organic hydrocarbon compounds that have only single chemical bonds in their chemical structure. Cycloalkanes consist of only carbon (C) and hydrogen (H) atoms and are saturated because there are no multiple C-C bonds to hydrogenate (add more hydrogen to). A general chemical formula for cycloalkanes would be $C_nH_2(n+1-g)$ where n = number of C atoms and g = number of rings in the molecule. Cycloalkanes with a single ring are named analogously to their normal alkane counterpart of the same carbon count : cyclopropane, cyclobutane, cyclopentane, cyclohexane, etc. The larger cycloalkanes, with greater than 20 carbon atoms are typically called cycloparaffins.

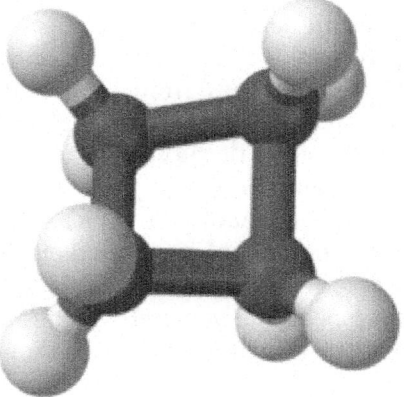

Fig. : Ball-and-stick model of cyclobutane.

Cycloalkanes are classified into small, common, medium, and large cycloalkanes, where cyclopropane and cyclobutane are the small ones, cyclopentane, cyclohexane, cycloheptane are the common ones, cyclooctane through cyclotridecane are the medium ones, and the rest are the larger ones.

Nomenclature

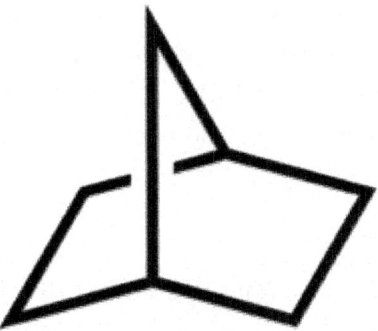

Fig. : Norbornane (also called *bicyclo[2.2.1] heptane*).

The naming of polycyclic alkanes such as bicyclic alkanes and spiro alkanes is more complex, with the base name indicating the number of carbons in the ring system, a prefix indicating the number of rings (*e.g.*, "bicyclo"), and a numeric prefix before that indicating the number of carbons in each part of each ring, exclusive of vertices. For instance, a bicyclooctane that consists of a six-member ring and a four-member ring, which share two adjacent carbon atoms that form a shared edge, is [4.2.0]-bicyclooctane. That part of the six-member ring, exclusive of the shared edge has 4 carbons. That part of the four-member ring, exclusive of the shared edge, has 2 carbons. The edge itself, exclusive of the two vertices that define it, has 0 carbons.

There is more than one convention (method or nomenclature) for the naming of compounds, which can be confusing for those who are just learning, and inconvenient for those who are well rehearsed in the older ways. For beginners it is best to learn IUPAC nomenclature from a source that is up to date, because this system is constantly being revised. In the example [4.2.0]-bicyclooctane would be written bicyclo [4.2.0] octane to fit the conventions for IUPAC naming. It has then got room for an additional numerical prefix if there is the need to include details of other attachments to the molecule such as chlorine or a methyl group. Another convention for the naming of compounds is the **common name**, which is a shorter name and it gives less information about the compound. An example of a common name is terpineol, the name of which can tell us only that it is an alcohol (because the suffix 'ol' is in the name) and it should then have a hydroxide (OH) group attached to it.

An example of the IUPAC method is given in the image to the right. In this example the base name is listed first, which indicates the total number of carbons in both rings including the carbons making up the shared edge (*e.g.*, heptane, which means *hept* or 7 carbons, and *ane*, which indicates only single bonding between carbons). Then in front of the base name is the numerical prefix, which lists the number of carbons in each ring, excluding the carbons that are shared by each ring, plus the number of carbons on the bridge between the rings. In this case there are two rings with two carbons each and a single bridge with one carbon, excluding the carbons shared by it and the other two rings. There is a total of three numbers and they are listed in descending order separated by dots, thus : [2.2.1].

Before the numerical prefix is another prefix indicating the number of rings (*e.g.*, "bicyclo"). Thus, the name is bicyclo [2.2.1] heptane.

The group of cycloalkanes are also known as **naphthenes**.

Properties

Cycloalkanes are similar to alkanes in their general physical properties, but they have higher boiling points, melting points, and densities than alkanes. This is due to stronger London forces because the ring shape allows for a larger area of contact. Containing only C-C and C-H bonds, unreactivity of cycloalkanes with little or no ring strain are comparable to non-cyclic alkanes.

Ring Strain

The carbon atoms in cycloalkanes are sp^3 hybridized and are therefore a deviation from the ideal tetrahedral bond angles of 109°28′. This causes an increase in potential energy and an overall destabilizing effect. Eclipsing of hydrogen atoms is an important destabilizing effect, as well. The **strain energy** of a cycloalkane is the theoretical increase in energy caused by the compound's geometry, and is calculated by comparing the experimental standard enthalpy change of combustion of the cycloalkane with the value calculated using average bond energies.

Ring strain is highest for cyclopropane, in which the carbon atoms form a triangle and therefore have 60 degree C-C-C bond angles. There are also three pairs of eclipsed hydrogens. The ring strain is calculated to be around 120 kJ/mol.

Cyclobutane has the carbon atoms in a puckered square with approximately 90-degree bond angles; "puckering" reduces the eclipsing interactions between hydrogen atoms. Its ring strain is therefore slightly less, at around 110 kJ/mol.

For a theoretical planar cyclopentane the C-C-C bond angles would be 108 degrees, very close to the measure of the tetrahedral angle. Actual cyclopentane molecules are puckered, but this changes only the bond angles slightly so that angle strain is relatively small. The eclipsing interactions are also reduced, leaving a ring strain of about 25 kJ/mol.

In cyclohexane the ring strain and eclipsing interactions are negligible because the puckering of the ring allows ideal tetrahedral bond angles to be achieved. As well, in the most stable *chair form* of cyclohexane, axial hydrogens on adjacent carbon atoms are pointed in opposite directions, virtually eliminating eclipsing strain.

After cyclohexane, the molecules are unable to take a structure with no ring strain, resulting in an increase in strain energy, which peaks at 9 carbons (around 50 kJ/mol). After that, strain energy slowly decreases until 12 carbon atoms, where it drops significantly; at 14, another significant drop occurs and the strain is on a level comparable with 10 kJ/mol. After 14 carbon atoms, sources disagree on what happens to ring strain, some indicating that it increases steadily, others saying that it disappears entirely. However, bond angle strain and eclipsing strain are an issue only for smaller rings.

Reactions

The simple and the bigger cycloalkanes are very stable, like alkanes, and their reactions, for example, radical chain reactions, are like alkanes.

The small cycloalkanes — in particular, cyclopropane — have a lower stability due to Baeyer strain and ring strain. They react similarly to alkenes, though they do not react in electrophilic addition, but in nucleophilic aliphatic substitution. These reactions are ring-opening reactions or ring-cleavage reactions of alkyl cycloalkanes. Cycloalkanes can be formed in a Diels-Alder reaction followed by a catalytic hydrogenation.

Chapter 16

AMINE

Primary amine	Secondary amine	Tertiary amine
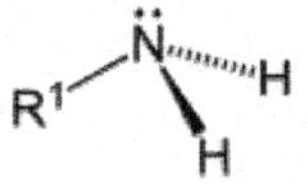 $R^1 \overset{\overset{\displaystyle ..}{N}}{\diagdown} \,^{\cdots\cdots} H$ H	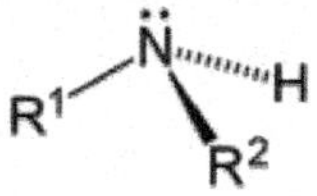 $R^1 \overset{\overset{\displaystyle ..}{N}}{\diagdown} \,^{\cdots\cdots} H$ R^2	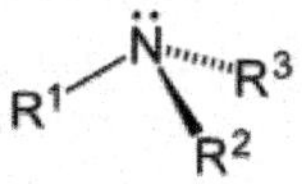 $R^1 \overset{\overset{\displaystyle ..}{N}}{\diagdown} \,^{\cdots\cdots} R^3$ R^2

Amines are organic compounds and functional groups that contain a basic nitrogen atom with a lone pair. Amines are derivatives of ammonia, wherein one or more hydrogen atoms have been replaced by a substituent such as an alkyl or aryl group. Important amines include amino acids, biogenic amines, trimethylamine, and aniline. Inorganic derivatives of ammonia are also called amines, such as chloramine ($NClH_2$).

Compounds with the nitrogen atom attached to a carbonyl of the structure R–CO–NR'R" are called amides and have different chemical properties from amines.

CLASSES OF AMINES

An aliphatic amine has no aromatic ring attached directly to the nitrogen atom. Aromatic amines have the nitrogen atom connected to an aromatic ring as in the various anilines. The aromatic ring decreases the alkalinity of the amine, depending on its substituents. The presence of an amine group strongly increases the reactivity of the aromatic ring, due to an electron-donating effect.

Amines are organized into four sub-categories :

• *Primary Amines :* Primary amines arise when one of three hydrogen atoms in ammonia is replaced by an alkyl or aromatic. Important primary alkyl amines include methylamine, ethanolamine (2-aminoethanol), and the buffering agent tris, while primary aromatic amines include aniline.

• *Secondary Amines :* Secondary amines have two organic substituents (alkyl, aryl or both) bound to N together with one hydrogen (or no hydrogen if one of the substituent bonds is double). Important representatives include dimethylamine and methylethanolamine, while an example of an aromatic amine would be diphenylamine.

* *Tertiary Amines :* In tertiary amines, all three hydrogen atoms are replaced by organic substituents. Examples include trimethylamine, which has a distinctively fishy smell or triphenylamine.

* *Cyclic Amines :* Cyclic amines are either secondary or tertiary amines. Examples of cyclic amines include the 3-member ring aziridine and the six-membered ring piperidine. N-methylpiperidine and N-phenylpiperidine are examples of cyclic tertiary amines.

It is also possible to have four organic substituents on the nitrogen. These species are not amines but are quaternary ammonium cations and have a charged nitrogen center. Quaternary ammonium salts exist with many kinds of anions.

NAMING CONVENTIONS

Amines are named in several ways. Typically, the compound is given the prefix "amino-" or the suffix : "-amine". The prefix "N-" shows substitution on the nitrogen atom. An organic compound with multiple amino groups is called a diamine, triamine, tetraamine and so forth.

Systematic names for some common amines :

Lower amines are named with the suffix-*amine*.

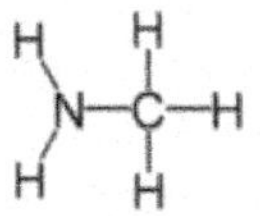

Higher amines have the prefix *amino* as a functional group. IUPAC however does not recommend this convention, but prefers the alkanamine form, *e.g.* pentan-2-amine.

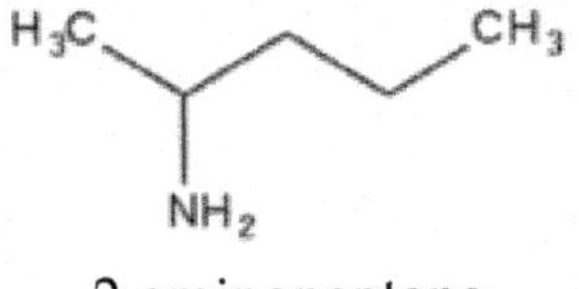

methylamine

2-aminopentane

(or sometimes : *pent-2-yl-amine* or *pentan-2-amine*)

PHYSICAL PROPERTIES

Hydrogen bonding significantly influences the properties of primary and secondary amines. Thus the boiling point of amines is higher than those of the corresponding phosphines, but generally lower than those of the corresponding alcohols. For example, methylamine and ethylamine are gases under standard conditions, whereas the corresponding methyl alcohol and ethyl alcohols are liquids. Gaseous amines possess a characteristic ammonia smell, liquid amines have a distinctive "fishy" smell.

Also reflecting their ability to form hydrogen bonds, most aliphatic amines display some solubility in water. Solubility decreases with the increase in the number of carbon atoms. Aliphatic amines display significant solubility in organic solvents, especially polar organic solvents. Primary amines react with ketones such as acetone.

The aromatic amines, such as aniline, have their lone pair electrons conjugated into the benzene ring, thus their tendency to engage in hydrogen bonding is diminished. Their boiling points are high and their solubility in water is low.

Chirality

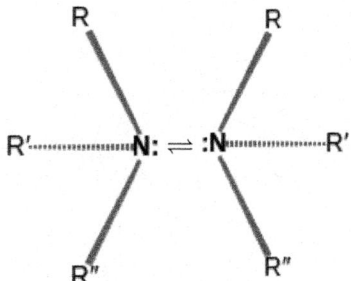

Fig. : Inversion of an amine. The pair of dots represents the lone electron pair on the nitrogen atom.

Amines of the type NHRR' and NRR'R" are chiral : the nitrogen atom bears four substituents counting the lone pair. The energy barrier for the inversion of the stereo-center is relatively low, *e.g.*, ~7 kcal/mol for a trialkylamine. The interconversion of the stereoisomers has been compared to the inversion of an open umbrella into a strong wind. Because of this low barrier, amines such as NHRR' cannot be resolved optically and NRR'R" can only be resolved when the R, R', and R" groups are constrained in cyclic structures such as aziridines. Quaternary ammonium salts with four distinct groups on the nitrogen are capable of exhibiting optical activity.

Properties as Bases

Like ammonia, amines are bases. Compared to alkali metal hydroxides, amines are weaker. The basicity of amines depends on :

1. The electronic properties of the substituents (alkyl groups enhance the basicity, aryl groups diminish it).
2. Steric hindrance offered by the groups on nitrogen.
3. The degree of solvation of the protonated amine.

The nitrogen atom features a lone electron pair that can bind H^+ to form an ammonium ion R_3NH^+. The water solubility of simple amines is largely due to hydrogen bonding between protons in the water molecules and these lone electron pairs.

• Inductive effect of alkyl groups

Ions of compound	K_b
Ammonia NH_3	$1.8 \cdot 10^{-5}$ M
Propylamine $CH_3CH_2CH_2NH_2$	$4.7 \cdot 10^{-4}$ M
Isopropylamine $(CH_3)_2CHNH_2$	$3.4 \cdot 10^{-4}$ M
Methylamine CH_3NH_2	$4.4 \cdot 10^{-4}$ M
Dimethylamine $(CH_3)_2NH$	$5.4 \cdot 10^{-4}$ M
Trimethylamine $(CH_3)_3N$	$5.9 \cdot 10^{-5}$ M

+I effect of alkyl groups raises the energy of the lone pair of electrons, thus elevating the basicity. Thus the basicity of an amine may be expected to increase with the number of alkyl groups on the amine. However, there is no strict trend in this regard, as basicity is also governed by other factors. The increase in Kb from methylamine to dimethylamine may be attributed to +I effect; however, there is a decrease from dimethylamine to trimethyl amine due to the predominance of steric hindrance offered by the three methyl groups to the approaching Brönsted acid.

- Mesomeric effect of aromatic systems

Ions of compound	K_b
Ammonia NH_3	$1.8 \cdot 10^{-5}$ M
Aniline $C_6H_5NH_2$	$3.8 \cdot 10^{-10}$ M
4-Methylaniline $4\text{-}CH_3C_6H_4NH_2$	$1.2 \cdot 10^{-9}$ M
2-Nitroaniline	$1.5 \cdot 10^{-15}$ M
3-Nitroaniline	$2.8 \cdot 10^{-13}$ M
4-Nitroaniline	$9.5 \cdot 10^{-14}$ M

-M effect of aromatic ring delocalises the lone pair of electrons on nitrogen into the ring, resulting in decreased basicity. Substituents on the aromatic ring, and their positions relative to the amine group may also considerably alter basicity.

The solvation of protonated amines changes upon their conversion to ammonium compounds. Typically salts of ammonium compounds exhibit the following order of solubility in water : primary ammonium (RNH_3^+) > secondary ammonium ($R_2NH_2^+$) > tertiary ammonium (R_3NH^+). Quaternary ammonium salts usually exhibit the lowest solubility of the series.

In sterically hindered amines, as in the case of trimethylamine, the protonated form is not well-solvated. For this reason the parent amine is less basic than expected. In the case of aprotic polar solvents (like DMSO and DMF), wherein the extent of solvation is not as high as in protic polar solvents (like water and methanol), the basicity of amines is almost solely governed by the electronic factors within the molecule.

SYNTHESIS

Alkylation

The most industrially significant amines are prepared from ammonia by alkylation with alcohols :

$ROH + NH_3 \rightarrow RNH_2 + H_2O$

These reactions require catalysts, specialized apparatus, and additional purification measures since the selectivity can be problematic. The same amines can be prepared by treatment of Haloalkanes with ammonia and amines :

$RX + 2\ R'NH_2 \rightarrow RR'NH + [RR'NH_2]X$

Such reactions, which are most useful for alkyl iodides and bromides, are rarely employed because the degree of alkylation is difficult to control.

Reductive Routes

Via the process of hydrogenation, nitriles are reduced to amines using hydrogen in the presence of a nickel catalyst. Reactions are sensitive acidic or alkaline conditions, which can cause hydrolysis of-CN group. $LiAlH_4$ is more commonly employed for the reduction of nitriles on the laboratory scale. Similarly, $LiAlH_4$ reduces amides to amines. Many amines are produced from aldehydes and ketones via reductive amination, which can either proceed catalytically or stoichiometrically.

Aniline ($C_6H_5NH_2$) and its derivatives are prepared by reduction of the nitroaromatics. In industry, hydrogen is the preferred reductant, whereas in the laboratory, tin and iron are often employed.

Specialized Methods

Many laboratory methods exist for the preparation of amines, many of these methods being rather specialized.

Reaction name	Substrate	Comment
Gabriel synthesis	organohalide	reagent : potassium phthalimide
Staudinger reduction	Azide	This reaction also takes place with a reducing agent such as lithium aluminium hydride.
Schmidt reaction	carboxylic acid	
Aza-Baylis–Hillman reaction	imine	Synthesis of allylic amines
Hofmann degradation	amide	This reaction is valid for preparation of primary amines only. Gives good yields of primary amines uncontaminated with other amines.
Hofmann elimination	Quaternary ammonium salt	upon treatment with strong base
Amide reduction	amide	

(Contd...)

(Contd...)

Reaction name	Substrate	Comment
Nitrile reduction	nitriles	either accomplished with reducing agents or by electrosynthesis
Reduction of nitro compounds	nitro compounds	can be accomplished with elemental zinc, tin or iron with an acid.
Amine alkylation	haloalkane	
Delepine reaction	organohalide	reagent hexamine
Buchwald–Hartwig reaction	aryl halide	specific for aryl amines
Menshutkin reaction	tertiary amine	reaction product a quaternary ammonium cation
Hydroamination	alkenes and alkynes	
Oxime reduction	oximes	
Leuckart reaction	ketones and aldehydes	reductive amination with formic acid and ammonia via an imine intermediate
Hofmann–Löffler reaction	haloamine	
Eschweiler–Clarke reaction	amine	reductive amination with formic acid and formaldehyde *via* an imine intermediate.

REACTIONS

Alkylation, Acylation, and Sulfonation

Aside from their basicity, the dominant reactivity of amines is their nucleophilicity. Most primary amines are good ligands for metal ions to give co-ordination complexes. Amines are alkylated by alkyl halides. Acyl chlorides and acid anhydrides react with primary and secondary amines to form amides (the "Schotten–Baumann reaction").

Similarly, with sulfonyl chlorides, one obtains sulfonamides. This transformation, known as the Hinsberg reaction, is a chemical test for the presence of amines.

Because amines are basic, they neutralize acids to form the corresponding ammonium salts R_3NH^+. When formed from carboxylic acids and primary and secondary amines, these salts thermally dehydrate to form the corresponding amides.

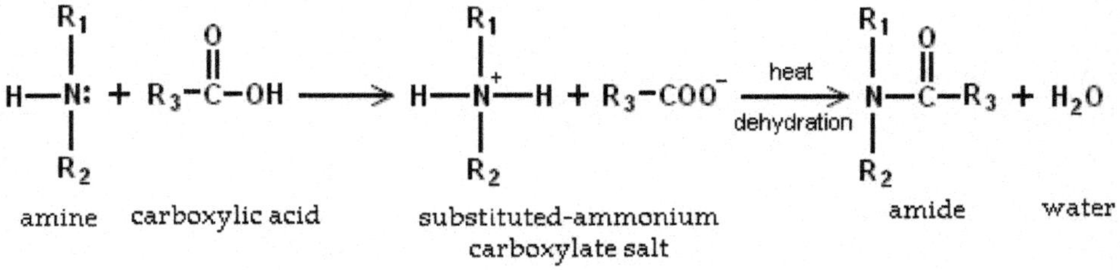

| amine | carboxylic acid | substituted-ammonium carboxylate salt | amide | water |

Diazotization

Amines react with nitrous acid to give diazonium salts. The alkyl diazonium salts are of little synthetic importance because they are too unstable. The most important members are derivatives of aromatic amines such as aniline ("phenylamine") (A = aryl or naphthyl) :

$$ANH_2 + HNO_2 + HX \rightarrow AN_2^+X^- + 2\,H_2O$$

Anilines and naphthylamines form more stable diazonium salts, which can be isolated in the crystalline form. Diazonium salts undergo a variety of useful transformations involving replacement of the N_2 group with anions. For example, cuprous cyanide gives the corresponding nitriles :

$$AN_2^+ + Y^- \rightarrow AY + N_2$$

Aryldiazonium couple with electron-rich aromatic compounds such as a phenol to form azo compounds. Such reactions are widely applied to the production of dyes.

Conversion to imines

Imine formation is an important reaction. Primary amines react with ketones and aldehydes to form imines. In the case of formaldehyde (R' = H), these products typically exist as cyclic trimers.

$$RNH_2 + R'_2C=O \rightarrow R'_2C=NR + H_2O$$

Reduction of these imines gives secondary amines :

$$R'_2C=NR + H_2 \rightarrow R'_2CH-NHR$$

Similarly, secondary amines react with ketones and aldehydes to form enamines :

$$R_2NH + R'(R''CH_2)C=O \rightarrow R''CH=C(NR_2)R' + H_2O$$

Overview

An overview of the reactions of amine is given below :

Reaction name	Reaction product	Comment
Amine alkylation	amines	degree of substitution increases
Schotten–Baumann reaction	amide	Reagents : acyl chlorides, acid anhydrides
Hinsberg reaction	Sulfonamides	Reagents : sulfonyl chlorides
Amine-carbonyl condensation	imines	
Organic oxidation	nitroso com-pounds	Reagent : peroxymonosulfuric acid
Organic oxidation	diazonium salt	Reagent : nitrous acid
Zincke reaction	Zincke aldehyde	reagent pyridinium salts, with primary and secondary amines
Emde degradation	tertiary amine	reduction of quaternary ammonium cations
Hofmann–Martius re-arrangement	aryl substituted anilines	
Von Braun reaction	Organocyanamide	By cleavage (tertiary amines only) with cyanogen bromide
Hofmann elimination	Alkene	proceeds by β-elimination of less hindered carbon
Cope reaction	Alkene	Similar to Hofmann elimination
carbylamine reaction	Isonitrile	(primary amines only)
Hoffmann's mustard oil test	Isothiocyanate	CS_2 and $HgCl_2$ are used. Thiocyanate smells like mustard.

BIOLOGICAL ACTIVITY

Amines are ubiquitous in biology. The breakdown of amino acids releases amines, famously in the case of decaying fish which smell of trimethylamine. Many neurotransmitters are amines, including epinephrine, norepinephrine, dopamine, serotonin, and histamine. Protonated amino groups ($-NH_3^+$) are the most common positively charged moieties in proteins, specifically in the amino acid lysine. The anionic polymer DNA is typically bound to various amine-rich proteins. Additionally, the terminal charged primary ammonium on lysine forms salt bridges with carboxylate groups of other amino acids in polypeptides, which is one of the primary influences on the three-dimensional structures of proteins.

APPLICATION OF AMINES

Dyes

Primary aromatic amines are used as a starting material for the manufacture of *azo* dyes. It reacts with nitrous acid to form diazonium salt, which can undergo coupling reaction to form azo compound. As azo-compounds are highly coloured, they are widely used in dyeing industries, such as :

- Methyl orange
- Direct brown 138
- Sunset yellow FCF
- Ponceau.

Drugs

Many drugs are designed to mimic or to interfere with the action of natural amine neurotransmitters, exemplified by the amine drugs :

- Chlorpheniramine is an antihistamine that helps to relieve allergic disorders due to cold, hay fever, itchy skin, insect bites and stings.
- Chlorpromazine is a tranquillizer that sedates without inducing sleep. It is used to relieve anxiety, excitement, restlessness or even mental disorder.
- Ephedrine and phenylephrine, as amine hydrochlorides, are used as decongestants.
- Amphetamine, methamphetamine, and methcathinone are psychostimulant amines that are listed as controlled substances by the US DEA.
- Amitriptyline, imipramine, lofepramine and clomipramine are tricyclic antidepressants and tertiary amines.
- Nortriptyline, desipramine, and amoxapine are tricyclic anti-depressants and secondary amines. (The tricyclics are grouped by the nature of the final amine group on the side chain.)
- Substituted tryptamines and phenethylamines are key basic structures for a large variety of psychedelic drugs.
- Opiate analgesics such as morphine, codeine, and heroin are tertiary amines.

Gas Treatment

Aqueous monoethanolamine (MEA), diglycolamine (DGA), diethanolamine (DEA), diisopropanolamine (DIPA) and methyldiethanolamine (MDEA) are widely used industrially for removing carbon dioxide (CO_2) and hydrogen sulfide (H_2S) from natural gas and refinery process streams. They may also be used to remove CO_2 from combustion gases/flue gases and may have potential for abatement of greenhouse gases. Related processes are known as sweetening.

SAFETY

Low molecular weight amines, such as ethylamine, are toxic, and some are easily absorbed through the skin. Many higher molecular weight amines are, biologically, highly active.

BIOGENIC AMINE

A **biogenic amine** is a biogenic substance with one or more amine groups.

Examples

Some prominent examples of biogenic amines include :

- Histamine — a substance derived from the amino acid histidine that acts as a neurotransmitter mediating arousal and attention, as well as a pro-inflammatory signal released from mast cells in response to allergic reactions or tissue damage. Histamine is also an important stimulant of HCl secretion by the stomach through histamine H_2 receptors.
- Serotonin — a central nervous system neurotransmitter derived from the amino acid tryptophan involved in regulating mood, sleep, appetite, and sexuality.
- The three catecholamine neurotransmitters :
  - Norepinephrine (noradrenaline) — a neurotransmitter involved in sleep and wakefulness, attention, and feeding behaviour, as well as a stress hormone released by the adrenal glands that regulates the sympathetic nervous system.
  - Epinephrine (adrenaline) — an adrenal stress hormone, as well as a neurotransmitter present at lower levels in the brain.
  - Dopamine-a neurotransmitter involved in motivation, reward, addiction, behavioural reinforcement, and co-ordination of bodily movement.
- The trace amines :
  - 3-Iodothyronamine — a metabolite of the thyroid hormones, and has been hypothesized to be the primary endogenous ligand for the trace amine-associated receptor 1 (TAAR1).
  - Tryptamine — a monoamine alkaloid found in trace amounts in the brains of mammals, and believed to play a role as a neuromodulator or neurotransmitter.
  - Tyramine — a substance that is found in many common foods, and is associated with increased blood pressure and headaches.
  - As well as others such as dimethyltryptamine (DMT), phenethylamine, and octopamine and the *meta*-substituted positional isomers of octopamine and tyramine.

Physiological Importance

There is a distinction between endogenous and exogenous biogenic amines. Endogenous amines are produced in many different tissues (for example : adrenaline in adrenal medulla or histamine in mast cells and liver). The amines are transmitted locally or via the blood system. The exogenous amines are directly absorbed from food in the intestine. Alcohol can increase the absorption rate. Monoamine oxidase (MAO) breaks down biogenic amines and prevents excessive resorption.

MAO inhibitors (MAOIs) are also used as medications for the treatment of depression to prevent MAO from breaking down amines important for positive mood.

ACID-BASE EXTRACTION

Acid-base extraction is a procedure using sequential liquid–liquid extractions to purify acids and bases from mixtures based on their chemical properties.

Acid-base extraction is routinely performed during the work-up after chemical syntheses and for the isolation of compounds and natural products like alkaloids from crude extracts. The product is largely free of neutral and acidic or basic impurities. It is not possible to separate chemically similar acids or bases using this simple method.

Theory

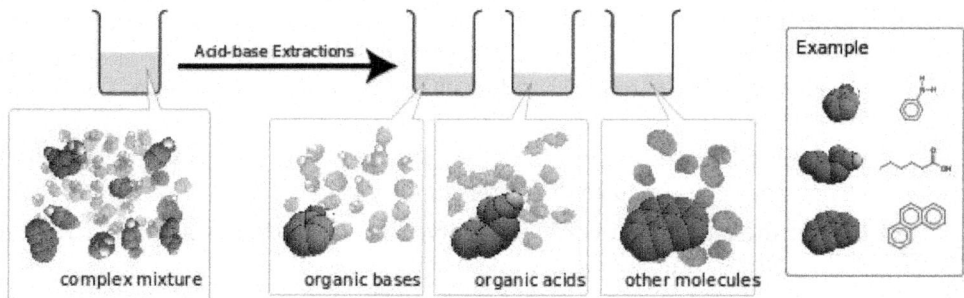

The fundamental theory behind this technique is that salts, which are ionic, tend to be water-soluble while neutral molecules tend not to be.

The addition of an acid to a mixture of an organic base and acid will result in the acid remaining uncharged, while the base will be protonated. If the organic acid, such as a carboxylic acid, is sufficiently strong, its self-ionization can be suppressed by the added acid.

Conversely, the addition of a base to a mixture of an organic acid and base will result in the base remaining uncharged, while the acid is deprotonated to give the corresponding salt. Once again, the self-ionization of a strong base is suppressed by the added base.

The acid-base extraction procedure can also be used to separate very-weak acids from stronger acids and very-weak bases from stronger bases as long as the difference of their pK_a (or pK_b) constants is large enough. Examples are :

- Very weak acids with phenolic OH groups like phenol, 2-naphthol, or 4-hydroxyindole (pK_a around 10) from stronger acids like benzoic acid or sorbic acid (pK_a around 4–5)
- Very weak bases like caffeine or 4-nitroaniline (pK_b around 13–14) from stronger bases like mescaline or dimethyltryptamine (pK_b around 3–4).

Usually the pH is adjusted to a value roughly between the pK_a (or pK_b) constants of the compounds to be separated. Weak acids like citric acid, phosphoric acid, or diluted sulfuric acid are used for moderately acidic pH values and hydrochloric acid or more concentrated sulfuric acid is used for strongly acidic pH values. Similarly, weak bases like ammonia or sodium bicarbonate ($NaHCO_3$) are used for moderately basic pH values while stronger bases like potassium carbonate (K_2CO_3) or sodium hydroxide (NaOH) are used for strongly alkaline conditions.

Technique

Usually, the mixture is dissolved in a suitable solvent such as dichloromethane or diethyl ether (ether), and poured into a separating funnel. An aqueous solution of the acid or base is added, and the pH of the aqueous phase is adjusted to bring the compound of interest into its required form. After shaking and allowing for phase separation, the phase containing the compound of interest is collected. The procedure is then repeated with this phase at the opposite pH range. The order of the step is not important and the process can be repeated to increase the separation. However, it is often convenient to have the compound dissolved in the organic phase after the last step, so that the evaporation of the solvent yields the product.

Limitations

The procedure works only for acids and bases with a large difference in solubility between their charged and their uncharged form. The procedure does not work for :

- Zwitterions with acidic and basic functional groups in the same molecule, *e.g.* glycine which tend to be water soluble at most pH.
- Very lipophilic amines that do not easily dissolve in the aqueous phase in their charged form, *e.g.* triphenylamine and trihexylamine.
- Very lipophilic acids that do not easily dissolve in the aqueous phase in their charged form, *e.g.* fatty acids.
- Lower amines like ammonia, methylamine, or triethanolamine which are miscible or significantly soluble in water at most pH.
- Hydrophilic acids like acetic acid, citric acid, and most inorganic acids like sulfuric acid or phosphoric acid.

Alternatives

Alternatives to acid-base extraction including :

- Filtering the mixture through a plug of silica gel or alumina — charged salts tend to remain strongly adsorbed to the silica gel or alumina.
- Ion exchange chromatography can separate acids, bases, or mixtures of strong and weak acids and bases by their varying affinities to the column medium at different pH.

AMINE GAS TREATING

Amine gas treating, also known as **gas sweetening** and **acid gas removal**, refers to a group of processes that use aqueous solutions of various alkylamines (commonly referred to simply as amines) to remove hydrogen sulfide (H_2S) and carbon dioxide (CO_2) from gases. It is a common unit process used in refineries, and is also used in petro-chemical plants, natural gas processing plants and other industries.

Processes within oil refineries or chemical processing plants that remove hydrogen sulfide are referred to as "sweetening" processes because the odour of the processed products is improved by the absence of hydrogen sulfide. An alternative to the use of amines involves membrane technology. Membranes are attractive since no reagents are consumed.

Many different amines are used in gas treating :

- Diethanolamine (DEA)
- Monoethanolamine (MEA)
- Methyldiethanolamine (MDEA)
- Diisopropanolamine (DIPA)
- Aminoethoxyethanol (Diglycolamine) (DGA).

The most commonly used amines in industrial plants are the alkanolamines DEA, MEA, and MDEA. These amines are also used in many oil refineries to remove sour gases from liquid hydrocarbons such as liquified petroleum gas (LPG).

Description of a Typical Amine Treater

Gases containing H_2S or both H_2S and CO_2 are commonly referred to as *sour gases* or *acid gases* in the hydrocarbon processing industries.

The chemistry involved in the amine treating of such gases varies somewhat with the particular amine being used. For one of the more common amines, monoethanolamine (MEA) denoted as RNH_2, the chemistry may be expressed as :

$$RNH_2 + H_2S \Leftrightarrow RNH+$$

$$3 + SH^-$$

A typical amine gas treating process includes an **absorber** unit and a **regenerator** unit as well as accessory equipment. In the absorber, the down-flowing amine solution absorbs H_2S and CO_2 from the up-flowing sour gas to produce a sweetened gas stream (*i.e.*, a gas free of hydrogen sulfide and carbon dioxide) as a product and an amine solution rich in the absorbed acid gases. The resultant "rich" amine is then routed into the regenerator (a stripper with a re-boiler) to produce regenerated or "lean" amine that is recycled for reuse in the absorber. The stripped overhead gas from the regenerator is concentrated H_2S and CO_2. In oil refineries, that stripped gas is mostly H_2S, much of which often comes from a sulfur-removing process called hydrodesulfurization. This H_2S-rich stripped gas stream is then usually routed into a Claus process to convert it into elemental sulfur. In fact, the vast majority of the 64,000,000 metric tons of sulfur produced worldwide in 2005 was by-product sulfur from refineries and other hydrocarbon processing plants. Another sulfur-removing process is the WSA Process which recovers sulfur in any form as concentrated sulfuric acid. In some plants, more than one amine absorber unit may share a common regenerator unit.

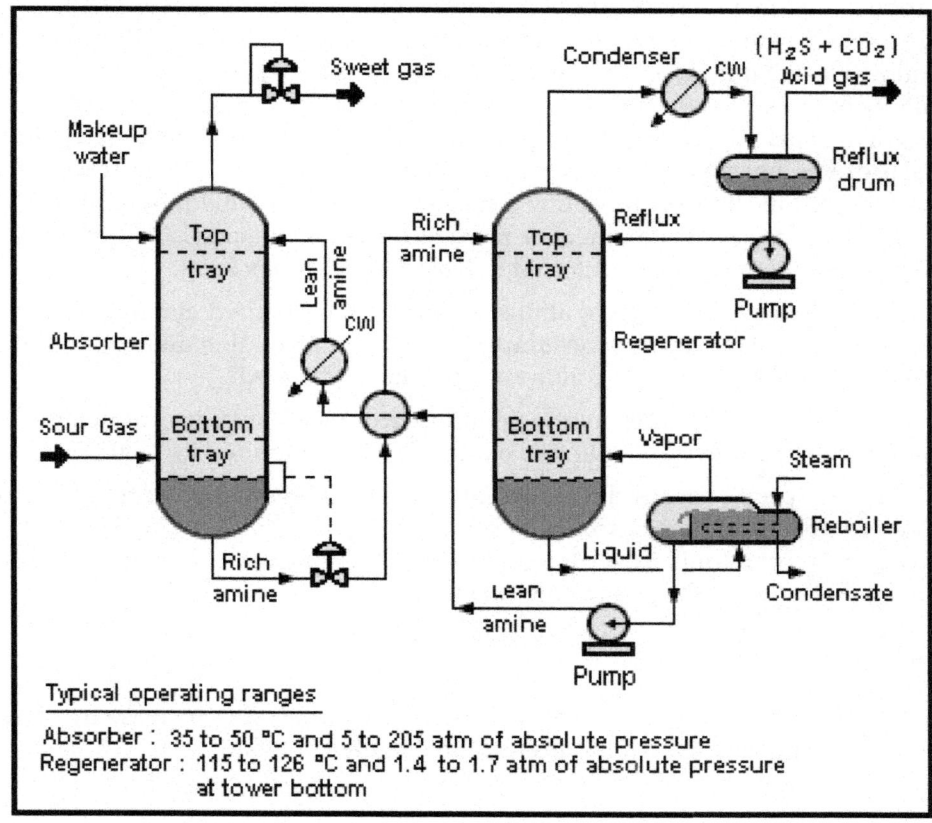

Typical operating ranges

Absorber : 35 to 50 °C and 5 to 205 atm of absolute pressure
Regenerator : 115 to 126 °C and 1.4 to 1.7 atm of absolute pressure
           at tower bottom

Process flow diagram of a typical amine treating process used in petroleum refineries, natural gas processing plants and other industrial facilities.

The amine concentration in the absorbent aqueous solution is an important parameter in the design and operation of an amine gas treating process. Depending on which one of the following four amines the unit was designed to use and what gases it was designed to remove, these are some typical amine concentrations, expressed as weight per cent of pure amine in the aqueous solution :

- *Monoethanolamine* : About 20% for removing H_2S and CO_2, and about 32% for removing only CO_2.
- *Diethanolamine* : About 20 to 25% for removing H_2S and CO_2
- *Methyldiethanolamine* : About 30 to 55%% for removing H_2S and CO_2
- *Diglycolamine* : About 50% for removing H_2S and CO_2.

The choice of amine concentration in the circulating aqueous solution depends upon a number of factors and may be quite arbitrary. It is usually made simply on the basis of experience. The factors involved include whether the amine unit is treating raw natural gas or petroleum refinery by-product gases that contain relatively low concentrations of both H_2S and CO_2 or whether the unit is treating gases with a high percentage of CO_2 such as the off-gas from the steam reforming process used in ammonia production or the flue gases from power plants. Both H_2S and CO_2 are acid gases and hence corrosive to carbon steel. However, in an amine treating unit, CO_2 is the stronger acid of the two. H_2S forms a film of iron sulfide on the surface of the steel that acts to protect the steel. When treating gases with a high percentage of CO_2, corrosion inhibitors are often used and that permits the use of higher concentrations of amine in the circulating solution. Another factor involved in choosing an amine concentration is the relative solubility of H_2S and CO_2 in the selected amine. For more information about selecting the amine concentration, the reader is referred to Kohl and Nielsen's book.

The choice of the type of amine will affect the required circulation rate of amine solution, the energy consumption for the regeneration and the ability to selectively remove either H_2S alone or CO_2 alone if desired.

The current emphasis on removing CO_2 from the flue gases emitted by fossil fuel power plants has led to much interest in using amines for that purpose.

In the specific case of the industrial synthesis of ammonia, for the steam reforming process of hydrocarbons to produce gaseous hydrogen, amine treating is one of the commonly used processes for removing excess carbon dioxide in the final purification of the gaseous hydrogen.

aMDEA

Activated MDEA or aMDEA uses piperazine as a catalyst to increase the speed of the reaction with CO_2. It has been commercially successful.

Chapter 17

NOVEL FIELD TEST EQUIPMENT FOR LITHIUM-ION BATTERIES IN HYBRID ELECTRICAL VEHICLE APPLICATIONS

Pontus Svens[1,2,*], Johan Lindstrom[1], Olle Gelin[1], Marten Behm[2] and Goran Lindbergh[2]

[1]   Scania CV AB, SE-151 87, Sodertalje, Sweden; E-Mails: johan.lindstrom@scania.com (J.L.); olle.gelin@scania.com (O.G.)

[2]   School of Chemical Science and Engineering, Department of Chemical Engineering and Technology, Applied Electrochemistry, KTH Royal Institute of Technology, SE-100 44, Stockholm, Sweden; E-Mails: behm@kth.se (M.B.); gnli@kth.se (G.L.)

*   Author to whom correspondence should be addressed; E-Mail: ponsvens@kth.se; Tel.: +46-8-553-51661; Fax: +46-8-553-82841.

ABSTRACT

Lifetime testing of batteries for hybrid-electrical vehicles (HEV) is usually performed in the lab, either at the cell, module or battery pack level. Complementary field tests of battery packs in vehicles are also often performed. There are, however, difficulties related to field testing of battery-packs. Some examples are cost issues and the complexity of continuously collecting battery performance data, such as capacity fade and impedance increase. In this paper, a novel field test equipment designed primarily for lithium-ion battery cell testing is presented. This equipment is intended to be used on conventional vehicles, not hybrid vehicles, as a cheaper and faster field testing method for batteries, compared to full scale HEV testing. The equipment emulates an HEV environment for the tested battery cell by using real time vehicle sensor information and the existing starter battery as load and source. In addition to the emulated battery cycling, periodical capacity and pulse testing capability are implemented as well. This paper begins with presenting some background information about hybrid electrical vehicles and describing the limitations with today's HEV battery testing. Furthermore, the functionality of the test equipment is described in detail and, finally, results from verification of the equipment are presented and discussed.

Keywords

Battery testing; hybrid electrical vehicle (HEV); lithium ion battery

1. INTRODUCTION

During the last 50 years vehicle emissions have been reduced significantly [1,2]. Further emission reductions related to the internal combustion engine, ICE, are becoming more and more difficult to achieve. One approach to further reduce the emissions is to hybridize vehicles by introducing an additional energy converter and energy storage. Even if this technology is considered to be relatively new, it has a long history. For example, the first hybrid passenger car, the Lohner-Porsche, was built in 1900 and used lead acid batteries as energy storage [3]. There are also several publications available regarding vehicle hybridization, some published already in the 1970s [4]. It was, however, not until the launch of the first generation Toyota Prius hybrid passenger car in 1997 that public interest for this technology increased considerably [5]. Today, even trucks and buses are subject to hybridization, and this paper will discuss field testing of HEV-batteries within this segment.

A well-known problem when using batteries as energy storage in HEVs is the limited and relatively unpredictable battery lifetime. The ageing process for lithium-ion batteries in particular is very complex and is influenced by several factors, such as for example temperature and depth of discharge (DOD). Because of this, extensive battery testing is required from the vehicle manufacturer's side to be able to guarantee battery lifetime for HEV-customers. In the early stage of HEV-battery development, there is often a desire to do performance and ageing testing on several batteries available on the market. Hence, measurements are initially performed on the battery cell level in the laboratory, followed by battery pack laboratory testing when the number of interesting candidates has been reduced. Finally, field testing is performed on a limited population of vehicles, and seldom over a time frame that gives significant information about battery ageing before the product is launched. Consequently, vehicle manufacturers typically refer to laboratory ageing data when predicting HEV-battery lifetime.

For this reason, we see a need for test equipment that can be used on conventional vehicles, making the tested battery cell experience the same environmental and cycling conditions as the battery pack in an HEV. In this paper, we describe a novel cost-effective and simplified method of performing that kind of field testing on battery cells intended for HEV usage. The main purpose with the concept is the possibility to compare battery testing in laboratory with field testing in an early HEV-battery development phase. With this test equipment, field testing of batteries is possible to perform already in parallel with the battery cell laboratory screening phase. Since this test equipment does not require HEV-vehicles or full-scale battery packs, it is less costly, less complex and provides battery ageing information faster than full scale HEV-vehicle tests. With this test equipment, it is possible to measure battery performance periodically on board the vehicle during a lifetime test, thus providing detailed data regarding performance degradation over time. This kind of performance testing is usually only done in the beginning and at the end of conventional battery pack field tests [6,7]. By using this test method in parallel with laboratory measurements, it is possible to acquire a better understanding of differences in battery ageing between lab and field tests. Further more, even if it is possible to cycle batteries in laboratory using recorded data from vehicle and with similar climate changes and vibration conditions as in field, it is too complicated and expensive to perform in practice. Another example of application for the test equipment is comparing different hybrid strategies regarding battery ageing. This can be performed by placing several test equipments on one or several vehicles, outfitted with same batteries but different hybrid strategies. Quantitative measurements are also a possible application for this concept. By collecting data from a large fleet of vehicles operating in a specific geographical area (for example inner city buses), and using the same batteries, it should be possible to obtain a statistical battery lifetime distribution for this specific vehicle population. Combining those results with battery ageing results from laboratory tests on a large population of the same type of battery cells would provide information about battery lifetime distribution regarding the HEV that are operating in that geographic area. Finally, since vehicle performance data is stored in parallel with the battery parameters, drive cycle analysis can be performed, making it possible to relate battery ageing to different driving patterns [8-10].

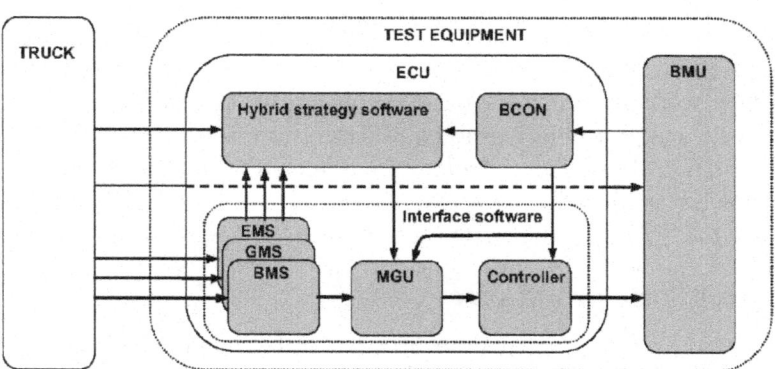

Figure 1. Schematic view describing the interaction between the truck and the test equipment.

The development goal was to create a test equipment that could be used on a conventional vehicles, making the tested battery cell experience the same environmental and cycling conditions as a battery pack in an HEV. The concept is based on using the starter battery in the vehicle both as load and source when charging and discharging the tested battery cell. When charging the tested battery cell, current is drawn from the vehicle starter battery and the opposite during discharge. The cycling current is derived using sensor data available in a conventional vehicle, for example accelerator and brake pedal positions. The test equipment is designed for single battery cell testing and handles battery cell voltages between 1.5 and 4.5 V. Further more, the equipment is designed to work with a 24 V-system that is common in heavy-duty vehicles, but it could be adapted to other electrical systems, such as the 12 V-system in passenger cars. The complete test equipment was verified on a Scania hybrid truck [11], showing that a battery cell cycled with the test equipment can experience the same cycle conditions as a HEV-battery pack. The battery cell chosen for the verification was the AHR32113 from A123 Systems.

2. SYSTEM DESCRIPTION

2.1. Overview

The test equipment will be described in detail in this section following the layout in Figure 1, beginning with the interface to the truck and finishing with the assembly and mounting. Figure 1 shows an overview of the test equipment and the interaction with the truck system. The arrows show the communication between the different parts. The test equipment layout is both hardware- and system-wise divided into two parts: the battery management unit (BMU) and the electronic control unit (ECU), as shown in Figure 1. The arrows indicate the communication flow.

2.2. Electronic Control Unit, ECU

The electronic control unit, ECU, is a vehicle-specific unit from Scania CV AB [11]. It contains a MPC5554 microprocessor from Freescale that is well suited for the complex software functions used in this application [12]. The strategy when creating the test equipment was to handle the communication to the vehicle with the ECU and handle all battery cell interactions with the BMU. Consequently, the hybrid vehicle specific software was solely implemented in the ECU, making it possible to use the BMU together with other hybrid vehicle software systems as well.

2.2.1. Hybrid Strategy Software

The hybrid strategy software from a real hybrid vehicle (in our case a Scania hybrid truck) was chosen as the foundation for the ECU-software development. The hybrid strategy software makes decisions about how to apply the available electrical motor (EM) torque in a hybrid vehicle into the drive train, based on information about vehicle and battery status, *e.g.*, vehicle speed and acceleration, and battery SOC and temperature. For example, using the EM for faster and smoother gear shifting has a higher priority than using it for propulsion of the vehicle. The Scania hybrid strategy software was implemented into the test equipment as an important part of the complete ECU-software.

2.2.2. Interface Software

Since hybrid specific components are missing in conventional vehicles, software had to be developed with the purpose of emulating the missing components. The signals needed are generated using vehicle sensor information available in conventional trucks. This part of the ECU software, the interface software, is divided into four parts that emulate different functions in a hybrid truck. In addition to those parts, the controller software which handles the toggling between normal mode and test mode is included in the interface software. The four parts in the interface software that emulates hybrid functions feed the hybrid strategy software with information needed to decide how to cycle the battery cell. Those parts will be described in detail in the following sections.

2.2.3. Engine Management Software, EMS

When propelling a Scania hybrid truck, the EM is used together with the ICE, and the level of EM-power used during propulsion is determined by the hybrid strategy software, using vehicle system information that is already available in conventional trucks. However, when engine braking is used, the hybrid strategy software is expecting a signal that is not generated in a conventional truck. Engine braking occurs when the accelerator pedal is released or at small brake pedal deviations. When engine braking is activated in a hybrid truck, EM-braking is prioritized before other braking systems which will be explained further later on. This functionality is handled by the EMS software by using information about vehicle speed and accelerator and brake pedal positions to generate the signal needed for the hybrid strategy software to calculate the correct corresponding EM torque.

2.2.4. Gearbox Management Software, GMS

The gearbox in a Scania hybrid truck is in principal a robotized manual gearbox with a clutch. Since the presence of an EM in the driveline would slow down the gear shifting speed, the EM is actively used for compensating this. The effect on the battery when using the EM for gear shifting is current peaks that might have an impact on ageing. Hence, it was essential to implement this effect in the software. The signal needed for the hybrid strategy software to calculate the corresponding EM torque is generated by the GMS software based on information about present gear and gear shifting occurrences.

2.2.5. Brake Management Software, BMS

A conventional truck has four possible ways of braking: engine, exhaust, retarder and mechanical braking [13]. In a hybrid truck, it is also possible to use the EM for braking. When using the brake pedal or the retarder lever in a hybrid truck, the vehicle brake system sends a request for braking to the hybrid strategy software and, based on the battery status and the priority list, the possible amount of braking power by using the EM is calculated by the hybrid strategy software. Since EM-braking does not exist on a conventional truck, this functionality is handled by the BMS-software. By using information about the brake pedal position and the retarder lever position, the signal needed for the hybrid strategy software to calculate the corresponding EM torque is generated. Since engine and exhaust braking is related to the engine system, it is handled by the EMS software, not the BMS software as explained earlier.

2.2.6. Motor-Generator Unit Software, MGU

The EM torque signal sent out from the interface software is converted to a corresponding current by the MGU software. The relationship between torque, rotational speed and output power for the EM was obtained from the manufacturer in form of a lookup table. The conversion is done by using the lookup table together with present shaft speed and battery voltage.

2.2.7. Controller Software

The controller software is working as a switch that in normal operation forwards the signals coming from the MGU to the BMU, but instead forwards the signals from the BCON to the BMU in test mode.

2.2.8. Battery Condition Software, BCON

In a hybrid vehicle, the battery pack usually has a separate electronic control unit that monitors the individual cell voltages and cell temperatures, calculates the battery state of charge (SOC) and manages the charge balancing of the cells. In the test equipment, the SOC-calculations and temperature regulation is managed by the battery condition (BCON) software in the ECU, as well as the test mode functionality that will be described later. One important part of this software is the SOC-management and hence, this functionality will be described more in detail.

The allowed SOC-window for a battery in a hybrid vehicle application depends on the cell chemistry and the hybrid strategy conditions. The choice of SOC-window is set as a parameter in the BCON software. To establish that the battery SOC-value always is within the allowed region, it is necessary to accurately calculate SOC. A common way of determine SOC is by integrating the current over time, also called coulomb counting. This method requires a well known battery capacity (Q_{max}) according to Equation (1):

$$SOC = SOC_{Init} - 100 \frac{\int_0^t I(t)\,dt}{Q_{max}} \tag{1}$$

A problem with this method is that current measurement errors are also integrated over time, causing a drift of the measured SOC-value compared to the actual value. By using complementary determination methods, such as correlating SOC to the open circuit voltage (OCV), the drift of SOC can be reduced. Both methods described above are implemented in the BCON-software. The OCV calibration is implemented by using a lookup table with SOC-values and correlating

OCV-values at different temperatures. The lookup table for the battery cell used during the test equipment development was created using data obtained from galvanostatic titration technique (GITT) measurements in laboratory [14]. A plot describing the OCV *vs.* SOC at 25 °C is shown in Figure 2.

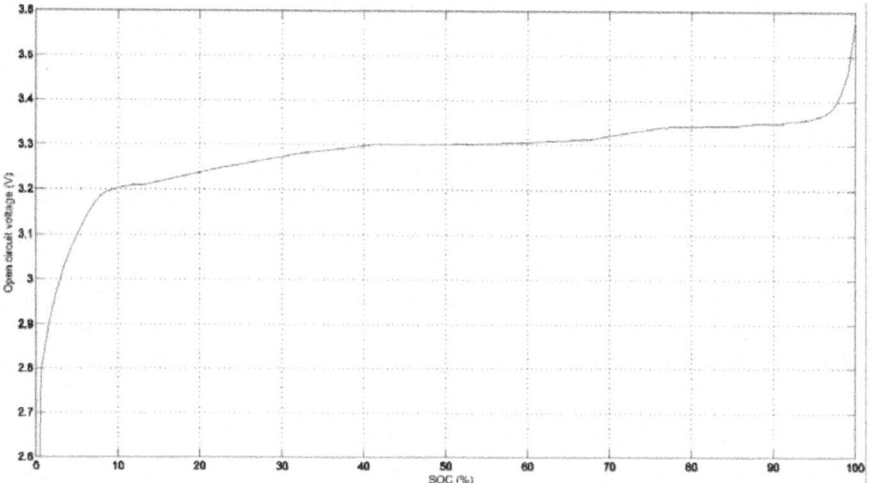

Figure 2. Open circuit voltage *vs.* SOC at 25 °C for the battery cell used in the system validation.

The OCV-calibration procedure in the BCON-software is designed to be performed every time the voltage is stable during a predefined period. The OCV-measurement procedure checks the cell voltage stability by calculating the difference between two consecutive average voltages, performing integration over two predefined periods as shown in Equation (2):

$$\Delta U = \frac{\int_0^{t_1} U(t)\, dt}{t_1} - \frac{\int_0^{t_2} U(t)\, dt}{t_2 - t_1} \tag{2}$$

The times t_1 and t_2 will depend on battery chemistry and have to be found empirically by assigning a ΔU-value that is considered acceptable. This ΔU value has to be set low enough to give a stable measurement, but also high enough to get a signal that is larger than the measurement noise. If the stability test succeeds, the measured voltage is used to receive a new correlating SOC value from the lookup table. The stability check is performed according to a predefined periodicity during normal operation. The LFP/Graphite battery chemistry has a region where the OCV is very flat over a wide SOC-region, approximately between 40% and 70% (Figure 2). In this region, it is undesirable to use the lookup table for SOC-determination. To solve this problem, the BCON software is designed with the possibility of excluding OCV-measurement in a selected voltage span. In this case, the SOC estimation is solely performed by coulomb counting. The OCV-measurement procedure will also be executed if the cell voltage frequently reaches the limits during normal operation. This should not occur in a well tuned test system but if it happens, the voltage stability check described earlier is performed. If this occurs when the OCV is in a region where voltage calibration is prohibited, the battery will be discharged during a predefined period with a predefined current until the voltage is outside that region. When a stable OCV is measured, a new SOC-value is obtained from the lookup table.

When performing lifetime tests on batteries with flat voltage *vs.* SOC characteristics, the cycling region is usually inside the flat SOC-region where it is undesirable to perform OCV-measurements to calibrate SOC. To avoid SOC drift in this case, it is however possible to use OCV-calibration with a lower frequency than normal. Since the battery will be discharged before the OCV-calibration in this case, it is essential to find a frequency that will not affect the ageing of the tested battery during a complete lifetime test. The SOC-calibration that anyway will be performed during the periodic performance tests may be sufficient in those cases. One way of improve the SOC-estimation method would be to make it possible to estimate the errors from the current and voltage measurements and compensate for this when calculating SOC. This could be implemented in the ECU software by introducing a Kalman-filtering algorithm [15].

The test mode functionality implemented in the BCON software makes it possible to check the battery condition during a field test without having to dismount the test equipment and bringing the battery cell into the lab. The two available test functions are described below.

A capacity test comprising a constant current (CC) discharge is available, making it possible to continuously measure the capacity fade during a lifetime test. When starting a capacity test, the battery cell will initially be charged to its maximum capacity by a constant current, constant voltage procedure (CCCV), followed by the constant current discharge until the cell voltage reaches the lower limit. When the capacity test has finished, the battery capacity is calculated and stored, and the SOC-algorithm is updated with the present SOC-value and the new capacity.

In addition to the capacity test, a pulse test is also available, making it possible to measure changes in cell resistance. When entering the pulse test mode, the battery cell is first discharged to the lower voltage limit with a constant current. After this, the pulse test will be performed at a predefined number of SOC-values. The internal resistance will subsequently be calculated according to the EUCAR High Voltage HEV Traction Battery Test Procedure [16]. If a voltage limit is reached during the pulse test, the BMU automatically enters a constant voltage (CV) state and the test is continued until the end. The periodicity of the tests is programmable and if the battery cell is placed in a temperature controlled device, it is possible to perform the tests at a controlled temperature.

2.3. Battery Management Unit, BMU

The main parts of the BMU hardware are the power electronics in form of a step-up, step-down DC/DC-converter, the microprocessor and the memory card, all placed on the same printed circuit board (PCB). Figure 3 shows a schematic view of the BMU hardware, illustrating that several inputs and outputs are available.

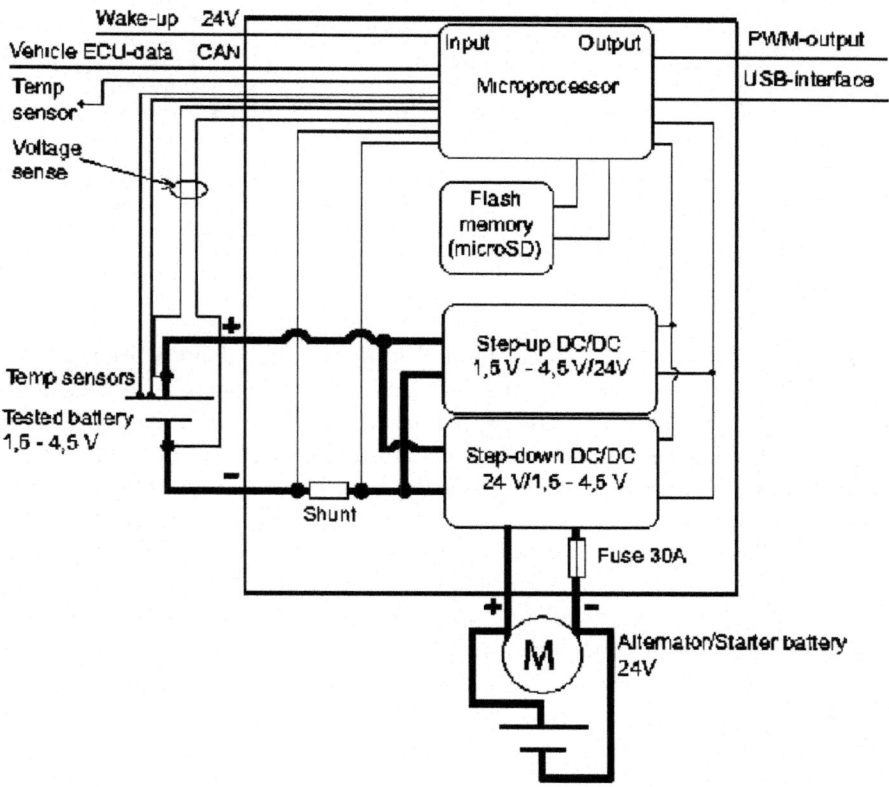

Figure 3. BMU schematic.

The BMU is activated using the wake-up port and the cycling parameters (current, min. and max. voltages) that are sent from the ECU are received through the vehicle ECU-data input port. The communication between the ECU and the BMU is done through a vehicle specific communication network, called Controller Area Network (CAN) [17]. Three NTC-temperature sensors can be connected to the BMU and a pulse width modulated (PWM) output signal is available. An USB-interface is available for transferring data to a computer. The complete BMU was manufactured by the company Elektronikkonsult AB, Sweden [18]. The component and manufacturing costs for one BMU is estimated to be less than 2000 euro.

2.3.1. Power Electronics

The power electronics is essentially a two-quadrant, four-phase DC/DC converter with high efficiency [19]. The two quadrant design permits bidirectional power flow (positive and negative currents) that allows the battery cell to be both charged and discharged (Figure 4).

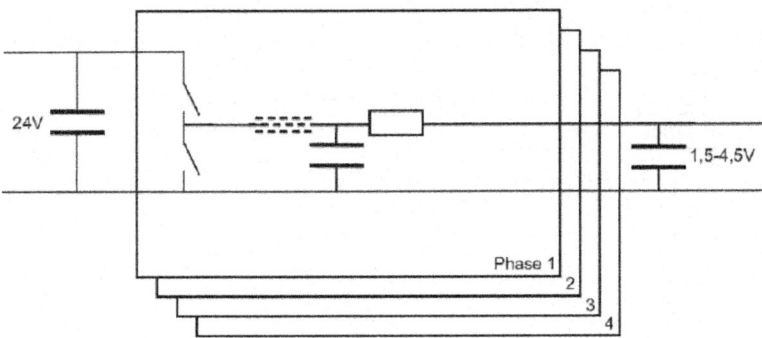

Figure 4. Simplified schematic of the DC/DC-converter.

The DC/DC converter is controlled and monitored by a microprocessor on the PCB, which in turn is controlled by the ECU via the CAN-interface. The charge/discharge current is split into four equal phases that are synchronous and shifted 90 degrees from each other to minimize disturbances. When the four current leads are connected together at the battery cell, switching disturbances are theoretically cancelled out. Each current phase provides a maximum of ±40 A and is controlled by the microprocessor via a regulator circuit. This two quadrant converter, also referred to as a half bridge converter, will take current from the vehicle 24 V-system when charging, and deliver current to it when discharging the battery cell. This approach requires that the vehicle electrical system has the capacity to both deliver and receive the needed current without disturbing vehicle performance. In worst case, (max current into a discharged battery) the internal power dissipation from the complete BMU is typically 40 W, which gives an average efficiency of about 85%.

2.3.2. Microprocessor

The microprocessor used in the BMU is the STM32F103ZE from STMicroelectronics [20]. This microprocessor has 512 kB Flash and 64 kB RAM memory, 72 MHz maximum processor speed and 124 available I/O (Input/Output) pins. This microprocessor comes with extensive support for several I/O devices and it has the capability of communicating via both CAN and USB.

2.3.3. Sensors

The voltage measurement of the battery cell is implemented as a differential measurement. The possibility to measure three additional voltages (one differential measurement per phase) is also implemented in the hardware. This allows testing of up to four battery cells in parallel by making the four phases independent with a software modification. The voltage of the tested cell is measured and recorded with a sample rate of 10 Hz and an accuracy of ±0.01 V. The input voltage from the truck side is also measured but not stored since this information only is used to detect low or high truck voltage.

The cycling current is controlled by the ECU via the CAN-interface. The maximum current level is ±160 A, the sum of the currents in all phases. Four current sensors are used since there are four current phases, and each current is measured separately with a frequency of 10 kHz. An average current is calculated to obtain a stable reading each millisecond and this value is stored with a sample rate of 10 Hz. The measurement accuracy is ±0.1 A.

Three negative-temperature-coefficient (NTC) thermistors can be used. During development and verification of the test equipment, the RH16-3H103FB NTC thermistor from Mitsubishi was chosen [21]. The measurement range is with this temperature sensor -40 °C to +80 °C. The measurement periodicity is 0.1 Hz and the accuracy ±0.5 °C.

One analogue output is available and this output can for example be used for powering a thermal management system. The output signal is a PWM signal controlled by the microprocessor. The available power can be adjusted between 0 W and about 280 W, depending on the truck voltage level since the maximum current from the output is 10 A. The power can be adjusted in the range 0 to 100%, with a resolution of 0.5% (*ca.* 1.4 W).

Table 1. Parameters stored on the memory card.

Parameter	Data Type	Description
Accumulated positive energy throughput	Calculated	Charge power integrated over time (Wh)
Accumulated negative energy throughput	Calculated	Discharge power integrated over time (Wh)
Battery capacity	Calculated	Data from the capacity test (Ah)
Current	Raw	Battery cell current (A)
Voltage	Raw	Battery cell voltage (V)
Temperature	Raw	Battery cell temperature (°C)
Ambient temperature	Raw	Air temperature inside the casing (°C)
SOC	Calculated	SOC-value (%)
Electric machine torque	Calculated	Simulated EM torque data from the MGU (Nm)
Vehicle speed	Raw	Vehicle speed – data from CAN (kph)
EM rotational speed	Raw	Input shaft speed – data from CAN (rpm)
Torque request info	Raw	Total torque request – data from the hybrid strategy software (Nm)

2.3.4. Memory Card

It is possible to store battery current and voltage as well as vehicle sensor data on a memory card on the BMU. The memory card slot on the test equipment is compatible with microSD and microSDHC memory cards and the data is saved to the memory card in a binary format to save space. When a test mode is activated, a separate file with those measurements will be generated to make it easier to access battery performance data during evaluation. Since storing all available vehicle sensor data would acquire too large space on the memory card, a choice of what to store had to be made. To make the upcoming data analysis more convenient, some key parameters were chosen to be directly calculated by the ECU software and stored. Table 1 shows a short description of the parameters chosen to be stored.

2.3.5. Layout

Figure 4 shows a picture of the BMU hardware. The connections to the battery cell and the 24 V electrical system are located at the top of the PCB in the figure. The CAN and USB-connector and the sensor inputs are located at the bottom of the figure, as well as the Wake-up input and PWM-output. The position of the memory card slot and the microprocessor are also marked in the figure.

Figure 5. BMU hardware.

2.4. Assembly

The BMU PCB, sensor and power cables (including fuses) and battery housing including battery, are placed in an aluminium casing, as seen in Figure 6. The BMU PCB is connected to the casing via a heat bridge to ensure cooling for the power transistors. To be able to resemble a true HEV environment in the best possible way for the battery, some kind of thermal management system is desirable. As discussed earlier, the ECU has an output that can be used for controlling a climate unit.

(a)   (b)

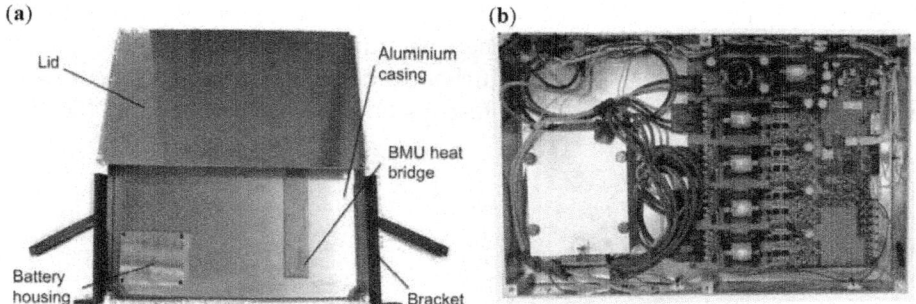

Figure 6. (a) Test equipment casing including a heat bridge for the BMU, battery housing, brackets and lid, looking from above, with the lid slide off; (b) Assembly of the test equipment components, except the ECU, in the aluminium casing.

The thermal management system implemented for the verification of the test equipment consisted of power resistors built into an aluminum housing surrounding the battery (Figure 7).

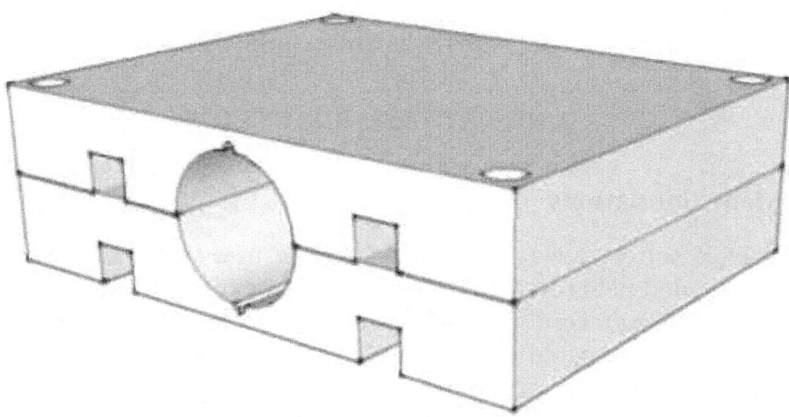

Figure 7. Battery housing with place for battery and temperature sensors (circular opening), and power resistors (square openings).

Since the battery housing is placed in contact with the aluminum casing, the battery is also indirectly cooled by the air flow around the casing. Implementing active cooling would have been possible but had added complexity and cost to the system. The battery cell temperature should be possible to keep in a predefined region by using active heating and passive cooling for tests performed in climates with moderate temperatures. However, when performing tests in warmer climates, some kind of active cooling should be considered.

2.5. Mounting

The test equipment was mounted on the frame of a Scania truck when the system verification tests were performed (Figure 8). The ECU was placed in the cabin, close to the connection point for the CAN-communication. This was done to avoid unnecessarily long cables and hence minimize the risk of collecting disturbing signals. In this way, eventual disturbances on the CAN-path between the ECU and the BMU would only affect the test system, not other systems in the truck.

Figure 8. Test equipment (encircled) mounted on a Scania truck.

3. RESULTS AND DISCUSSION

3.1. System Simulations

The ECU software was verified separately in a simulation environment based on recorded field test data from a hybrid truck. The current from the ECU was generated using recorded vehicle parameters from a test drive with a hybrid vehicle. This was compared to the corresponding recorded current through the hybrid vehicle battery pack during the same test drive. The results are presented in Figure 9.

It is seen from the two plots in Figure 9 that the simulated current in general follows the same pattern as the current recorded from the hybrid truck. There are however small differences in the pattern that mainly can be explained by two factors: (1) no current limitation was applied on the tested battery while it was implemented on the HEV battery pack current and, (2) the battery cell were of a different type compared to the ones used in the hybrid truck. Since the EM in an HEV requests power from the battery pack, it is expected that the current level is dependent on the current-voltage characteristics for the specific battery type. In our case a LFP/Graphite-battery was used in the test equipment and NiMH-batteries in the hybrid truck.

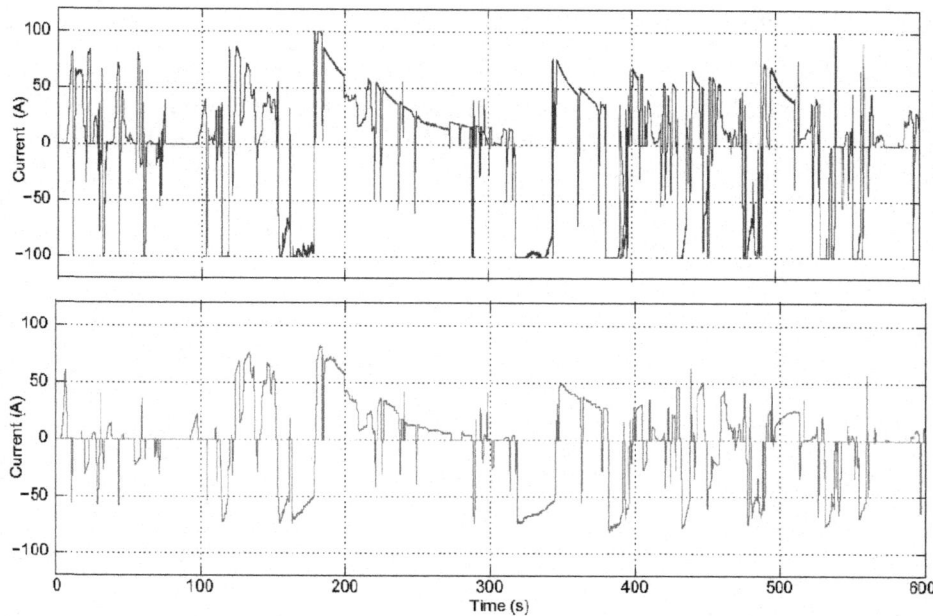

Figure 9. Comparison between ECU-generated current (upper plot) and recorded battery pack current (lower plot). Positive currents correspond to discharge.

3.2. Test Mode Verification

A verification of the capacity test was performed. The left plot in Figure 10 shows the discharge curve performed with a constant current of 3.6 A until the voltage reached the lower limit of 2 V.

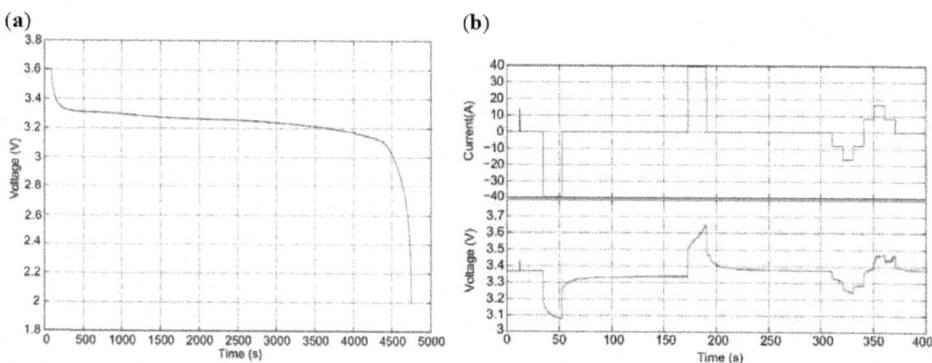

Figure 10. (a) Results from capacity test at 1 °C (3.6 A) and 23 °C; (b) Pulse test verification results: the upper plot shows the current pulse and the lower plot shows the corresponding voltage. The SOC at start was 60% and the temperature was 23 °C.

The pulse test sequence was verified after first adjusting the SOC of the battery cell to 60%. The plots at the right in Figure 10 show the current and corresponding voltage during the pulse test. The test mode verification fulfilled the expectations.

3.3. System Verification

Field verification of the complete test equipment was done on a Scania hybrid truck. By using a hybrid truck for the verification, it was possible to compare the measured current through the tested battery cell with the actual current flowing through the hybrid battery pack. The result from the test run is presented in Figure 11.

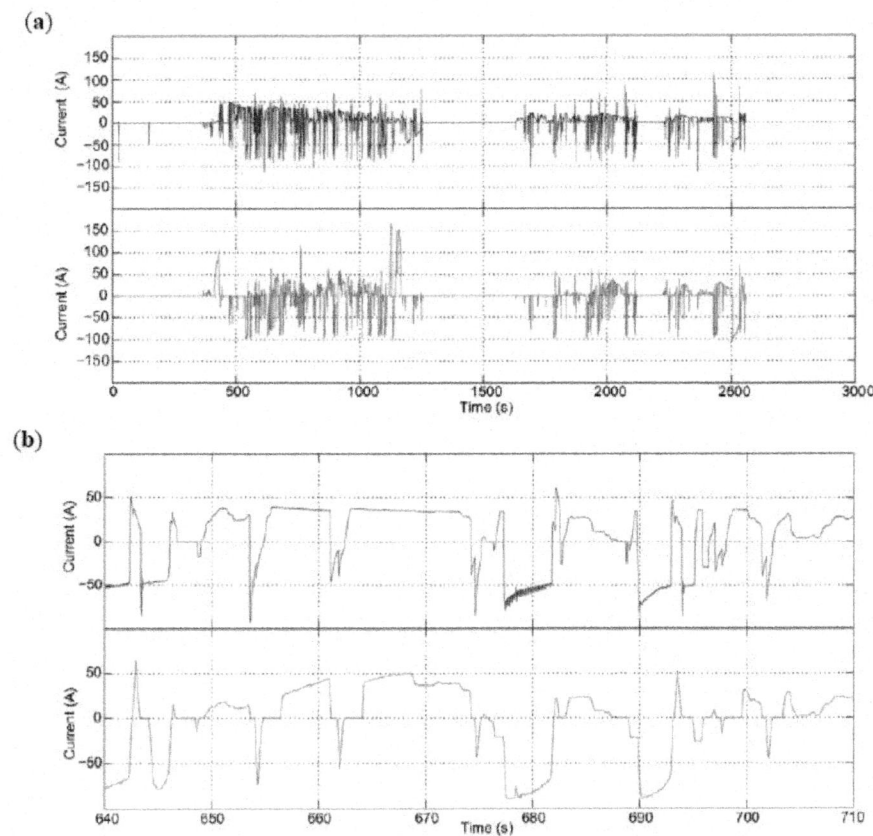

Figure 11. Comparison between test equipment battery current and hybrid truck battery pack current during a test run. (a) Current through the tested battery cell and current through the hybrid truck battery pack, respectively; (b) Magnification of plots in (a) between 640 and 710 seconds.

By zooming in between 640 and 710 seconds, it is possible to see the similarity between the current through the tested battery cell and the HEV battery current (Figure 11b). A comparison of the charge that passed through the tested battery cell compared to the hybrid vehicle battery pack during the test run indicates that the both cells were cycled similarly; only 3% less charge had passed through the tested battery compared to the hybrid vehicle battery pack during the around 40 minutes long test period. This is a satisfying result, considering that the hybrid truck batteries were of different type than the tested cell. The results also confirm that the emulated hybrid vehicle functions are influencing the tested battery as supposed. For example, current peaks due to gear changes are present at 654, 662, 675 and 702 seconds in Figure 11b, both for the tested battery and the hybrid vehicle battery. The conclusion from the verification results is that the test equipment has a most promising potential to be used as a hybrid vehicle emulator for batteries placed on conventional vehicles.

4. CONCLUSIONS

It is in this paper shown that the described novel test equipment has the ability to perform hybrid-battery cell testing on conventional vehicles by emulating an HEV-environment. The test equipment was successfully designed to be used on conventional vehicles (in our case heavy duty trucks). The core of the test equipment, the battery management unit (BMU), was designed to handle the battery cycling, as well as temperature control and measurement data storage. A vehicle specific ECU was used as an interface between the test equipment and the vehicle. This ECU was programmed with software that generates a signal that controls the BMU and the hence current through the battery. This was done by using vehicle sensor data received from the vehicle specific communication network, the CAN-BUS. The test equipment was successfully validated both in laboratory and on a hybrid truck. From the validation on vehicle, the comparison between current through the tested battery cell and current through the hybrid truck battery pack showed satisfying similarity. The idea of isolating a cell when performing field testing has both cost and test time benefits compared to full scale HEV testing. However, by doing this, some parameters that can influence battery ageing are excluded compared to full scale tests. For example, SOC-level differences and temperature differences between cells in a pack will not be addressed. Hence, this test method should be considered to be a complement to laboratory testing rather than a substitute to field testing. In other words, it should be seen as a link between laboratory testing and HEV field testing. Another aspect with this test equipment is the possibility to separate cycle ageing from calendar ageing by placing one additional cell in the battery casing. Both cells will hence experience the same environmental conditions while only one is cycled. One area of improvement for the test equipment is the SOC-estimation. One way of improving this would be to introduce Kalman-filtering. The component and manufacturing costs for the test equipment, excluding the ECU, is estimated to be in the order of 2000 euro, which should be considered low in this context. Results from lifetime testing of battery cells in field using this test equipment will be the subject of further publications.

Acknowledgements

The authors would like to send their gratitude to the company Elektronik-konsult AB, especially Anders Ohlsson, Ulf Soderberg, Erik Hansson and Vidar Wernoe, for their excellent work with the BMU-part of the test equipment. We also want to give Johannes Slettengren at Scania CV AB our appreciation for all his help with the ECU software during the design phase. Finally, we would like to send a big thanks to all supportive people at Scania CV AB, helping out with different tasks during the project. The project was financed by the Swedish Energy Agency and Scania CV AB.

REFERENCES

1. Beyea, J.; Stellman, S.D.; Hatch, M.; Gammon, M.D. Airborne emissions from 1961 to 2004 of Benzo a pyrene from US vehicles per km of travel based on tunnel studies. *Environ. Sci. Technol.*

2. Zamboni, G.; Capobianco, M.; Daminelli, E. Estimation of road vehicle exhaust emissions from 1992 to 2010 and comparison with air quality measurements in Genoa, Italy. *Atmos. Environ.*

3. *Porsche Automobile Holding SE: Lohner-Porsche: The First Hybrid Vehicle*, 2011. Available online: http://www.porsche.com/usa/aboutporsche/porscheandenvironment/hybrid/lohner/ (accessed on 16 February 2011).

4. Bumby, J.R. Hybrid electric vehicle—development and future prospects. *Futures* **1978**, *10*, 438–442.

5. Thornton, E. Japan's hybrid cars. *BusinessWeek*, December 15, 1997, pp. 108–110.

6. Karner, D.; Francfort, J. Hybrid and plug-in hybrid electric vehicle performance testing by the US Department of Energy Advanced Vehicle Testing Activity. *J. Power Sources* **2007**, *174*, 69–75.

7. Francfort J.E. *Hybrid Electric Vehicle and Lithium Polymer NEV Testing*; Idaho National Laboratory: Idaho Falls, ID, USA, 2005. Available online: http://avt.inel.gov/pdf/hev/hev_power_sourcepaper.pdf (accessed on 16 February 2011).

8. Dubarry, M. Analysis of electric vehicle usage of a Hyundai Santa Fe fleet in Hawaii. *J. Asian Electr. Veh.* **2005**, *3*, 657–663.

9. Dubarry, M.; Svoboda, V.; Hwu, R.; Liaw, B.Y. A roadmap to understand battery performance in electric and hybrid vehicle operation. *J. Power Sources* **2007**, *174*, 366–372.

10. Liaw, B.Y.; Dubarry, M. From driving cycle analysis to understanding battery performance in real-life electric hybrid vehicle operation. *J. Power Sources* **2007**, *174*, 76–88.

11. *Scania CV AB*, 2011. Available online: http://www.scania.com/ (accessed on 16 February 2011).

12. *Freescale: 32-bit Embedded Controller MPC5554*, 2005. Available online: http://www.freescale.com/files/32bit/doc/fact_sheet/MPC5554FS.pdf (accessed on 16 February 2011).

13 Fitch, J.W. *Motor Truck Engineering Handbook*, 4th ed.; SAE International: Warrendale, PA, USA, 1993.

14 Weppner, W.; Huggins, R.A. Determination of kinetic-parameters of mixed-conducting electrodes and application to system LI3SB. *J. Electrochem. Soc.* **1977**, *124*, 1569–1578.

15 Plett, G.L. Extended kalman filtering for battery management systems of LiPB-based HEV battery packs—Part 1. background. *J. Power Sources* **2004**, *134*, 252–261.

16. *EUCAR Traction Battery Working Group: High Voltage HEV Traction Battery Test Procedure*; EUCAR: Brussels, Belgium, 2005.

17. *J1939, Recommended Practice for a Serial Control and Communications Vehicle Network*; SAE International: Warrendale, PA, USA, 2009.

18. *Elektronikkonsult AB*, 2011. Available online: http://www.elektronikkonsult.se/home.aspx (accessed on 16 February 2011).

19. Luo, F.L. *Essential DC/DC Converters*; Taylor & Francis Group: Abingdon, UK, 2006.

20. *STMicroelectronics: High-Density Performance Line ARM-Based 32-bit MCU with 256 to512 KB Flash, USB, CAN, 11 Timers, 3 ADCs, 13 Communication Interfaces*, 2009. Available online: http://

www.st.com/internet/com/TECHNICAL_RESOURCES/TECHNICAL_LITERATURE/DA TASHEET/CD00191185.pdf (accessed on 16 February 2011).

21. *Mitsubishi Materials Corporation: On Board Thermistor Series*; Tokyo, Japan, 2009. Available online: http://www.mmc.co.jp/adv/dev/japan/pdf/Division_PDF/70.pdf (accessed on 16 February 2011).

This page left intentionally blank.

INDEX